# SEMA SIMAI Springer Series

Volume 10

More information about this series at http://www.springer.com/series/10532

Sergio Amat • Sonia Busquier

Editors

# Advances in Iterative Methods for Nonlinear Equations

 Springer

*Editors*
Sergio Amat
Departamento de Matemática
  Aplicada y Estadística Universidad
  Politécnica de Cartagena
Cartagena, Spain

Sonia Busquier
Departamento de Matemática
  Aplicada y Estadística Universidad
  Politécnica de Cartagena
Cartagena, Spain

ISSN 2199-3041
SEMA SIMAI Springer Series
ISBN 978-3-319-81845-0
DOI 10.1007/978-3-319-39228-8

ISSN 2199-305X (electronic)

ISBN 978-3-319-39228-8 (eBook)

# Contents

# Chapter 1
# Introduction

**Sergio Amat**

This book is devoted to the approximation of nonlinear equations using iterative methods. This area, as a subfield of Applied Mathematics, remains an active branch of research. Many problems in Computational Sciences and other disciplines can be stated in the form of a nonlinear equation or system using mathematical modeling. In particular, a large number of problems in Applied Mathematics and Engineering are solved by finding the solutions to these equations.

Study on the convergence of iterative methods usually centers on two main issues: semilocal and local convergence analysis. The former is based on the information around an initial point, and amounts to giving convergence conditions guaranteeing the convergence of the iterative process, while the latter is based on the information around a solution to find estimates of the radii of the convergence balls. Selection of the most efficient and robust iterative method for a given problem is crucial to guarantee good approximation results. The order of convergence, the computational cost, and the stability, including the dynamics, are properties that we can use to select the iterative method to be used.

This volume includes nine contributions relating to definition of the methods and to their analysis, including convergence, efficiency, robustness, dynamics, and applications.

The first chapter revisits some advances in the definition and analysis of Steffensen-type methods. The most important property of these methods is the absence of the computation of any derivative while nevertheless preserving the order of convergence of the Newton-type methods. In particular, we can approximate not only differentiable equations, but also non-differentiable ones.

S. Amat (✉)

Departamento de Matemática Aplicada y Estadística, Universidad Politécnica de Cartagena, Cartagena, Spain

e-mail: sergio.amat@upct.es

© Springer International Publishing Switzerland 2016

S. Amat, S. Busquier (eds.), *Advances in Iterative Methods for Nonlinear Equations*, SEMA SIMAI Springer Series 10, DOI 10.1007/978-3-319-39228-8_1

We review some of the recent proposals and some of the convergence theories, including the extension of these methods to the Banach space setting.

The second chapter is devoted to the Newton method for convex optimization. The mathematical analysis uses the notions of generalized Lipschitz conditions and majorizing sequences. Semilocal convergence analysis of (GNA) using L-average conditions is presented. Convex-majorant conditions are used for the semilocal convergence. Numerical examples to illustrate theoretical results, as well as favorable comparisons with earlier studies, are included.

In the third chapter, we are concerned with the problem of presenting a unified local convergence analysis of the inexact Newton method with relative residual error tolerance in order to approximate a singularity of a differentiable vector field on a complete Riemannian manifold setting. Using a combination of a majorant and a center majorant function, we present a local convergence analysis with the following advantages over earlier works using only a majorant function: a larger radius of convergence, more precise estimates on the distances involved, and a larger uniqueness ball. These advantages are obtained under the same computational cost of the functions and parameters involved. Special cases demonstrate the theoretical result as well as the advantages over earlier studies presented in the chapter.

A survey on the existing techniques utilized to design optimal iterative schemes for solving nonlinear equations is presented in the fourth chapter. Attention is focused on procedures that use some evaluations of the derivative of the nonlinear function. After introducing elementary concepts, the methods are classified according to the optimal order reached; some general families of arbitrary order are presented as well. Later, techniques of complex dynamics are introduced, as this is a resource recently used by many authors in order to classify and compare iterative methods of the same order of convergence. Finally, numerical tests are given to show the performance of some of the mentioned procedures, and conclusions are stated.

In the fifth chapter, we present, starting from the Kantorovich theory for Newton's method, two variants of the classic Newton-Kantorovich study that guarantee the semilocal convergence of the method for solving nonlinear equations.

The sixth chapter deals with the enlargement of the region of convergence of Newton's method for solving nonlinear equations defined in Banach spaces. We have used a homotopy method to obtain approximate zeros of the considered function. The novelty in our approach is the establishment of new convergence results based on a Lipschitz condition with an L-average for the involved operator. In particular, semilocal convergence results, as well as local convergence results, are obtained.

In the seventh chapter, we introduce a family of iterative processes in Banach spaces with an $R$-order of convergence of at least three. It includes the most popular iterative processes as particular cases. We also include their algorithms in series development. A study of the general semilocal convergence theorem for the new family is presented, and we also include information about the existence and uniqueness of solution, and a result on the a priori error estimates that leads to the third R-order of convergence of the iterative processes of the family. We analyze the accessibility domain for the family so defined. We define a hybrid iterative method that uses Newton's method as predictor and iterative processes of the family as correctors, so that it takes advantage of the domain of parameters of the predictor method and the speed of convergence of the corrector method. So, from the same starting points of Newton's method, predictor-corrector iterative methods converge with the same rate of convergence as iterative processes of the family. Finally, a new uni-parametric family of multi-point iterations is constructed in Banach spaces to solve nonlinear equations. The semilocal convergence and the $R$-order of convergence of the new iterations are analyzed under Kantorovich-type conditions.

The relaxed Newton's method modifies the classical Newton's method with a parameter in such a way that when it is applied to a polynomial with multiple roots and we take as parameter one of these multiplicities, the order of convergence to the related multiple root is increased. For polynomials of degree three or higher, the relaxed Newton's method may possess extraneous attracting (or even super-attracting) cycles. The eighth chapter presents some algorithms and implementations that allow us to compute the measure (area or probability) of the basin of a $p$-cycle when it is taken in the Riemann sphere. We quantify the efficiency of the relaxed Newton's method by computing, up to a given precision, the measure of the different attracting basins of non-repelling cycles. In this way, we can compare the measure of the basins of the ordinary fixed points (corresponding to the polynomial roots) with the measure of the basins of the point at infinity, and the basins of other non-repelling $p$-cyclic points for $p > 1$.

The aim of the ninth chapter is to provide an overview of theoretical results and numerical tools in some iterative schemes to approximate solutions of nonlinear equations. We examine the concept of iterative methods and their local order of convergence, numerical parameters that allow us to assess the order, and the development of inverse operators (derivative and divided differences). We also provide a detailed study of a new computational technique to analyze efficiency. Finally, we end the chapter with a discussion on adaptive arithmetic to accelerate computations.

This book will appeal to researchers whose field of interest is related to nonlinear problems and equations, and their approximation.

# Chapter 2
# An Overview on Steffensen-Type Methods

S. Amat, S. Busquier, Á.A. Magreñán, and L. Orcos

**Abstract** In this chapter we present an extensive overview of Steffensen-type methods. We first present the real study of the methods and then we present the complex dynamics related this type of methods applied to different polynomials. We also provide an extension to Banach space settings and an application to a Boundary Value Problem. We finish this chapter with contributions to this matter made by other authors.

## 2.1 Introduction

One of the most studied problems in Numerical Analysis is the approximation of nonlinear equations. A powerful tool is the use of iterative methods. It is well-known that Newton's method,

$$x_0 \in \Omega, \qquad x_n = x_{n-1} - [F'(x_{n-1})]^{-1} F(x_{n-1}), \quad n \in \mathbb{N},$$

is one of the most used iterative methods to approximate the solution $x^*$ of $F(x) = 0$. The quadratic convergence and the low operational cost of Newton's method ensure that it has a good computational efficiency.

If we are interesting in methods without using derivatives, then Steffensen-type methods will be a good alternative. These methods only compute divided differences

S. Amat • S. Busquier
Departamento de Matemática Aplicada y Estadística, Universidad Politécnica de Cartagena, Cartagena, Spain
e-mail: sergio.amat@upct.es; sonia.busquier@upct.es

Á.A. Magreñán (✉)
Escuela Superior de Ingeniería y Tecnología, Universidad Internacional de La Rioja (UNIR), C/Gran Vía 41, 26005 Logroño (La Rioja), Spain
e-mail: alberto.magrenan@unir.net; alberto.magrenan@gmail.com

L. Orcos
Facultad de Educación, Universidad Internacional de La Rioja (UNIR), C/Gran Vía 41, 26005 Logroño (La Rioja), Spain
e-mail: lara.orcos@unir.net

© Springer International Publishing Switzerland 2016
S. Amat, S. Busquier (eds.), *Advances in Iterative Methods for Nonlinear Equations*, SEMA SIMAI Springer Series 10, DOI 10.1007/978-3-319-39228-8_2

5

and can be used for nondifferentiable problems. Moreover, they have the same order of convergence than the Newton-type methods. For instance, if the evaluation of $F'(x)$ at each step of Newton's method is approximated by a divided difference of first order $[x, x + F(x); F]$, we will obtain the known method of Steffensen,

$$x_0 \in \Omega, \qquad x_n = x_{n-1} - [x_{n-1}, x_{n-1} + F(x_{n-1}); F]^{-1} F(x_{n-1}), \quad n \in \mathbb{N},$$

which has quadratic convergence and the same computational efficiency as Newton's method. Recall that a bounded linear operator $[x, y; F]$ from $X$ into $X$ is called divided difference of first order for the operator $F$ on the points $x$ and $y$ if $[x, y; F](x - y) = F(x) - F(y)$. Moreover, if $F$ is Fréchet differentiable, then $F'(x) = [x, x; F]$.

The organization of the paper is as follows. We start in Sect. 2.2 with the study of scalar equations. We present in Sect. 2.2.1.1 some convergence analysis and some dynamical aspects of the methods. Some numerical experiments and the dynamics associated to the previous analysis is presented in Sect. 2.2.1.2. In Sect. 2.4, we study the extension of these schemes to a Banach space setting and give some semilocal convergence analysis. Finally, some numerical experiments, including differentiable and nondifferentiable operators, are presented in Sect. 2.5. Finally, other contributions are reported in Sect. 2.6.

## 2.2 The Real Case

Steffensen's method is a root-finding method [39], similar to Newton's method, named after Johan Frederik Steffensen. It is well know that Steffensen's method also achieves quadratic convergence for smooth equations, but without using derivatives as Newton's method does. In this section, we recall the convergence analysis for semismooth equations that is less popular.

### 2.2.1 Semismooth Equations

In [9, 31] the definition of semismooth functions is extended to nonlinear operators. We say that $F : \mathbb{R}^n \to \mathbb{R}^n$ is semismooth at $x$ if $F$ is locally Lipschitz at $x$ and the following limit

$$\lim_{V \in \partial F(x+th')h' \to h, t \downarrow 0} Vh'$$

exists for any $h \in \mathbb{R}^n$, where $\partial F$ is the generalized Jacobian defined,

$$\partial F(x) = conv \partial_B F(x)$$

Most nonsmooth equations involve semismooth operators at practice [32]. We say that $F$ is strongly semismooth at $x$ if $F$ is semismooth at $x$ and for any $V \in \partial F(x+h), h \to 0, Vh - F'(x; h) = O\left(\|h\|^2\right)$.

For $n = 1$, we denote by $\delta F(x, y)$ the divided differences of the form:

$$\delta F(x, y) = \frac{F(x) - F(y)}{x - y}.$$

For the convergence analysis we will need the following result.

**Lemma 1** *Suppose that $F$ is semismooth at $x^*$ and denote the lateral derivatives of $F$ at $x^*$ by*

$$d^- = -F'\left(x^*-\right), \ d^+ = F'\left(x^*+\right)$$

*then*

$$d^- - \delta F(u, v) = o(1) \ u \uparrow x^*, v \uparrow x^*,$$
$$d^+ - \delta F(u, v) = o(1) \ u \downarrow x^*, v \downarrow x^*.$$

*Moreover if $F$ is strongly semismooth at $x^*$, then*

$$d^- - \delta F(u, v) = O\left(|u - x^*| + |v - x^*|\right) \ u, v < x^*,$$
$$d^- - \delta F(u, v) = O\left(|u - x^*| + |v - x^*|\right) \ u, v > x^*.$$

#### 2.2.1.1 A Modification of Steffensen's Method and Convergence Analysis

The classical Steffensen's method can be written as

$$x_{n+1} = x_n - \delta F(x_n, x_n + F(x_n))^{-1} F(x_n).$$

Our iterative procedure would be considered as a new approach based in a better approximation to the derivative $F'(x_n)$ from $x_n$ and $x_n + F(x_n)$ in each iteration. It takes the following form

$$x_{n+1} = x_n - \delta F(x_n, \tilde{x}_n)^{-1} F(x_n) \tag{2.1}$$

where $\tilde{x}_n = x_n + \alpha_n |F(x_n)|$.

These parameters $\alpha_n \in \mathbb{R}$ will be a control of the good approximation to the derivative. Theoretically, if $\alpha_n \to 0$, then

$$\delta F(x_n, \tilde{x}_n) \to F'(x_n).$$

In order to control the stability in practice, but having a good resolution at every iteration, the parameters $\alpha_n$ can be computed such that

$$tol_c << |\alpha_n|F(x_n)|F(x_n)| \leq tol_u,$$

where $tol_c$ is related with the computer precision and $tol_u$ is a user's free parameter.

As the classical Steffensen's method the modification (2.1) needs two evaluations of the function in each iteration and it is quadratically convergent in the smooth case. In the next theorem, we prove that the iterative method (2.1) is quadratically convergent for strongly semismooth equations as well.

**Theorem 1** *Suppose that $F$ is semismooth at a solution $x^*$ of $F(x) = 0$. If $d^-$ and $d^+$ are nonzero, then the algorithm (2.1) is well defined in a neighborhood of $x^*$ and converges to $x^*$ Q-superlinearly. Furthermore, if $F$ is strongly semismooth at $x^*$, the converge to $x^*$ is Q-quadratic.*

*Proof* We may choose $x_0$ sufficiently close to $x^*$ (and/or $\alpha_n$ sufficiently small) such that we have either $x_0, \tilde{x}_0 > x^*$ or $x_0, \tilde{x}_0 < x^*$. According to Lemma 1 is well defined for $k = 0$. It is easy to check that

$$|\tilde{x}_n - x_n| = O(|F(x_n)^2|) = O(|x_n - x^*|^2).$$

Then from Lemma 1,

$$\delta F(x_n, \tilde{x}_n) = \delta F(x_n, x^*) + o(1) = d^+ + o(1) \; (or \; d^- + o(1)).$$

Thus,

$$\begin{aligned}
|x_{n+1} - x^*| &= |x_n - x^* - \delta F(x_n, \tilde{x}_n)^{-1} F(x_n)| \\
&\leq |\delta F(x_n, \tilde{x}_n)^{-1}| \, |F(x_n) - F(x^*) - \delta F(x_n, \tilde{x}_n)(x_n - x^*)| \\
&\leq |\delta F(x_n, \tilde{x}_n)^{-1}| \, |\delta F(x_n, x^*) - \delta F(x_n, \tilde{x}_n)| \, |x_n - x^*| \\
&= o(|x_n - x^*|).
\end{aligned}$$

And we obtain superlinear convergence of $\{x_n\}$. If $F$ is strongly semismooth at $x^*$, we may prove similarly the Q-quadratic convergence of $\{x_n\}$. $\qquad \square$

At practice, this modified Steffensen's method will present some advantages. Firstly, since in general $\delta F(x_n, \tilde{x}_n)$ is a better approximation to the derivative $F'(x_n)$ than $\delta F(x_n, x_n + \epsilon|F(x_n)|F(x_n))$ the convergence will be faster (the first iterations will be better). Secondly, the size of the neighborhood can be higher, that is, we can consider worse starting points $x_0$ (taking $\alpha_0$ sufficiently small), as we will see at the numerical experiments. Finally, if we consider $\epsilon$ sufficiently small in order to obtain similar results at the first iterations and solving the above mentioned disadvantages, then some numerical stability problems will appear at the next iterations.

See [32] and its references for more details on this topic.

### 2.2.1.2  Numerical Experiments and Conclusions

In order to show the performance of the modified Steffensen's method, we have compared it with the classical Steffensen's method and the modified secant's type method proposed in [34]. We consider $tol_u = 10^{-8} >> tol_c$. We have tested on several semismooth equations. Now, we present one.

We consider

$$F(x) = \begin{cases} k(x^4 + x) & \text{if } x < 0 \\ -k(x^3 + x) & \text{if } x \geq 0 \end{cases} \tag{2.2}$$

where $k$ is a real constant.

For $x_0 = 0.1$ and $k = 1$, all the iterative method are Q-quadratically convergent, see Table 2.1. Nevertheless, for $\epsilon$ small the method proposed in [34] has problems with the last iterations. If we consider a stop criterium in order to avoid this problems then we would not be arrived to the convergence. However, our scheme converges without stability pathologies.

If we consider now $x_0 = 1$ and $k = 10$, the classical Steffensen's method and the modified secant method with $\epsilon = 1$ have problems of convergence, in fact they need 258 and 87,174 iterations to converge respectively, see Table 2.2.

The other schemes obtain similar results as before, see Table 2.3.

Finally, in Tables 2.4 and 2.5 we take different initials guesses and different values of $k$. In these tables, we do not write the results for Steffensen's and for $\epsilon = 1$ because in all cases the method do not converge after $10^6$ iterations. On the other hand, if $\epsilon$ is not small enough the convergence is slow, but if it is too small stability

**Table 2.1** Error, Eq. (2.2) $k = 1, x_0 = 0.1$

| Iter. | Steff. | $\epsilon = 1$ | $\epsilon = 10^{-4}$ | $\epsilon = 10^{-8}$ | $tol_u = 10^{-8}$ |
|---|---|---|---|---|---|
| 1 | $1.38e - 03$ | $3.62e - 04$ | $2.99e - 04$ | $2.99e - 04$ | $2.99e - 04$ |
| 2 | $5.09e - 11$ | $5.19e - 14$ | $1.72e - 13$ | $5.21e - 09$ | $2.26e - 14$ |
| 3 | $0.00e + 00$ | $0.00e + 00$ | $NaN$ | $NaN$ | $0.00e + 00$ |

**Table 2.2** Iterations and error, Eq. (2.2) $k = 10$, $x_0 = 1$

| Steff. | $\epsilon = 1$ |
|---|---|
| 256 $1.00e - 02$ | 87172 $3.22e - 02$ |
| 257 $1.41e - 06$ | 87173 $3.42e - 05$ |
| 258 $0.00e + 00$ | 87174 $0.00e + 00$ |

**Table 2.3** Error, Eq. (2.2) $k = 10, x_0 = 1$

| Iter. | $\epsilon = 10^{-4}$ | $\epsilon = 10^{-8}$ | $tol_u = 10^{-8}$ |
|---|---|---|---|
| 5 | $1.26e - 02$ | $1.13e - 02$ | $1.13e - 02$ |
| 6 | $7.49e - 07$ | $4.79e - 07$ | $4.92e - 07$ |
| 7 | $9.08e - 14$ | $7.84e - 09$ | $0.00e + 00$ |
| 8 | $8.32e - 15$ | $NaN$ | |
| 9 | $NaN$ | | |

**Table 2.4** Final iteration and
error, Eq. (2.2), $x_0 = 1$

| $k$ | $\epsilon = 10^{-4}$ | $\epsilon = 10^{-8}$ | $tol_u = 10^{-8}$ |
|---|---|---|---|
| $10^3$ | 50870 $3.88e - 13$ | 14 $3.68e - 15$ | 12 $0.00e + 00$ |
| $10^4$ | $> 10^6$ | 17 $0.00e + 00$ | 15 $0.00e + 00$ |
| $10^6$ | $> 10^6$ | $> 10^6$ | 20 $0.00e + 00$ |
| $10^8$ | $> 10^6$ | $> 10^6$ | 26 $0.00e + 00$ |
| $10^{16}$ | $> 10^6$ | $> 10^6$ | 47 $0.00e + 00$ |

**Table 2.5** Final iteration and
error, Eq. (2.2), $k = 1$

| $x_0$ | $\epsilon = 10^{-4}$ | $\epsilon = 10^{-8}$ | $tol_u = 10^{-8}$ |
|---|---|---|---|
| 4 | 13 $4.20e - 13$ | 9 $6.14e - 09$ | 9 $0.00e + 00$ |
| 8 | 457095 $1.27e - 13$ | 11 $2.42e - 09$ | 12 $0.00e + 00$ |
| 16 | $> 10^6$ | 20 $3.46e - 09$ | 14 $0.00e + 00$ |
| 32 | $> 10^6$ | $> 10^6$ | 16 $0.00e + 00$ |

problems appear, as we said before. Our iterative method gives goods results in all the cases.

## 2.3 Dynamics

In the last years many authors has been studied the dynamics of iterative methods [7, 8, 13, 14, 27]. This classical methods require the computation of the inverse of derivatives which is well known that it can involves a very high computational cost, so other authors have worked in developing tools in order to study nondifferentiable methods [28] and studying the dynamics them [10, 15, 26].

We begin the study with the modification of the following classical iterative methods:

1. *Newton*

$$x_{n+1} = x_n - \frac{f(x_n)}{f'(x_n)}.$$

2. *Two-steps*

$$y_n = x_n - \frac{f(x_n)}{f'(x_n)},$$

$$x_{n+1} = y_n - \frac{f(y_n)}{f'(x_n)}.$$

3. *Chebyshev*

$$x_{n+1} = x_n - \left(1 + \frac{1}{2}L_f(x_n)\right)\frac{f(x_n)}{f'(x_n)}.$$

4. *Halley*

$$x_{n+1} = x_n - \left( \frac{1}{1 - \frac{1}{2}L_f(x_n)} \right) \frac{f(x_n)}{f'(x_n)},$$

where

$$L_f(x) = \frac{f(x)f''(x)}{f'(x)^2}.$$

We denote by $[\cdot, \cdot; f]$ and $[\cdot, \cdot, \cdot; f]$ the first and the second divided difference of the function $f$.

Our modify Steffensen-type methods associated to the above schemes write:

1. *Modify Steffensen*

$$x_{n+1} = x_n - \frac{f(x_n)}{[x_n, x_n + \alpha_n f(x_n); f]}.$$

2. *Modify Steffensen-Two-steps*

$$y_n = x_n - \frac{f(x_n)}{[x_n - \alpha_n f(x_n), x_n + \alpha_n f(x_n); f]},$$

$$x_{n+1} = y_n - \frac{f(y_n)}{[x_n - \alpha_n f(x_n), x_n + \alpha_n f(x_n); f]}.$$

3. *Modify Steffensen-Chebyshev*

$$x_{n+1} = x_n - \left( 1 + \frac{1}{2}L_f(x_n) \right) \frac{f(x_n)}{[x_n - \alpha_n f(x_n), x_n + \alpha_n f(x_n); f]}.$$

4. *Modify Steffensen-Halley*

$$x_{n+1} = x_n - \left( \frac{1}{1 - \frac{1}{2}\mathcal{L}_f(x_n)} \right) \frac{f(x_n)}{[x_n - \alpha_n f(x_n), x_n + \alpha_n f(x_n); f]},$$

where

$$\mathcal{L}_f(x) = \frac{f(x)[x_n - \alpha_n f(x_n), x_n, x_n + \alpha_n f(x_n); f]}{[x_n - \alpha_n f(x_n), x_n + \alpha_n f(x_n); f]^2}.$$

These methods depend, in each iteration, of some parameters $\alpha_n$. These parameters are a control of the good approximation to the derivatives. In order to control the accuracy and stability in practice, the $\alpha_n$ can be computed such that

$$tol_c << \frac{tol_u}{2} \leq ||\alpha_n f(x_n)|| \leq tol_u,$$

where $tol_c$ is related with the computer precision and $tol_u$ is a free parameter for the user.

The classical Steffensen-type methods use $\alpha_n = 1$.

In this section we compare the dynamics of the above methods to introduce the benefits of using the parameters $\alpha_n$. In the experiments we have taken $tol_u = 10^{-6}$.

We approximate the roots of polynomials. We use different colored painting regions of convergence of each root and dark violet is used for no convergence.

We include only the examples for $p(z) = z^3 - 1$ but similar conclusions are obtained for other examples.

The clear conclusion is that the good approximation of the derivatives (for instance using the parameters $\alpha_n$) is crucial to remain the characteristic of the basins of attraction. The classical Steffensen-type methods ($\alpha_n = 1$) have smaller basins of attraction and great regions of no convergence (Figs. 2.1, 2.2, 2.3 and 2.4).

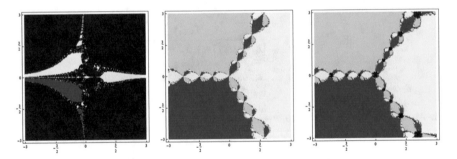

**Fig. 2.1** Basins of attraction for $p(z) = z^3 - 1$. *Left*: Steffensen's method, *Middle*: Newton's method, *Right*: modified Steffensen's method

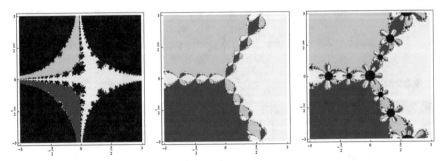

**Fig. 2.2** Basins of attraction for $p(z) = z^3 - 1$. *Left*: two-steps Steffensen's method, *Middle*: two-steps Newton's method, *Right*: modified two-step Steffensen's method

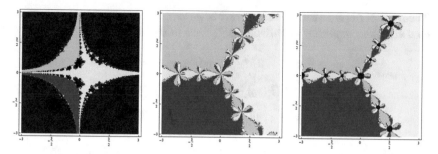

**Fig. 2.3** Basins of attraction for $p(z) = z^3 - 1$. *Left*: Chebyshev-Steffensen's method, *Middle*: Chebyshev's method, *Right*: modified Chebyshev-Steffensen's method

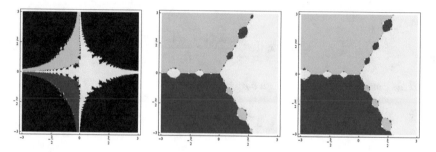

**Fig. 2.4** Basins of attraction for $p(z) = z^3 - 1$. *Left*: Halley-Steffensen's method, *Middle*: Halley's method, *Right*: modified Halley-Steffensen's method

## 2.4 Extension to Banach Space Setting

We only consider the case of second order methods, but similar results can be found for higher order methods.

### 2.4.1 Convergence Analysis

We consider both type of equations: $F(x) = x$ and the usual $F(x) = 0$.

First of all, we must recall the expression of the method for fixed point type equations:

$$x_{n+1} = x_n + (I - [F(x_n), x_n; F])^{-1}(F(x_n) - x_n). \tag{2.3}$$

**Theorem 2** *Let B be an open convex set of a Banach space X. Let $F : B \subset X \to X$ be a nonlinear operator, with divided difference in $B \subset X$. Let $x_0$ be such that*

$$||F(x_0) - x_0|| \leq a_0 \qquad (2.4)$$

$$||(I - [\alpha_0(F(x_0) - x_0) + x_0, x_0; F])^{-1}|| \leq b_0 \qquad (2.5)$$

$$||[x', x''; F] - [y', y''; F]|| \leq k \cdot (||x' - y'|| - ||x'' - y''||) \qquad (2.6)$$

*para todo $x', x'', y', y''$ en $S_0 = \{x : ||x - x_0|| \leq \max(a_0, 2a_0b_0)\}$. Si $S_0 \subset B$, $\alpha_n < 2\alpha_{n-1}$ ($\alpha_n \in (0, 1]$, $\forall n$) y $h = 2ka_0b_0(a_0 + b_0) \leq \frac{1}{2}$ then, the sequence $\{x_n\}$ given by (2.3) is well defined and converges to a fixed point of $F(x)$. Moreover, $x^*$ belong to the ball*

$$||x - x_0|| \leq \frac{a_0b_0(1 - \sqrt{1 - 2h_0})}{h_0}, \qquad (2.7)$$

*and the convergence radius is give by*

$$||x_n - x^*|| \leq \frac{a_0b_0(2h_0)^{2^n}}{2^n h_0}. \qquad (2.8)$$

*Finally, if condition (2.6) is held in $||x - x_0|| \leq a_0 + \frac{a_0b_0(1+\sqrt{1-2h_0})}{h_0} = a_0 + M_0$ the fixed point $x^*$ in unique in the ball $||x - x_0|| < M_0$.*

The basic hypothesis given in the previous theorem is that the divided difference of $F$ was Lipchitz in any ball in a neighbourhood of the initial iteration, in particular the Fréchet derivative of $F$ exists. In some recent works [20–22] (for secant methods), Hernández and Rubio relax these hypotheses and they only suppose that the divided difference satisfy that

$$||[x, y; F] - [v, w; F]|| \leq \omega(||x - v||, ||y - w||), \qquad x, y, v, w \in B$$

where $\omega : \mathbb{R}_+ \times \mathbb{R}_+ \to \mathbb{R}_+$ is a nondecreasing continuous function in both components.

In the next theorem, we will extend that theory to our method

$$x_{n+1} = x_n - ([x_n, x_n + \alpha_n F(x_n); F])^{-1} F(x_n) \qquad (2.9)$$

in order to solve the equation $F(x) = 0$.

**Theorem 3** *Let X be a Banach space. Let B be an open convex subset of X and let suppose that there exists a divided difference of first order of G such that*

$$||[x, y; F] - [v, w; F]|| \leq \omega(||x - v||, ||y - w||), \qquad x, y, v, w \in B$$

where $\omega : \mathbb{R}_+ \times \mathbb{R}_+ \rightarrow \mathbb{R}_+$ is a nondecreasing continuous function in both components. Let $\alpha_n$ be such that $\|\alpha_n G(x_n)\| \leq tol_u$.

Let $x_0 \in B$ and let suppose that

1) $\|\Gamma_0^{-1} := [x_0, x_0 + \alpha_0 G(x_0); F]^{-1}\| \leq \beta$.
2) $\|\Gamma_0^{-1} G(x_0)\| \leq \eta$.
3) Let $m = \beta \omega(\eta, tol_u)$. Let us suppose that

$$t\left(1 - \frac{m}{1 - \beta\omega(t, t + 2tol_u)}\right) - \eta = 0 \tag{2.10}$$

has a minimum positive root which we call R.

If $\beta\omega(R, R + 2tol_u) < 1$, $M := \frac{m}{1 - \beta\omega(R, R + 2tol_u)} < 1$ y $\overline{B(x_0, R)} \subset B$ then, the sequence given by (2.9) is well defined, belongs to $\overline{B(x_0, R)}$ and converges to the unique solution of $F(x) = 0$ in $\overline{B(x_0, R)}$.

## 2.5 Application to Boundary Value Problems

We consider the following boundary problem

$$y''(t) = f(t, y(t), y'(t)), \qquad y(a) = \alpha, \qquad y(b) = \beta, \tag{2.11}$$

choose a discretization of $[a, b]$ with $N$ subintervals,

$$t_j = a + \frac{T}{N}j, \qquad T = b - a, \qquad j = 0, 1, \ldots, N,$$

and propose the use of the multiple shooting method for solving it. First, in each interval $[t_j, t_{j+1}]$, we compute the function $y(t; s_0, s_1, \ldots, s_{j-1})$ recursively, by solving the initial value problems

$$y''(t) = f(t, y(t), y'(t)), \qquad y(t_j) = y(t_j; s_0, s_1, \ldots, s_{j-1}), \qquad y'(t_j) = s_j,$$

whose solution is denoted by $y(t; s_0, s_1, \ldots, s_j)$.

To approximate a solution of problem (2.11), we approximate a solution of the nonlinear system of equations $F(s) = 0$, where $F : \mathbb{R}^N \longrightarrow \mathbb{R}^N$ and

$$\begin{cases} F_1(s_0, s_1, \ldots, s_{N-1}) = s_1 - y'(t_1; s_0) \\ F_2(s_0, s_1, \ldots, s_{N-1}) = s_2 - y'(t_2; s_0, s_1) \\ \vdots \\ F_{N-1}(s_0, s_1, \ldots, s_{N-1}) = s_{N-1} - y'(t_{N-1}; s_0, s_1, \ldots, s_{N-2}) \\ F_N(s_0, s_1, \ldots, s_{N-1}) = \beta - y(t_N; s_0, s_1, s_{N-2}, s_{N-1}). \end{cases}$$

For this, we consider Steffensen's method and method (2.9) and compare their numerical performance. In our study, we consider the usual divided difference of first order. So, for $\mathbf{u}, \mathbf{v} \in \mathbb{R}^N$, such that $\mathbf{u} \neq \mathbf{v}$, we consider $[\mathbf{u}, \mathbf{v}; F] = \left([\mathbf{u}, \mathbf{v}; F]_{ij}\right)_{i,j=1}^N \in \mathcal{L}(\mathbb{R}^N, \mathbb{R}^N)$, where

$$[\mathbf{u}, \mathbf{v}; F]_{ij} = \frac{1}{u_j - v_j} \left( F_i(u_1, \ldots, u_j, v_{j+1}, \ldots, v_N) - F_i(u_1, \ldots, u_{j-1}, v_j, \ldots, v_N) \right).$$

For the initial slope $\mathbf{s}_0 = \left(s_0^0, s_1^0, \ldots, s_{N-1}^0\right)$, to apply Steffensen's method and method (2.9), we consider

$$\begin{cases} s_0^0 = \dfrac{\beta - \alpha}{b - a} = \dfrac{y(t_N) - y(t_0)}{t_N - t_0}, \\[2mm] s_1^0 = \dfrac{y(t_N) - y(t_1; s_0)}{t_N - t_1}, \\[2mm] s_2^0 = \dfrac{y(t_N) - y(t_2; s_0, s_1)}{t_N - t_2}, \\[2mm] \quad \vdots \\[2mm] s_{N-1}^0 = \dfrac{y(t_N) - y(t_{N-1}; s_0, s_1, \ldots, s_{N-2})}{t_N - t_{N-1}}. \end{cases}$$

In particular, to show the performance of method (2.9), we consider the following boundary value problem:

$$y''(t) = y(t) \left( y'(t)^2 + \cos^2 t \right), \qquad y(0) = -1, \qquad y(1) = 1.$$

In this case, we have $T = 1$ and consider three iterations of the schemes for $N = 2, 3$ and four subintervals in the multiple shooting method. The exact solution is obtained with ND-Solve of MATHEMATICA taking $y'(0) = 0.6500356840546128$ in order to have a trustworthy error for values near to $10^{-15}$ (tolerance in double precision).

In Tables 2.6, 2.7, 2.8, 2.9 and 2.10, we observe that Steffensen's method obtains poor results. Notice that when $N$ decreases (or the interval increases), the initial guess is less closer to the solution. This is the reason of the improvements of method (2.9) proposed in this work. For the worst case, $N = 2$, Steffensen's method diverges. And, for $N = 3, 4$, we observe clearly the second order of the methods, as well as the best performance of method (2.9).

**Table 2.6** Method (2.9), $a = 0, b = 10^{-3}; N = 2$

| $n$ | $\|F(s_n)\|_\infty$ | $\|y(t) - y_n\|_\infty$ | $\|y'(t) - y'_n\|_\infty$ |
|---|---|---|---|
| 1 | $1.190\ldots \times 10^{-1}$ | $1.190\ldots \times 10^{-1}$ | $9.634\ldots \times 10^{-2}$ |
| 2 | $6.292\ldots \times 10^{-3}$ | $6.292\ldots \times 10^{-3}$ | $6.297\ldots \times 10^{-3}$ |
| 3 | $1.680\ldots \times 10^{-5}$ | $1.680\ldots \times 10^{-5}$ | $1.772\ldots \times 10^{-5}$ |

**Table 2.7** Method (2.9), $a = 0, b = 10^{-3}; N = 3$

| $n$ | $\|F(\mathbf{s}_n)\|_\infty$ | $\|y(t) - y_n\|_\infty$ | $\|y'(t) - y'_n\|_\infty$ |
|---|---|---|---|
| 1 | $9.839\ldots \times 10^{-2}$ | $2.041\ldots \times 10^{-2}$ | $8.140\ldots \times 10^{-2}$ |
| 2 | $7.274\ldots \times 10^{-4}$ | $1.342\ldots \times 10^{-4}$ | $5.189\ldots \times 10^{-4}$ |
| 3 | $1.445\ldots \times 10^{-8}$ | $1.169\ldots \times 10^{-8}$ | $1.808\ldots \times 10^{-8}$ |

**Table 2.8** Steffensen's method; $N = 3$

| $n$ | $\|F(\mathbf{s}_n)\|_\infty$ | $\|y(t) - y_n\|_\infty$ | $\|y'(t) - y'_n\|_\infty$ |
|---|---|---|---|
| 1 | $2.665\ldots \times 10^{-1}$ | $2.527\ldots \times 10^{-1}$ | $3.865\ldots \times 10^{-1}$ |
| 2 | $1.893\ldots \times 10^{-2}$ | $1.893\ldots \times 10^{-2}$ | $1.965\ldots \times 10^{-2}$ |
| 3 | $6.407\ldots \times 10^{-4}$ | $1.669\ldots \times 10^{-4}$ | $4.999\ldots \times 10^{-4}$ |

**Table 2.9** Method (2.9), $a = 0, b = 10^{-3}; N = 4$

| $n$ | $\|F(\mathbf{s}_n)\|_\infty$ | $\|y(t) - y_n\|_\infty$ | $\|y'(t) - y'_n\|_\infty$ |
|---|---|---|---|
| 1 | $4.680\ldots \times 10^{-2}$ | $2.484\ldots \times 10^{-2}$ | $6.815\ldots \times 10^{-2}$ |
| 2 | $2.331\ldots \times 10^{-5}$ | $8.762\ldots \times 10^{-5}$ | $2.334\ldots \times 10^{-5}$ |
| 3 | $3.636\ldots \times 10^{-9}$ | $3.636\ldots \times 10^{-9}$ | $4.208\ldots \times 10^{-9}$ |

**Table 2.10** Steffensen's method; $N = 4$

| $n$ | $\|F(\mathbf{s}_n)\|_\infty$ | $\|y(t) - y_n\|_\infty$ | $\|y'(t) - y'_n\|_\infty$ |
|---|---|---|---|
| 1 | $1.215\ldots \times 10^{-1}$ | $5.081\ldots \times 10^{-2}$ | $1.593\ldots \times 10^{-1}$ |
| 2 | $6.728\ldots \times 10^{-3}$ | $2.532\ldots \times 10^{-3}$ | $5.457\ldots \times 10^{-3}$ |
| 3 | $1.052\ldots \times 10^{-5}$ | $3.891\ldots \times 10^{-6}$ | $6.043\ldots \times 10^{-6}$ |

## 2.6 Other Contributions

Finally, we introduce briefly some recent contributions.

- In [17] the authors study the convergence of a Newton-Steffensen type method for solving nonlinear equations introduced by Sharma [37]. Under simplified assumptions regarding the smoothness of the nonlinear function, they show that the q-convergence order of the iterations is 3. Moreover, they show that if the nonlinear function maintains the same monotony and convexity on an interval containing the solution, and the initial approximation satisfies the Fourier condition, then the iterations converge monotonically to the solution. They also obtain a posteriori formulas for controlling the errors.
- Based on Steffensen's method, the paper [23] derives a one-parameter class of fourth-order methods for solving nonlinear equations. In the proposed methods, an interpolating polynomial is used to get a better approximation to the derivative of the given function. Each member of the class requires three evaluations of the given function per iteration. Therefore, this class of methods has efficiency index which equals 1.587.
- For solving nonlinear equations, the paper [33] suggests a second-order parametric Steffensen-like method, which is derivative free and only uses two evaluations of the function in one step. A variant of the Steffensen-like method which is still derivative free and uses four evaluations of the function to achieve cubic

convergence is also presented. Moreover, a fast Steffensen-like method with super quadratic convergence and a fast variant of the Steffensen-like method with super cubic convergence are proposed by using a parameter estimation. The error equations and asymptotic convergence constants are obtained for the discussed methods.

- In [34], a parametric variant of Steffensen-secant method and three fast variants of Steffensen-secant method for solving nonlinear equations are suggested. They achieve cubic convergence or super cubic convergence for finding simple roots by only using three evaluations of the function per step. Their error equations and asymptotic convergence constants are deduced. Modified Steffensen's method and modified parametric variant of Steffensen-secant method for finding multiple roots are also discussed.

- In [36], a family of fourth-order Steffensen-type two-step methods is constructed to make progress in including Ren-Wu-Bi's methods [23] and Liu-Zheng-Zhao's method [Z. Liu, Q. Zheng, P. Zhao, A variant of Steffensen's method of fourth-order convergence and its applications, Appl. Math. Comput. 216 (2010) 1978–1983.] as its special cases. Its error equation and asymptotic convergence constant are deduced. The family provides the opportunity to obtain derivative-free iterative methods varying in different rates and ranges of convergence.

- In [11], a family of Steffensen-type methods of fourth-order convergence for solving nonlinear smooth equations is suggested. In the proposed methods, a linear combination of divided differences is used to get a better approximation to the derivative of the given function. Each derivative-free member of the family requires only three evaluations of the given function per iteration. Therefore, this class of methods has efficiency index equal to 1.587. The new class of methods agrees with this conjecture.

- A new derivative-free iterative method for solving nonlinear equations with efficiency index equal to 1.5651 is presented in [18].

- In the paper [12], by approximating the derivatives in the well known fourth-order Ostrowski's method and in a sixth-order improved Ostrowski's method by central difference quotients, we obtain new modifications of these methods free from derivatives. The authors prove the important fact that the methods obtained preserve their convergence orders 4 and 6, respectively, without calculating any derivatives.

- The authors of [19] present a modification of Steffensen's method as a predictor-corrector iterative method, so that they can use Steffensen's method to approximate a solution of a nonlinear equation in Banach spaces from the same starting points from which Newton's method converges. They study the semilocal convergence of the predictor-corrector method by using the majorant principle.

- A derivative free method for solving nonlinear equations of Steffensen's type is presented in [17]. Using a self-correcting parameter, calculated by using Newton's interpolatory polynomial of second degree, the $R$-order of convergence is increased from 2 to 3. This acceleration of the convergence rate is attained without any additional function calculations, which provides a very high computational efficiency of the proposed method.

- The paper [38] proposes two classes of three-step without memory iterations based on the well known second-order method of Steffensen. Per computing step, the methods from the developed classes reach the order of convergence eight using only four evaluations, while they are totally free from derivative evaluation. Hence, they agree with the optimality conjecture of Kung-Traub for providing multi-point iterations without memory.
- In [40], based on some known fourth-order Steffensen-type methods, we present a family of three-step seventh-order Steffensen-type iterative methods for solving nonlinear equations and nonlinear systems. For nonlinear systems, a development of the inverse first-order divided difference operator for multivariable function is applied to prove the order of convergence of the new methods.

  Other related works can be found in [1–6, 16, 24, 25, 29, 30, 35, 41–43].

**Acknowledgements** The research has been partially funded by UNIR Research (http://research. unir.net), Universidad Internacional de La Rioja (UNIR, http://www.unir.net), under the Research Support Strategy 3 [2015–2017], Research Group: MOdelación Matemática Aplicada a la INgeniería (MOMAIN), by the Grant SENECA 19374/PI/14 and by the project MTM2014-52016-C2-1-P of the Spanish Ministry of Economy and Competitiveness.

# References

1. Alarcón, V., Amat, S., Busquier, S., López, D.J.: A Steffensen's type method in Banach spaces with applications on boundary-value problems. J. Comput. Appl. Math. **216**, 243–250 (2008)
2. Amat, S., Busquier, S.: Convergence and numerical analysis of a family of two-step Steffensen's methods. Comput. Math. Appl. **49**, 13–22 (2005)
3. Amat, S., Busquier, S.: A two-step Steffensen's method under modified convergence conditions. J. Math. Anal. Appl. **324**, 1084–1092 (2006)
4. Amat, S., Busquier, S.: On a Steffensen's type method and its behavior for semismooth equations. Appl. Math. Comput. **177**, 819–823 (2006)
5. Amat, S., Blanda, J., Busquier, S.: A Steffensen type method with modified functions. Riv. Mat. Univ. Parma **7**, 125–133 (2007)
6. Amat, S., Bermúdez, C., Busquier, S., Mestiri, D.: A family of Halley-Chebyshev iterative schemes for non-Fréchet differentiable operators. J. Comput. Appl. Math. **228**, 486–493 (2009)
7. Amat, S., Busquier, S., Magreñán, Á.A.: Reducing chaos and bifurcations in Newton-type methods. Abstr. Appl. Anal. **2013**, 10 pp. (2013). Article ID 726701. http://dx.doi.org/10.1155/2013/726701
8. Argyros, I.K., Magreñán, Á.A.: On the convergence of an optimal fourth-order family of methods and its dynamics. Appl. Math. Comput. **252**, 336–346 (2015)
9. Chen, X., Qi, L., Sun, D.: Global and superlinear convergence of the smoothing Newton method and its application to general box constrained variational inequalities. Math. Comput. **67**, 519–540 (1998)
10. Chicharro, F., Cordero, A., Gutiérrez, J.M., Torregrosa, J.R.: Complex dynamics of derivative-free methods for nonlinear equations. Appl. Math. Comput. **219**, 7023–7035 (2013)
11. Cordero, A., Torregrosa, J.R.: A class of Steffensen type methods with optimal order of convergence. Appl. Math. Comput. **217**, 7653–7659 (2011)
12. Cordero, A., Hueso, J.L., Martínez, E., Torregrosa, J.R.: Steffensen type methods for solving nonlinear equations. J. Comput. Appl. Math. **236**(12), 3058–3064 (2012)

13. Cordero, A., García-Maimó, J., Torregrosa, J.R., Vassileva, M.P., Vindel, P.: Chaos in King's iterative family. Appl. Math. Lett. **26**, 842–848 (2013)
14. Cordero, A., Torregrosa, J.R., Vindel, P.: Dynamics of a family of Chebyshev-Halley type methods. Appl. Math. Comput. **219**, 8568–8583 (2013)
15. Cordero, A., Soleymani, F., Torregrosa, J.R., Shateyi, S.: Basins of attraction for various Steffensen-type methods. J. Appl. Math. **2014**, 17 pp. (2014). Article ID 539707. http://dx.doi.org/10.1155/2014/539707
16. Dehghan, M., Hajarian, M.: Some derivative free quadratic and cubic convergence iterative formulas for solving nonlinear equations. Comput. Appl. Math. **29**(1), 19–30 (2010)
17. Džunić, J., Petković, M.S.: A cubically convergent Steffensen-like method for solving nonlinear equations. Appl. Math. Lett. **25**, 1881–1886 (2012)
18. Eftekhari, T.: A new sixth-order Steffensen-type iterative method for solving nonlinear equations. Int. J. Anal. **2014**, 5 pp. (2014). Article ID 685796
19. Ezquerro, J.A., Hernández, M.A., Romero, N., Velasco, A.I.: On Steffensen's method on Banach spaces. J. Comput. Appl. Math. **249**, 9–23 (2013)
20. Hernández, M.A., Rubio, M.J.: The secant method and divided differences Hölder continuous. Appl. Math. Comput. **124**, 137–149 (2001)
21. Hernández, M.A., Rubio, M.J.: Semilocal convergence of the secant method under mild convergence conditions of differentiability. Comput. Math. Appl. **44**(3/4), 277–285 (2002)
22. Hernández, M.A., Rubio, M.J., Ezquerro, J.A.: Secant-like methods for solving nonlinear integral equations of the Hammerstein type. J. Comput. Appl. Math. **115**(1–2), 245–254 (2000)
23. Hongmin, R., Qingbiao, W., Weihong, B.: A class of two-step Steffensen type methods with fourth-order convergence. Appl. Math. Comput. **209**, 206–210 (2009)
24. Jain, P.: Steffensen type methods for solving non-linear equations. Appl. Math. Comput. **194**, 527–533 (2007)
25. Kung, H.T., Traub, J.F.: Optimal order of one-point and multipoint iteration. J. Assoc. Comput. Math. **21**, 634–651 (1974)
26. Lotfi, T., Magreñán, Á.A., Mahdiani, K., Rainer, J.J.: A variant of Steffensen–King's type family with accelerated sixth-order convergence and high efficiency index: dynamic study and approach. Appl. Math. Comput. **252**, 347–353 (2015)
27. Magreñán, Á.A.: Different anomalies in a Jarratt family of iterative root-finding methods. Appl. Math. Comput. **233**, 29–38 (2014)
28. Magreñán, Á.A.: A new tool to study real dynamics: the convergence plane. Appl. Math. Comput. **248**, 215–224 (2014)
29. Păvăloiu, I., Cătinas, E.: On a Newton-Steffensen type method. Appl. Math. Lett. **26**, 659–663 (2013)
30. Petković, M.S., Ilić, S., Džunić, J.: Derivative free two-point methods with and without memory for solving nonlinear equations. Appl. Math. Comput. **217**, 1887–1895 (2010)
31. Qi, L., Sun, J.: A nonsmooth version of Newton's method. Math. Program. **58**, 353–367 (1993)
32. Qi, L., Sun, J.: A survey of some nonsmooth equations and smoothing Newton methods. In: Progress in Optimization. Applied Optimization, vol. 30, pp. 121–146. Kluwer Academic Publishers, Dordrecht (1999)
33. Quan, Z., Jing, W., Peng, Z., Li, Z.: A Steffensen-like method and its higher-order variants. Appl. Math. Comput. **214**, 10–16 (2009)
34. Quan, Z., Peng, Z., Li, Z., Wenchao, M.: Variants of Steffensen-secant method and applications. Appl. Math. Comput. **216**, 3486–3496 (2010)
35. Quan, Z., Jingya, L., Fengxi, H.: An optimal Steffensen-type family for solving nonlinear equations. Appl. Math. Comput. **217**, 9592–9597 (2011)
36. Quan, Z., Peng, Z., Fengxi, H.: A family of fourth-order Steffensen-type methods with the applications on solving nonlinear ODEs. Appl. Math. Comput. **217**, 8196–8203 (2011)
37. Sharma, J.R.: A composite third order Newton-Steffensen method for solving nonlinear equations. Appl. Math. Comput. **169**, 242–246 (2005)

38. Soleymani, F., Karimi Vanani, S.: Optimal Steffensen-type methods with eighth order of convergence. Comput. Math. Appl. **62**, 4619–4626 (2011)
39. Steffensen, I.F.: Remarks on iteration. Skand. Aktuarietidskr. **16**, 64–72 (1933)
40. Xiaofeng, W., Tie, Z.: A family of Steffensen type methods with seventh-order convergence. Numer. Algorithm **62**, 429–444 (2013)
41. Zheng, Q., Wang, C.C., Sun, G.Q.: A kind of Steffensen method and its third-order variant. J. Comput. Anal. Appl. **11**, 234–238 (2009)
42. Zheng, Q., Li, J., Huang, F.: An optimal Steffensen-type family for solving nonlinear equations. Appl. Math. Comput. **217**, 9592–9597 (2011)
43. Zhu, X.: Modified Chebyshev-Halley methods free from second derivative. Appl. Math. Comput. **203**, 824–827 (2008)

# Chapter 3
# Newton's Method for Convex Optimization

Ioannis K. Argyros and Daniel González

**Abstract** In this chapter we deal with the convex optimization problem (COP). Using the generalized-Newton's algorithm (GNA) we generate a sequence that converges to a solution of the COP. We use weak-center and weak Lipschitz-type conditions in our semilocal convergence analysis leading to a finer convergence analysis than in earlier studies. Numerical examples where earlier sufficient convergence conditions are not satisfied but our conditions are satisfied are also presented in this chapter.

## 3.1 Newton's Method

The iterative method

$$x_{n+1} = x_n - \frac{f(x_n)}{f'(x_n)} \quad \text{for each} \quad n = 0, 1, 2, \ldots,$$

where $x_0$ is an initial guess and $f$ is a differentiable real-valued function defined on an open subset of $R$ is the well-known Newton's method. This method was inaugurated by Sir Isaac Newton (1642–1727), who introduced it in 1669 for computing zeros of polynomials.

Let $F : \Omega \subset \mathcal{X} \to \mathcal{Y}$ be a Fréchet-differentiable operator, where $\mathcal{X}$ and $\mathcal{Y}$ are normed vector spaces and $\Omega$ is an open subset of $\mathcal{X}$. Then, Newton's method defined by

$$x_{n+1} = x_n - F'(x_n)^{-1}F(x_n), \quad n = 0, 1, 2, \ldots$$

I.K. Argyros (✉)
Department of Mathematical Sciences, Cameron University, Lawton, OK 73505, USA
e-mail: ioannisa@cameron.edu

D. González
Departamento de Matemática, Escuela Politécnica Nacional, Quito, Ecuador
e-mail: daniel.gonzalezs@epn.edu.ec

© Springer International Publishing Switzerland 2016
S. Amat, S. Busquier (eds.), *Advances in Iterative Methods for Nonlinear Equations*, SEMA SIMAI Springer Series 10, DOI 10.1007/978-3-319-39228-8_3

is used to approximate zeros of $F$ in $\Omega$, i. e., those points $x^* \in \Omega$ such that $F(x^*) = 0$.

Clearly, this iterative method can be realized only if all the points $x_k$, which are called the Newton iterates for operator $F$, remain in $\Omega$, and only if the derivatives $F'(x_k) \in L(\mathcal{X}, \mathcal{Y})$ are invertible for each $n = 0, 1, 2, \ldots$

In 1948, Kantorovich [27] published a semilocal theorem, since then called the Newton–Kantorovich theorem, which gives sufficient conditions guaranteeing that Newton's method converges to a zero of $F$ in $\Omega$. The assumptions of this theorem involve the values of the function $F$ and its derivative $F'$ at the initial guess $x_0$ and on the behavior of $F'$ in a neighborhood of $x_0$. Hence, all these assumptions are in principle verifiable a priori. Ever since, numerous papers have been published involving extensions of this theorem under various conditions. We refer the reader to the [6, 18, 19, 21, 28] and the references therein.

Many problems in Computational Sciences and other disciplines can be brought in form $F(x) = 0$, using Mathematical Modelling [19]. In particular, a large number of problems in Applied Mathematics and also in Engineering are solved by finding the solutions of certain equations of the form $F(x) = 0$. For example, dynamic systems are mathematically modelled by difference or differential equations, and their solutions usually represent the states of the systems. For the sake of simplicity, assume that a time invariant system is driven by the equation $\dot{x} = Q(x)$ (or some suitable operator $Q$), where $x$ is the state. Then, the equilibrium states are determined by solving on equation like $F(x) = 0$. Similar equations are used in the case of discrete systems. The unknowns of engineering equations can be functions (difference, differential and integral equations), vectors (systems of linear or nonlinear algebraic equations), or real or complex numbers (single algebraic equations with single unknowns). Iterations methods are also applied for solving optimization problems. In such cases, the iteration sequence converges to an optimal solution of the problem at hand. Except in special cases, the solutions of these equations cannot be found in closed form. That is why most commonly used solution methods for these equations are iterative. In particular, the practice of Numerical Analysis for finding such solutions is essentially connected to variants of Newton's method.

The study about convergence matter of iterative methods is usually centered on two types: semi-local and local convergence analysis. The semi-local convergence analysis is based on the information around an initial point, to give convergence conditions guaranteeing the convergence of the iterative process; while the local one is, based on the information around a solution to find estimates of the radii of the convergence balls.

The chapter is organized as follows: in Sect. 3.2, the background of the iterative method is presented as well as what we achieve in this chapter. Section 3.3 contains the notions of generalized Lipschitz conditions and the majorizing sequences for (GNA). In order for us to make the paper as self contained as possible, the notion of quasi-regularity is re-introduced (see, e.g., [29, 32]) in Sect. 3.4. Semilocal convergence analysis of (GNA) using $L$-average conditions is presented in Sect. 3.5. In Sect. 3.6, convex majorant conditions are used for the semilocal convergence of (GNA). Numerical examples to illustrate our theoretical results and favorable comparisons to earlier studies (see, e.g., [29, 33, 34, 39, 40]) are presented in Sect. 3.7. The chapter ends with a conclusion in Sect. 3.8.

## 3.2 Introduction

In this chapter we are concerned with the convex composite optimizations problem. Many problems in mathematical programming such as convex inclusion, minimax problems, penalization methods, goal programming, constrained optimization and other problems can be formulated like composite optimizations problem (see, e.g., [6, 18, 20, 23, 24, 33, 34, 38]).

Recently, in the elegant study by Li and Ng (see, e.g., [29]), the notion of quasi-regularity for $x_0 \in \mathbb{R}^l$ with respect to inclusion problem was used. This notion generalizes the case of regularity studied in the seminal paper by Burke and Ferris (see, e.g.,[20]) as well as the case when $d \to F'(x_0)d - \mathcal{C}$ is surjective. Relevant work can be found in the excellent studies by Giannessi, Moldovan and Pellegrini (see, e.g., [24, 33, 34]). This condition was inaugurated by Robinson in [36, 37] (see also, e.g., [6, 14, 15, 19]).

In this chapter we present a semilocal convergence analysis of Gauss–Newton method (see Algorithm (GNA) in Sect. 3.3). In [29], the convergence of (GNA) is based on the generalized Lipschitz conditions inaugurated by Wang [39, 40] (to be precised in Sect. 3.3). In [14], we presented a finer semilocal convergence analysis in a Banach space setting than in [39–43] for (GNM), with the advantages $(\mathcal{A})$: tighter error estimates on the distances involved and the information on the location of the solution is at least as precise. These advantages were obtained (under the same computational cost) using same or weaker hypotheses. Here, we provide the same advantages $(\mathcal{A})$ but for (GNA).

## 3.3 Generalized Lipschitz Conditions and Majorizing Sequences

The purpose of this section is to study the convex composite optimization problem

$$\min_{x \in \mathbb{R}^l} g(x) := h(F(x)), \tag{3.1}$$

where $h : \mathbb{R}^m \to \mathbb{R}$ is convex, $F : \mathbb{R}^l \to \mathbb{R}^m$ is Fréchet-differentiable operator and $m, l \in \mathbb{N}^\star$. The study of (3.1) is very important. On the one hand the study of (3.1) provides a unified framework for the development and analysis of algorithmic method and on the other hand it is a powerful tool for the study of first and second-order optimality conditions in constrained optimality (see, e.g., [6, 18, 20, 23, 24, 29, 33, 34, 38]). We assume that the minimum $h_{min}$ of the function

$h$ is attained. Problem (3.1) is related to

$$F(x) \in \mathcal{C}, \tag{3.2}$$

where

$$\mathcal{C} = \operatorname{argmin} h \tag{3.3}$$

is the set of all minimum points of $h$.

A semilocal convergence analysis for Gauss–Newton method (GNM) was presented using the popular algorithm (see, e.g., [6, 25, 29]):

---

**Algorithm (GNA) : $(\xi, \Delta, x_0)$**

Let $\xi \in [1, \infty[,\ \Delta \in ]0, \infty]$ and for each $x \in \mathbb{R}^l$, define $d_\Delta(x)$ by

$$d_\Delta(x) = \{d \in \mathbb{R}^l : \| d \| \leq \Delta,\ h(F(x) + F'(x)\, d) \leq h(F(x) + F'(x)\, d') \tag{3.4}$$
$$\text{for all } d' \in \mathbb{R}^l \text{ with } \| d' \| \leq \Delta\}.$$

Let also $x_0 \in \mathbb{R}^l$ be given. Having $x_0, x_1, \ldots, x_k$ $(k \geq 0)$, determine $x_{k+1}$ by:
If $0 \in d_\Delta(x_k)$, then STOP;
If $0 \notin d_\Delta(x_k)$, choose $d_k$ such that $d_k \in d_\Delta(x_k)$ and

$$\| d_k \| \leq \xi\, d(0, d_\Delta(x_k)). \tag{3.5}$$

Then, set $x_{k+1} = x_k + d_k$.

---

Here, $d(x, W)$ denotes the distance from $x$ to $W$ in the finite dimensional Banach space containing $W$. Note that the set $d_\Delta(x)$ $(x \in \mathbb{R}^l)$ is nonempty and is the solution of the following convex optimization problem

$$\min_{d \in \mathbb{R}^l, \|d\| \leq \Delta} h(F(x) + F'(x)\, d), \tag{3.6}$$

which can be solved by well known methods such as the subgradient or cutting plane or bundle methods (see, e.g., [26]).

Let $U(x, r)$ denote the open ball in $\mathbb{R}^l$ (or $\mathbb{R}^m$) centered at $x$ and of radius $r > 0$. By $\overline{U}(x, r)$ we denote its closure. Let $W$ be a closed convex subset of $\mathbb{R}^l$ (or $\mathbb{R}^m$). The negative polar of $W$ denoted by $W^\ominus$ is defined as

$$W^\ominus = \{z : \langle z, w \rangle \leq 0 \quad \text{for each} \quad w \in W\}. \tag{3.7}$$

We need the following notion of generalized Lipschitz condition due to Wang in [39, 40] (see also, e.g., [29]). From now on $L : [0, \infty[ \rightarrow ]0, \infty[$ (or $L_0$) denotes a nondecreasing and absolutely continuous function. Moreover, $\eta$ and $\alpha$ denote given positive numbers.

**Definition 3.3.1** Let $\mathcal{Y}$ be a Banach space and let $x_0 \in \mathbb{R}^l$. Let $G : \mathbb{R}^l \to \mathcal{Y}$. Then, $G$ is said to satisfy:

(a) The center $L_0$-average condition on $U(x_0, r)$, if

$$\| G(x) - G(x_0) \| \le \int_0^{\|x - x_0\|} L_0(u)\, du \quad \text{for all} \quad x \in U(x_0, r). \tag{3.8}$$

(b) The $L$-average Lipschitz condition on $U(x_0, r)$, if

$$\| G(x) - G(y) \| \le \int_{\|y - x_0\|}^{\|x - y\| + \|y - x_0\|} L(u)\, du \tag{3.9}$$

for all $x, y \in U(x_0, r)$ with $\| x - y \| + \| y - x_0 \| \le r$.

*Remark 3.3.1* It follows from (3.8) and (3.9) that if $G$ satisfies the $L$-average condition, then it satisfies the center $L$-Lipschitz condition, but not necessarily vice versa. We have that

$$L_0(u) \le L(u) \quad \text{for each} \quad u \in [0, r] \tag{3.10}$$

holds in general and $L/L_0$ can be arbitrarily large (see, e.g., [6, 18, 19]).

**Definition 3.3.2** Define majorizing function $\psi_\alpha$ on $[0, +\infty)$ by

$$\psi_\alpha(t) = \eta - t + \alpha \int_0^t L(u)\,(t - u)\, du \quad \text{for each} \quad t \ge 0 \tag{3.11}$$

and majorizing sequence $\{t_{\alpha,n}\}$ by

$$t_{\alpha,0} = 0, \quad t_{\alpha,n+1} = t_{\alpha,n} - \frac{\psi_\alpha(t_{\alpha,n})}{\psi_\alpha'(t_{\alpha,n})} \quad \text{for each} \quad n = 0, 1, \ldots \tag{3.12}$$

$\{t_{\alpha,n}\}$ was used in [29] as a majorizing sequence for $\{x_n\}$ generated by (GNA).

Sequence $\{t_{\alpha,n}\}$ can also be written equivalently for each $n = 1, 2, \ldots$ and $t_{\alpha,1} = 1$ as

$$t_{\alpha,n+1} = t_{\alpha,n} - \frac{\gamma_{\alpha,n}}{\psi_\alpha'(t_{\alpha,n})}, \tag{3.13}$$

where

$$\begin{aligned}
\gamma_{\alpha,n} &= \int_0^1 \int_{t_{\alpha,n-1}}^{t_{\alpha,n-1} + \theta\,(t_{\alpha,n} - t_{\alpha,n-1})} L(u)\, du\, d\theta\, (t_{\alpha,n} - t_{\alpha,n-1}) \\
&= \int_0^{t_{\alpha,n} - t_{\alpha,n-1}} L(t_{\alpha,n-1} + u)\,(t_{\alpha,n} - t_{\alpha,n-1} - u)\, du,
\end{aligned} \tag{3.14}$$

since (see (4.20) in [29])

$$\psi_\alpha(t_{\alpha,n}) = \frac{\gamma_{\alpha,n}}{\alpha} \quad \text{for each} \quad n = 1, 2, \ldots \tag{3.15}$$

From now on we show how our convergence analysis for (GNA) is finer than the one in [29]. Define a supplementary majorizing function $\psi_{\alpha,0}$ on $[0, +\infty)$ by

$$\psi_{\alpha,0}(t) = \eta - t + \alpha \int_0^t L_0(u)\,(t-u)\,du \quad \text{for each} \quad t \geq 0 \tag{3.16}$$

and corresponding majorizing sequence $\{s_{\alpha,n}\}$ by

$$s_{\alpha,0} = 0, \quad s_{\alpha,1} = \eta, \quad s_{\alpha,n+1} = s_{\alpha,n} - \frac{\beta_{\alpha,n}}{\psi'_{\alpha,0}(s_{\alpha,n})} \quad \text{for each} \quad n = 1, 2, \ldots,$$
$$\tag{3.17}$$

where $\beta_{\alpha,n}$ is defined as $\alpha_{\alpha,n}$ with $s_{\alpha,n-1}, s_{\alpha,n}$ replacing $t_{\alpha,n-1}, t_{\alpha,n}$, respectively.

The results concerning $\{t_{\alpha,n}\}$ are already in the literature (see, e.g., [6, 14, 29]), whereas the corresponding ones for sequence $\{s_{\alpha,n}\}$ can be derivated in an analogous way by simple using $\psi'_{\alpha,0}$ instead of $\psi'_\alpha$. First, we need some auxiliary results for the properties of functions $\psi_\alpha$, $\psi_{\alpha,0}$ and the relationship between sequences $\{s_{\alpha,n}\}$ and $\{t_{\alpha,n}\}$. The proofs of the next four lemmas involving the $\psi_\alpha$ function can be found in [29], whereas the proofs for function $\psi_{\alpha,0}$ are analogously obtained by simple replacing $L$ by $L_0$.

Let $r_\alpha > 0$, $b_\alpha > 0$, $r_{\alpha,0} > 0$ and $b_{\alpha,0} > 0$ be such that

$$\alpha \int_0^{r_\alpha} L(u)\,du = 1, \quad b_\alpha = \alpha \int_0^{r_\alpha} L(u)\,u\,du \tag{3.18}$$

and

$$\alpha \int_0^{r_{\alpha,0}} L_0(u)\,du = 1, \quad b_{\alpha,0} = \alpha \int_0^{r_{\alpha,0}} L_0(u)\,u\,du. \tag{3.19}$$

Clearly, we have that

$$b_\alpha < r_\alpha \tag{3.20}$$

and

$$b_{\alpha,0} < r_{\alpha,0}. \tag{3.21}$$

In view of (3.10), (3.18) and (3.19), we get that

$$r_\alpha \leq r_{\alpha,0} \tag{3.22}$$

and

$$b_\alpha \leq b_{\alpha,0}. \tag{3.23}$$

**Lemma 3.3.1** *Suppose that* $0 < \eta \leq b_\alpha$. *Then,* $b_\alpha < r_\alpha$ *and the following assertions hold:*

(i) $\psi_\alpha$ *is strictly decreasing on* $[0, r_\alpha]$ *and strictly increasing on* $[r_\alpha, \infty)$ *with* $\psi_\alpha(\eta) > 0$, $\psi_\alpha(r_\alpha) = \eta - b_\alpha \leq 0$, $\psi_\alpha(+\infty) \geq \eta > 0$.
(ii) $\psi_{\alpha,0}$ *is strictly decreasing on* $[0, r_{\alpha,0}]$ *and strictly increasing on* $[r_{\alpha,0}, \infty)$ *with* $\psi_{\alpha,0}(\eta) > 0$, $\psi_{\alpha,0}(r_{\alpha,0}) = \eta - b_{\alpha,0} \leq 0$, $\psi_{\alpha,0}(+\infty) \geq \eta > 0$. *Moreover, if* $\eta < b_\alpha$, *then* $\psi_\alpha$ *has two zeros, denoted by* $r_\alpha^\star$ *and* $r_\alpha^{\star\star}$, *such that*

$$\eta < r_\alpha^\star < \frac{r_\alpha}{b_\alpha} \eta < r_\alpha < r_\alpha^{\star\star} \tag{3.24}$$

*and if* $\eta = b_\alpha$, $\psi_\alpha$ *has an unique zero* $r_\alpha^\star = r_\alpha$ *in* $(\eta, \infty)$;
$\psi_{\alpha,0}$ *has two zeros, denoted by* $r_{\alpha,0}^\star$ *and* $r_{\alpha,0}^{\star\star}$, *such that*

$$\eta < r_{\alpha,0}^\star < \frac{r_{\alpha,0}}{b_{\alpha,0}} \eta < r_{\alpha,0} < r_{\alpha,0}^{\star\star},$$

$$r_{\alpha,0}^\star \leq r_\alpha^\star, \tag{3.25}$$

$$r_{\alpha,0}^{\star\star} \leq r_\alpha^{\star\star} \tag{3.26}$$

*and if* $\eta = b_{\alpha,0}$, $\psi_{\alpha,0}$ *has an unique zero* $r_{\alpha,0}^\star = r_{\alpha,0}$ *in* $(\eta, \infty)$.
(iii) $\{t_{\alpha,n}\}$ *is strictly monotonically increasing and converges to* $r_\alpha^\star$.
(iv) $\{s_{\alpha,n}\}$ *is strictly monotonically increasing and converges to its unique least upper bound* $s_\alpha^\star \leq r_{\alpha,0}^\star$.
(v) *The convergence of* $\{t_{\alpha,n}\}$ *is quadratic if* $\eta < b_\alpha$ *and linear if* $\eta = b_\alpha$.

**Lemma 3.3.2** *Let* $r_\alpha$, $r_{\alpha,0}$, $b_\alpha$, $b_{\alpha,0}$, $\psi_\alpha$, $\psi_{\alpha,0}$ *be as defined above. Let* $\overline{\alpha} > \alpha$. *Then, the following assertions hold:*

(i) *Functions* $\alpha \to r_\alpha$, $\alpha \to r_{\alpha,0}$, $\alpha \to b_\alpha$, $\alpha \to b_{\alpha,0}$ *are strictly decreasing on* $[0, \infty)$.
(ii) $\psi_\alpha < \psi_{\overline{\alpha}}$ *and* $\psi_{\alpha,0} < \psi_{\overline{\alpha},0}$ *on* $[0, \infty)$.
(iii) *Function* $\alpha \to r_\alpha^\star$ *is strictly increasing on* $I(\eta)$, *where* $I(\eta) = \{\alpha > 0 : \eta \leq b_\alpha\}$.
(iv) *Function* $\alpha \to r_{\alpha,0}^\star$ *is strictly increasing on* $I(\eta)$.

**Lemma 3.3.3** *Let* $0 \leq \lambda < \infty$. *Define functions*

$$\chi(t) = \frac{1}{t^2} \int_0^t L(\lambda + u)(t - u)\, du \quad \text{for all} \quad t \geq 0 \tag{3.27}$$

*and*

$$\chi_0(t) = \frac{1}{t^2} \int_0^t L_0(\lambda + u)(t - u)\, du \quad \text{for all} \quad t \geq 0. \tag{3.28}$$

*Then, functions $\chi$ and $\chi_0$ are increasing on $[0, \infty)$.*

**Lemma 3.3.4** *Define function*

$$g_\alpha(t) = \frac{\psi_\alpha(t)}{\psi'_\alpha(t)} \quad \text{for all} \quad t \in [0, r^\star_\alpha).$$

*Suppose $0 < \eta \leq b_\alpha$. Then, function $g_\alpha$ is increasing on $[0, r^\star_\alpha)$.*

Next, we shall show that sequence $\{s_{\alpha,n}\}$ is tighter than $\{t_{\alpha,n}\}$.

**Lemma 3.3.5** *Suppose that hypotheses of Lemma 3.3.1 hold and sequences $\{s_{\alpha,n}\}$, $\{t_{\alpha,n}\}$ are well defined for each $n = 0, 1, \ldots$ Then, the following assertions hold for $n = 0, 1, \ldots$*

$$s_{\alpha,n} \leq t_{\alpha,n}, \tag{3.29}$$

$$s_{\alpha,n+1} - s_{\alpha,n} \leq t_{\alpha,n+1} - t_{\alpha,n} \tag{3.30}$$

*and*

$$s^\star_\alpha = \lim_{n \to \infty} s_{\alpha,n} \leq r^\star_\alpha = t^\star_\alpha = \lim_{n \to \infty} t_{\alpha,n}. \tag{3.31}$$

*Moreover, if strict inequality holds in (3.10) so does in (3.29) and (3.30) for $n > 1$. Furthermore, the convergence of $\{s_{\alpha,n}\}$ is quadratic if $\eta < b_\alpha$ and linear if $L_0 = L$ and $\eta = b_\alpha$.*

*Proof* We shall first show using induction that (3.29) and (3.30) are satisfied for each $n = 0, 1, \ldots$ These estimates hold true for $n = 0, 1$, since $s_{\alpha,0} = t_{\alpha,0} = 0$ and $s_{\alpha,1} = t_{\alpha,1} = \eta$. Using (3.10), (3.13) and (3.17) for $n = 1$, we have that

$$s_{\alpha,2} = s_{\alpha,1} - \frac{\beta_{\alpha,1}}{\psi'_{\alpha,0}(s_{\alpha,1})} \leq t_{\alpha,1} - \frac{\gamma_{\alpha,1}}{\psi'_\alpha(t_{\alpha,1})} = t_{\alpha,2}$$

and

$$s_{\alpha,2} - s_{\alpha,1} = -\frac{\beta_{\alpha,1}}{\psi'_{\alpha,0}(s_{\alpha,1})} \leq -\frac{\gamma_{\alpha,1}}{\psi'_\alpha(t_{\alpha,1})} = t_{\alpha,2} - t_{\alpha,1},$$

since

$$-\psi'_{\alpha,0}(s) \leq -\psi'_\alpha(t) \quad \text{for each} \quad s \leq t. \tag{3.32}$$

Hence, estimate (3.29) holds true for $n = 0, 1, 2$ and (3.30) holds true for $n = 0, 1$. Suppose that

$$s_{\alpha,m} \leq t_{\alpha,m} \quad \text{for each} \quad m = 0, 1, 2, \ldots, k + 1$$

and

$$s_{\alpha,m+1} - s_{\alpha,m} \leq t_{\alpha,m+1} - t_{\alpha,m} \quad \text{for each} \quad m = 0, 1, 2, \ldots, k.$$

Then, we have that

$$s_{\alpha,m+2} = s_{\alpha,m+1} - \frac{\beta_{\alpha,m+1}}{\psi'_{\alpha,0}(s_{\alpha,m+1})} \leq t_{\alpha,m+1} - \frac{\gamma_{\alpha,m+1}}{\psi'_{\alpha}(t_{\alpha,m+1})} = t_{\alpha,m+2}$$

and

$$s_{\alpha,m+2} - s_{\alpha,m+1} = -\frac{\beta_{\alpha,m+1}}{\psi'_{\alpha,0}(s_{\alpha,m+1})} \leq -\frac{\gamma_{\alpha,m+1}}{\psi'_{\alpha}(t_{\alpha,m+1})} = t_{\alpha,m+2} - t_{\alpha,m+1}.$$

The induction for (3.29) and (3.30) is complete. Finally, estimate (3.31) follows from (3.30) by letting $n \to \infty$. The convergence order part for sequence $\{s_{\alpha,n}\}$ follows from (3.30) and Lemma 3.3.1-(v). The proof of Lemma 3.3.5 is complete.

*Remark 3.3.2* If $L_0 = L$, the results in Lemmas 3.3.1–3.3.5 reduce to the corresponding ones in [29]. Otherwise (i.e., if $L_0 < L$), our results constitute an improvement [see also (3.22)–(3.26)].

## 3.4 Background on Regularities

In order for us to make the study as self contained as possible, we mention some concepts and results on regularities which can be found in [29] (see also, e.g., [6, 19, 22, 30, 35, 39, 42]). For a set-valued mapping $T : \mathbb{R}^l \rightrightarrows \mathbb{R}^m$ and for a set $A$ in $\mathbb{R}^l$ or $\mathbb{R}^m$, we denote by

$$D(T) = \{x \in \mathbb{R}^l : Tx \neq \emptyset\}, \quad R(T) = \bigcup_{x \in D(T)} Tx,$$

$$T^{-1}y = \{x \in \mathbb{R}^l : y \in Tx\} \quad \text{and} \quad \| A \| = \inf_{a \in A} \| a \|.$$

Consider the inclusion

$$F(x) \in C, \tag{3.33}$$

where $C$ is a closed convex set in $\mathbb{R}^m$. Let $x \in \mathbb{R}^l$ and

$$d(x) = \{d \in \mathbb{R}^l : F(x) + F'(x) d \in C\}. \tag{3.34}$$

**Definition 3.4.1** Let $x_0 \in \mathbb{R}^l$.

(a) $x_0$ is quasi-regular point of (3.33) if there exist $R \in ]0, +\infty[$ and an increasing positive function $\beta$ on $[0, R[$ such that

$$d(x) \neq \emptyset \text{ and } d(0, d(x)) \leq \beta(\| x - x_0 \|) d(F(x), C) \text{ for all } x \in U(x_0, R). \tag{3.35}$$

$\beta(\| x - x_0 \|)$ is an "error bound" in determining how for the origin is away from the solution set of (3.33).

(b) $x_0$ is a regular point of (3.33) if

$$ker(F'(x_0)^T) \cap (C - F(x_0))^\ominus = \{0\}. \tag{3.36}$$

**Proposition 3.4.1 (See, e.g., [20])** *Let $x_0$ be a regular point of (3.33). Then, there are constants $R > 0$ and $\beta > 0$ such that (3.35) holds for $R$ and $\beta(\cdot) = \beta$. Therefore, $x_0$ is a quasi-regular point with the quasi-regular radius $R_{x_0} \geq R$ and the quasi-regular bound function $\beta_{x_0} \leq \beta$ on $[0, R]$.*

*Remark 3.4.1*

(a) $d(x)$ can be considered as the solution set of the linearized problem associated to (3.33)

$$F(x) + F'(x) d \in C. \tag{3.37}$$

(b) If $C$ defined in (3.33) is the set of all minimum points of $h$ and if there exists $d_0 \in d(x)$ with $\| d_0 \| \leq \Delta$, then $d_0 \in d_\Delta(x)$ and for each $d \in \mathbb{R}^l$, we have the following equivalence

$$d \in d_\Delta(x) \Longleftrightarrow d \in d(x) \Longleftrightarrow d \in d_\infty(x). \tag{3.38}$$

(c) Let $R_{x_0}$ denote the supremum of $R$ such that (3.35) holds for some function $\beta$ defined in Definition 3.4.1. Let $R \in [0, R_{x_0}]$ and $\mathcal{B}_R(x_0)$ denotes the set of function $\beta$ defined on $[0, R)$ such that (3.35) holds. Define

$$\beta_{x_0}(t) = \inf\{\beta(t) : \beta \in \mathcal{B}_{R_{x_0}}(x_0)\} \quad \text{for each} \quad t \in [0, R_{x_0}). \tag{3.39}$$

All function $\beta \in \mathcal{B}_R(x_0)$ with $\lim_{t \to R^-} \beta(t) < +\infty$ can be extended to an element of $\mathcal{B}_{R_{x_0}}(x_0)$ and we have that

$$\beta_{x_0}(t) = \inf\{\beta(t) : \beta \in \mathcal{B}_R(x_0)\} \quad \text{for each} \quad t \in [0, R). \tag{3.40}$$

$R_{x_0}$ and $\beta_{x_0}$ are called the quasi-regular radius and the quasi-regular function of the quasi-regular point $x_0$, respectively.

**Definition 3.4.2**

(a) A set-valued mapping $T : \mathbb{R}^l \rightrightarrows \mathbb{R}^m$ is convex if the following items hold

   (i) $Tx + Ty \subseteq T(x + y)$ for all $x, y \in \mathbb{R}^l$.
   (ii) $T\lambda x = \lambda Tx$ for all $\lambda > 0$ and $x \in \mathbb{R}^l$.
   (iii) $0 \in T0$.

(b) Let $T : \mathbb{R}^l \rightrightarrows \mathbb{R}^m$ a convex set-valued map. The norm of $T$ be defined by $\| T \| = \sup\limits_{x \in D(T)} \{\| Tx \| : \| x \| \leq 1\}$. If $\| T \| < \infty$, we say that $T$ is normed.

(c) For two convex set-valued mappings $T$ and $S : \mathbb{R}^l \rightrightarrows \mathbb{R}^m$, the addition and multiplication are defined by $(T + S)x = Tx + Sx$ and $(\lambda T)x = \lambda(Tx)$ for all $x \in \mathbb{R}^l$ and $\lambda \in \mathbb{R}$, respectively.

(d) Let $T : \mathbb{R}^l \rightrightarrows \mathbb{R}^m$, $C$ be closed convex in $\mathbb{R}^m$ and $x \in \mathbb{R}^l$. We define $T_x$ by

$$T_x d = F'(x)d - C \quad \text{for all} \quad d \in \mathbb{R}^l \tag{3.41}$$

and its inverse by

$$T_x^{-1} y = \{d \in \mathbb{R}^l : F'(x)d \in y + C\} \quad \text{for all} \quad y \in \mathbb{R}^m. \tag{3.42}$$

Note that if $C$ is a cone then $T_x$ is convex. For $x_0 \in \mathbb{R}^l$, if the Robinson condition (see, e.g., [36, 37]):

$$T_{x_0} \text{ carries } \mathbb{R}^l \text{ onto } \mathbb{R}^m \tag{3.43}$$

is satisfied, then $D(T_x) = \mathbb{R}^l$ for each $x \in \mathbb{R}^l$ and $D(T_{x_0}^{-1}) = \mathbb{R}^m$.

*Remark 3.4.2* Let $T : \mathbb{R}^l \rightrightarrows \mathbb{R}^m$.

(a) $T$ is convex $\iff$ the graph $Gr(T)$ is a convex cone in $\mathbb{R}^l \times \mathbb{R}^m$.
(b) $T$ is convex $\implies$ $T^{-1}$ is convex from $\mathbb{R}^m$ to $\mathbb{R}^l$.

**Lemma 3.4.1 (See, e.g., [36])** *Let $C$ be a closed convex cone in $\mathbb{R}^m$. Suppose that $x_0 \in \mathbb{R}^l$ satisfies the Robinson condition (3.43). Then we have the following assertions*

(i) $T_{x_0}^{-1}$ *is normed.*
(ii) *If $S$ is a linear operator from $\mathbb{R}^l$ to $\mathbb{R}^m$ such that $\| T_{x_0}^{-1} \| \| S \| < 1$, then the convex set-valued map $\bar{T} = T_{x_0} + S$ carries $\mathbb{R}^l$ onto $\mathbb{R}^m$. Furthermore, $\bar{T}^{-1}$ is normed and*

$$\| \bar{T}^{-1} \| \leq \frac{\| T_{x_0}^{-1} \|}{1 - \| T_{x_0}^{-1} \| \| S \|}.$$

The following proposition shows that condition (3.43) implies that $x_0$ is regular point of (3.33). Using the center $L_0$-average Lipschitz condition, we also estimate in Proposition 3.4.2 the quasi-regular bound function. The proof is given in an analogous way to the corresponding result in [29] by simple using $L_0$ instead of $L$.

**Proposition 3.4.2** *Let $C$ be a closed convex cone in $\mathbb{R}^m$, $x_0 \in \mathbb{R}^l$ and define $T_{x_0}$ as in (3.41). Suppose that $x_0$ satisfies the Robinson condition (3.43). Then we have the following assertions.*

(i) *$x_0$ is a regular point of (3.33).*
(ii) *If $F'$ satisfies the center $L_0$-average Lipschitz condition (3.8) on $U(x_0, R)$ for some $R > 0$. Let $\beta_0 = \| T_{x_0}^{-1} \|$ and let $R_{\beta_0}$ such that*

$$\beta_0 \int_0^{R_{\beta_0}} L_0(u) \, du = 1. \tag{3.44}$$

*Then the quasi-regular radius $R_{x_0}$, the quasi-regular bound function $\beta_{x_0}$ satisfy $R_{x_0} \geq \min\{R, R_{\beta_0}\}$ and*

$$\beta_{x_0}(t) \leq \frac{\beta_0}{1 - \beta_0 \int_0^t L_0(u) \, du} \quad \text{for each} \quad 0 \leq t < \min\{R, R_{\beta_0}\}. \tag{3.45}$$

*Remark 3.4.3* If $L_0 = L$, Proposition 3.4.2 reduces to the corresponding one in [29]. Otherwise, it constitutes an improvement [see (3.20)–(3.26)].

## 3.5 Semilocal Convergence Analysis for (GNA)

Assume that the set $C$ satisfies (3.3). Let $x_0 \in \mathbb{R}^l$ be a quasi-regular point of (3.3) with the quasi-regular radius $R_{x_0}$ and the quasi-regular bound function $\beta_{x_0}$ [i.e., see (3.39)]. Let $\xi \in [1, +\infty)$ and let

$$\eta = \xi \, \beta_{x_0}(0) \, d(F(x_0), C). \tag{3.46}$$

For all $R \in (0, R_{x_0}]$, we define

$$\alpha_0(R) = \sup \left\{ \frac{\xi \, \beta_{x_0}(t)}{\xi \, \beta_{x_0}(t) \int_0^t L_0(s) \, ds + 1} : \eta \leq t < R \right\}. \tag{3.47}$$

**Theorem 3.5.1** *Let $\xi \in [1, +\infty)$ and $\Delta \in (0, +\infty]$. Let $x_0 \in \mathbb{R}^l$ be a quasi-regular point of (3.3) with the quasi-regular radius $R_{x_0}$ and the quasi-regular bound function $\beta_{x_0}$. Let $\eta > 0$ and $\alpha_0(R)$ be given in (3.46) and (3.47), respectively.*

*Let $0 < R < R_{x_0}$, $\alpha \geq \alpha_0(R)$ be a positive constant and let $b_\alpha$, $r_\alpha$ be as defined in (3.18). Let $\{s_{\alpha,n}\}$ ($n \geq 0$) and $s_\alpha^\star$ be given by (3.17) and (3.31), respectively. Suppose that $F'$ satisfies the L-average Lipschitz and the center $L_0$-average Lipschitz conditions on $U(x_0, s_\alpha^\star)$. Suppose that*

$$\eta \leq \min\{b_\alpha, \Delta\} \quad and \quad s_\alpha^\star \leq R. \tag{3.48}$$

*Then, sequence $\{x_n\}$ generated by (GNA) is well defined, remains in $\overline{U}(x_0, s_\alpha^\star)$ for all $n \geq 0$ and converges to some $x^\star$ such that $F(x^\star) \in C$. Moreover, the following estimates hold for each $n = 1, 2, \ldots$*

$$\| x_n - x_{n-1} \| \leq s_{\alpha,n} - s_{\alpha,n-1}, \tag{3.49}$$

$$\| x_{n+1} - x_n \| \leq (s_{\alpha,n+1} - s_{\alpha,n}) \left( \frac{\| x_n - x_{n-1} \|}{s_{\alpha,n} - s_{\alpha,n-1}} \right)^2, \tag{3.50}$$

$$F(x_n) + F'(x_n)(x_{n+1} - x_n) \in C \tag{3.51}$$

*and*

$$\| x_{n-1} - x^\star \| \leq s_\alpha^\star - s_{\alpha,n-1}. \tag{3.52}$$

*Proof* By (3.48), (3.49) and Lemma 3.3.1, we have that

$$\eta \leq s_{\alpha,n} < s_\alpha^\star \leq R \leq R_{x_0}. \tag{3.53}$$

Using the quasi-regularity property of $x_0$, we get that

$$d(x) \neq \emptyset \text{ and } d(0, d(x)) \leq \beta_{x_0}(\| x - x_0 \|) d(F(x), C) \text{ for all } x \in U(x_0, R). \tag{3.54}$$

We first prove that the following assertion holds

$(\mathcal{T})$ (3.49) holds for all $n \leq k - 1 \Longrightarrow$ (3.50) and (3.51) hold for all $n \leq k$.

Denote by $x_k^\theta = \theta x_k + (1 - \theta) x_{k-1}$ for all $\theta \in [0, 1]$. Using (3.53), we have that $x_k^\theta \in U(x_0, s_\alpha^\star) \subseteq U(x_0, R)$ for all $\theta \in [0, 1]$. Hence, for $x = x_k$, (3.54) holds, i.e.,

$$d(x_k) \neq \emptyset \quad and \quad d(0, d(x_k)) \leq \beta_{x_0}(\| x_k - x_0 \|) d(F(x_k), C). \tag{3.55}$$

We have also that

$$\| x_k - x_0 \| \leq \sum_{i=1}^{k} \| x_i - x_{i-1} \| \leq \sum_{i=1}^{k} s_{\alpha,i} - s_{\alpha,i-1} = s_{\alpha,k} \tag{3.56}$$

and

$$\| x_{k-1} - x_0 \| \leq s_{\alpha,k-1} \leq s_{\alpha,k}. \tag{3.57}$$

Now, we prove that

$$\xi \, d(0, d(x_k)) \leq (s_{\alpha,k+1} - s_{\alpha,k}) \left( \frac{\| x_k - x_{k-1} \|}{s_{\alpha,k} - s_{\alpha,k-1}} \right)^2 \leq s_{\alpha,k+1} - s_{\alpha,k}. \tag{3.58}$$

We show the first inequality in (3.58). We denote by $A_k = \| x_{k-1} - x_0 \|$ and $B_k = \| x_k - x_{k-1} \|$. We have the following identity

$$\int_0^1 \int_{A_k}^{A_k + \theta B_k} L(u) \, du \, d\theta = \int_0^{B_k} L(A_k + u) \left( 1 - \frac{u}{B_k} \right) du. \tag{3.59}$$

Then, by the $L$-average condition on $U(x_0, s_\alpha^\star)$, (3.51) for $n = k - 1$ and (3.55)–(3.59), we get that

$$\begin{aligned}
\xi \, d(0, d(x_k)) &\leq \xi \, \beta_{x_0}(\| x_k - x_0 \|) \, d(F(x_k), \mathcal{C}) \\
&\leq \xi \, \beta_{x_0}(\| x_k - x_0 \|) \, \| F(x_k) - F(x_{k-1}) - F'(x_{k-1}) (x_k - x_{k-1}) \| \\
&\leq \xi \, \beta_{x_0}(\| x_k - x_0 \|) \int_0^1 \| (F'(x_k^\theta) - F'(x_{k-1})) (x_k - x_{k-1}) \, d\theta \| \\
&\leq \xi \, \beta_{x_0}(\| x_k - x_0 \|) \int_0^1 \int_{A_k}^{A_k + \theta B_k} L(u) \, du \, B_k \, d\theta \\
&\leq \xi \, \beta_{x_0}(\| x_k - x_0 \|) \int_0^{B_k} L(A_k + u) (B_k - u) \, du \\
&\leq \xi \, \beta_{x_0}(s_{\alpha,k}) \int_0^{B_k} L(s_{\alpha,k-1} + u) (B_k - u) \, du.
\end{aligned} \tag{3.60}$$

For simplicity, we denote $\varXi_{\alpha,k} := s_{\alpha,k} - s_{\alpha,k-1}$. By (3.49) for $n = k$ and Lemma 3.3.3, we have in turn that

$$\frac{\displaystyle\int_0^{B_k} L(s_{\alpha,k-1} + u) (B_k - u) \, du}{B_k^2} \leq \frac{\displaystyle\int_0^{\varXi_{\alpha,k}} L(s_{\alpha,k-1} + u) (\varXi_{\alpha,k} - u) \, du}{\varXi_{\alpha,k}^2}. \tag{3.61}$$

Thus, we deduce that

$$\xi \, d(0, d(x_k)) \leq \xi \, \beta_{x_0}(s_{\alpha,k}) \left( \int_0^{\varXi_{\alpha,k}} L(s_{\alpha,k-1} + u) (\varXi_{\alpha,k} - u) \, du \right) \left( \frac{B_k}{\varXi_{\alpha,k}} \right)^2. \tag{3.62}$$

Using (3.47) and (3.53), we obtain that

$$\frac{\xi \beta_{x_0}(s_{\alpha,k})}{\alpha_0(R)} \leq \left(1 - \alpha_0(R) \int_0^{s_{\alpha,k}} L_0(u)\, du\right)^{-1}. \tag{3.63}$$

Note that $\alpha \geq \alpha_0(R)$. By (3.9), we have that

$$\frac{\xi \beta_{x_0}(s_{\alpha,k})}{\alpha} \leq \left(1 - \alpha \int_0^{s_{\alpha,k}} L_0(u)\, du\right)^{-1} = -(\psi'_{\alpha,0}(s_{\alpha,k}))^{-1}. \tag{3.64}$$

By (3.12), (3.62)–(3.64), we deduce that the first inequality in (3.58) holds. The second inequality of (3.58) follows from (3.49). Moreover, by (3.48) and Lemma 3.3.5, we have that

$$\Xi_{\alpha,k+1} = -\psi'_{\alpha,0}(s_{\alpha,k})^{-1}\beta_{\alpha,k} \leq -\psi'_{\alpha,0}(t_{\alpha,0})\,\gamma_{\alpha,0} = -\psi'_{\alpha,0}(t_{\alpha,0})\,\psi_\alpha(t_{\alpha,0})$$
$$= \eta \leq \Delta.$$

Hence, (3.58) implies that $d(0, d(x_k)) \leq \Delta$ and there exists $d_0 \in \mathbb{R}^l$ with $\| d_0 \| \leq \Delta$ such that $F(x_k) + F'(x_k)\, d_0 \in \mathcal{C}$. By Remark 3.4.1, we have that

$$d_\Delta(x_k) = \{d \in \mathbb{R}^l : \| d \| \leq \Delta \text{ and } F(x_k) + F'(x_k)\, d \in \mathcal{C}\}$$

and

$$d(0, d_\Delta(x_k)) = d(0, d(x_k)).$$

We deduce that (3.51) holds for $n = k$ since $d_k = x_{k+1} - x_k \in d(x_k)$. We also have that

$$\| x_{k+1} - x_k \| \leq \xi\, d(0, d_\Delta(x_k)) = \xi\, d(0, d(x_k)).$$

Hence (3.39) holds for $n = k$ and assertion $(\mathcal{T})$ holds. It follows from (3.49) that $\{x_k\}$ is a Cauchy sequence in a Banach space and as such it converges to some $x^\star \in \overline{U}(x_0, s_\alpha^\star)$ (since $\overline{U}(x_0, s_\alpha^\star)$ is a closed set).

We use now mathematical induction to prove that (3.49)–(3.51) hold. By (3.46), (3.48) and (3.54), we have that $d(x_0) \neq \emptyset$ and

$$\xi\, d(0, d(x_0)) \leq \xi\, \beta_{x_0}(0)\, d(F(x_0), \mathcal{C}) = \eta \leq \Delta.$$

We also have that

$$\| x_1 - x_0 \| = \| d_0 \| \leq \xi\, d(0, d_\Delta(x_0)) \leq \xi \beta_{x_0}(0)\, d(F(x_0), \mathcal{C}) = \eta = \Xi_{\alpha,0}$$

and (3.49) holds for $n = 1$. By induction argument, we get that

$$\| x_{k+1} - x_k \| \le \Xi_{\alpha,k+1} \left( \frac{\| x_k - x_{k-1} \|}{\Xi_{\alpha,k}} \right)^2 \le \Xi_{\alpha,k+1}.$$

The induction is completed. That completes the proof of Theorem 3.5.1.

*Remark 3.5.1*

(a) If $L = L_0$, then Theorem 3.5.1 reduces to the corresponding ones in [29]. Otherwise, in view of (3.29)–(3.31), our results constitute an improvement. The rest of [29] paper is improved, since those results are corollaries of Theorem 3.5.1. For brevity, we leave this part to the motivated reader.

(b) In view of the proof of our Theorem 3.5.1, we see that sequence $\{r_{\alpha,n}\}$ given by

$$
\begin{aligned}
&r_{\alpha,0} = 0, \quad r_{\alpha,1} = \eta, \\
&r_{\alpha,2} = r_{\alpha,1} - \frac{\alpha \displaystyle\int_0^{r_{\alpha,1}-r_{\alpha,0}} L_0(r_{\alpha,0} + u)\,(r_{\alpha,1} - r_{\alpha,0} - u)\,du}{\psi'_{\alpha,0}(r_{\alpha,1})}, \\
&r_{\alpha,n+1} = r_{\alpha,n} - \frac{\alpha \displaystyle\int_0^{r_{\alpha,n}-r_{\alpha,n-1}} L(r_{\alpha,n-1} + u)\,(r_{\alpha,n} - r_{\alpha,n-1} - u)\,du}{\psi'_{\alpha,0}(r_{\alpha,n})}
\end{aligned}
\tag{3.65}
$$

for each $n = 2, 3, \dots$

is also a majorizing sequences for (GNA). Following the proof of Lemma 3.3.5 and under the hypotheses of Theorem 3.5.1, we get that

$$r_{\alpha,n} \le s_{\alpha,n} \le t_{\alpha,n}, \tag{3.66}$$

$$r_{\alpha,n+1} - r_{\alpha,n} \le s_{\alpha,n+1} - s_{\alpha,n} \le t_{\alpha,n+1} - t_{\alpha,n} \tag{3.67}$$

and

$$r_\alpha^\star = \lim_{n \to \infty} r_{\alpha,n} \le s_\alpha^\star \le r_\alpha^\star. \tag{3.68}$$

Hence, $\{r_{\alpha,n}\}$ and $\{s_{\alpha,n}\}$ are tighter majorizing sequences for $\{x_n\}$ than $\{t_{\alpha,n}\}$ used by Li et al. in [29]. Sequences $\{r_{\alpha,n}\}$ and $\{s_{\alpha,n}\}$ can converge under hypotheses weaker than the ones given in Theorem 3.5.1. Such conditions have already given by us for more general functions $\psi$ and in the more general setting of a Banach space as in [14, 15] (see also, e.g., [5, 7–9]). Therefore, here, we shall only refer to the popular Kantorovich case as an illustration. Choose $\alpha = 1$, $L(u) = L$ and $L_0(u) = L_0$ for all $u \ge 0$. Then, $\{t_{\alpha,n}\}$ converges under the famous for its simplicity and clarity Newton-Kantorovich hypothesis (see, e.g., [6, 28])

$$h = L\eta \le \frac{1}{2}. \tag{3.69}$$

$\{r_{\alpha,n}\}$ converges provided that (see, e.g., [11])

$$h_1 = L_1 \, \eta \le \frac{1}{2},$$  (3.70)

where

$$L_1 = \frac{1}{8} \left( L + 4 L_0 + (L^2 + 8 L_0 L)^{1/2} \right)$$

and $\{r_{\alpha,n}\}$ converges if (see, e.g., [16, 17])

$$h_2 = L_2 \, \eta \le \frac{1}{2},$$  (3.71)

where

$$L_2 = \frac{1}{8} \left( 4 L_0 + (L L_0 + 8 L_0^2)^{1/2} + (L_0 L)^{1/2} \right).$$

It follows by (3.69)–(3.71) that

$$h \le \frac{1}{2} \implies h_1 \le \frac{1}{2} \implies h_2 \le \frac{1}{2},$$  (3.72)

but not vice versa unless if $L_0 = L$. Moreover, we get that

$$\frac{h_1}{h} \longrightarrow \frac{1}{4}, \quad \frac{h_2}{h} \longrightarrow 0, \quad \frac{h_2}{h_1} \longrightarrow 0 \quad \text{as} \quad \frac{L_0}{L} \longrightarrow 0.$$

(c) There are cases when the sufficient convergence conditions developed in the preceding work are not satisfied. Then, one can use the modified Gauss Newton method (MGNM). In this case, the majorizing sequence proposed in [29] is given by

$$q_{\alpha,0} = 0, \quad q_{\alpha,n+1} = q_{\alpha,n} - \frac{\psi_\alpha(q_{\alpha,n})}{\psi'_\alpha(0)} \quad \text{for each} \quad n = 0, 1, \dots$$  (3.73)

This sequence clearly converges under the hypotheses of Theorem 3.5.1, so that estimates (3.49)–(3.52) hold with sequence $\{q_{\alpha,n}\}$ replacing $\{s_{\alpha,n}\}$. However, according to the proof of Theorem 3.5.1, the hypotheses on $\psi_{\alpha,0}$ can replace the corresponding ones on $\psi_\alpha$. Moreover, the majorizing sequence is given by

$$p_{\alpha,0} = 0, \quad p_{\alpha,n+1} = p_{\alpha,n} - \frac{\psi_\alpha(p_{\alpha,n})}{\psi'_{\alpha,0}(0)} \quad \text{for each} \quad n = 0, 1, \dots$$  (3.74)

Furthermore, we have that

$$\psi_{\alpha,0}(s) \leq \psi_{\alpha}(s) \quad \text{for each} \quad s \geq 0. \tag{3.75}$$

Hence, we clearly have that for each $n = 0, 1, \dots$

$$p_{\alpha,n} \leq q_{\alpha,n}, \tag{3.76}$$

$$p_{\alpha,n+1} - p_{\alpha,n} \leq q_{\alpha,n+1} - q_{\alpha,n} \tag{3.77}$$

and

$$p_{\alpha}^{\star} = \lim_{n \to \infty} p_{\alpha,n} \leq q_{\alpha}^{\star} = \lim_{n \to \infty} q_{\alpha,n}. \tag{3.78}$$

[Notice also the advantages (3.20)–(3.26)].

In the special case when functions $L_0$ and $L$ are constants and $\alpha = 1$, we have that the conditions on function $\psi_{\alpha}$ reduce to (3.69), whereas using $\psi_{\alpha,0}$ to

$$h_0 = L_0 \eta \leq \frac{1}{2}. \tag{3.79}$$

Notice that

$$\frac{h_0}{h} \longrightarrow 0 \quad \text{as} \quad \frac{L_0}{L} \longrightarrow 0. \tag{3.80}$$

Therefore, one can use (MGNM) as a predictor until a certain iterate $x_N$ for which the sufficient conditions for (GNM) are satisfied. Then, we use $x_N$ as the starting iterate for faster than (MGNM) method (GNM). Such an approach was used by the author in [4].

## 3.6    General Majorant Conditions

In this section, we provide a semilocal convergence analysis for (GNA) using more general majorant conditions than (3.8) and (3.9).

**Definition 3.6.1** Let $\mathcal{Y}$ be a Banach space, $x_0 \in \mathbb{R}^l$ and $\alpha > 0$. Let $G : \mathbb{R}^l \to \mathcal{Y}$ and $f_{\alpha} : [0, r[\to] - \infty, +\infty[$ be continuously differentiable. Then, $G$ is said to satisfy:

(a)  The center-majorant condition on $U(x_0, r)$, if

$$\| G(x) - G(x_0) \| \leq \alpha^{-1} (f_{\alpha}'(\| x - x_0 \|) - f_{\alpha}'(0)) \quad \text{for all } x \in U(x_0, r). \tag{3.81}$$

(b) The majorant condition on $U(x_0, r)$, if

$$\| G(x) - G(y) \| \le \alpha^{-1} (f_\alpha'(\| x - y \| + \| y - x_0 \|) - f_\alpha'(\| y - x_0 \|)) \quad (3.82)$$

for all $x, y \in U(x_0, r)$ with $\| x - y \| + \| y - x_0 \| \le r$.

Clearly, conditions (3.81) and (3.82) generalize (3.8) and (3.9), respectively, in [22] (see also, e.g., [5, 7, 9, 14, 15]) (for $G = F'$ and $\alpha = 1$). Notice that (3.82) implies (3.81) but not necessarily vice versa. Hence, (3.81) is not additional to (3.82) hypothesis. Define majorizing sequence $\{t_{\alpha,n}\}$ by

$$t_{\alpha,0} = 0, \quad t_{\alpha,n+1} = t_{\alpha,n} - \frac{f_\alpha(t_{\alpha,n})}{f_\alpha'(t_{\alpha,n})}. \quad (3.83)$$

Moreover, as in (3.47) and for $R > 0$, define (implicitly):

$$\alpha_0(R) := \sup_{\xi \le t < R} -\frac{\eta \, \beta_{x_0}(t)}{f_{\alpha_0(R)}'(t)}. \quad (3.84)$$

Next, we provide sufficient conditions for the convergence of sequence $\{t_{\alpha,n}\}$ corresponding to the ones given in Lemma 3.3.1.

**Lemma 3.6.1 (See, e.g., [18, 19, 22])** *Let $r > 0$, $\alpha > 0$ and $f_\alpha : [0, r) \to (-\infty, +\infty)$ be continuously differentiable. Suppose*

(i) *$f_\alpha(0) > 0, f_\alpha'(0) = -1$;*
(ii) *$f_\alpha'$ is convex and strictly increasing;*
(iii) *equation $f_\alpha(t) = 0$ has positive zeros. Denote by $r_\alpha^\star$ the smallest zero. Define $r_\alpha^{\star\star}$ by*

$$r_\alpha^{\star\star} = \sup\{t \in [r_\alpha^\star, r) : f_\alpha(t) \le 0\}. \quad (3.85)$$

*Then, sequence $\{t_{\alpha,n}\}$ is strictly increasing and converges to $r_\alpha^\star$. Moreover, the following estimates hold*

$$r_\alpha^\star - t_{\alpha,n} \le \frac{D^- f_\alpha'(r_\alpha^\star)}{-2f_\alpha'(r_\alpha^\star)} (r_\alpha^\star - t_{\alpha,n-1})^2 \quad \text{for each} \quad n = 1, 2, \ldots, \quad (3.86)$$

*where $D^- f'$ is the left directional derivative of $f$ (see, e.g., [6, 10, 18, 19]).*

We can show the following semilocal convergence result for (GNA) using generalized majorant conditions (3.81) and (3.82).

**Theorem 3.6.1** *Let $\xi \in [1, +\infty)$ and $\Delta \in (0, +\infty]$. Let $x_0 \in \mathbb{R}^l$ be a quasi-regular point of (3.3) with the quasi-regular radius $R_{x_0}$ and the quasi-regular bound function $\beta_{x_0}$. Let $\eta > 0$ and $\alpha_0(r)$ be given in (3.46) and (3.84). Let $0 < R < R_{x_0}$, $\alpha \ge \alpha_0(R)$*

be a positive constant and let $r_\alpha^\star$, $r_\alpha^{\star\star}$ be as defined in Lemma 3.6.1. Suppose that $F'$ satisfies the majorant condition on $U(x_0, r_\alpha^\star)$, conditions

$$\eta \leq \min\{r_\alpha^\star, \Delta\} \quad and \quad r_\alpha^\star \leq R \tag{3.87}$$

hold. Then, sequence $\{x_n\}$ generated by (GNA) is well defined, remains in $\overline{U}(x_0, r_\alpha^\star)$ for all $n \geq 0$ and converges to some $x^\star$ such that $F(x^\star) \in \mathcal{C}$. Moreover, the following estimates hold for each $n = 1, 2, \ldots$

$$\| x_n - x_{n-1} \| \leq t_{\alpha,n} - t_{\alpha,n-1}, \tag{3.88}$$

$$\| x_{n+1} - x_n \| \leq (t_{\alpha,n+1} - t_{\alpha,n}) \left( \frac{\| x_n - x_{n-1} \|}{t_{\alpha,n} - t_{\alpha,n-1}} \right)^2, \tag{3.89}$$

$$F(x_n) + F'(x_n)(x_{n+1} - x_n) \in \mathcal{C} \tag{3.90}$$

and

$$\| x_{n-1} - x^\star \| \leq r_\alpha^\star - t_{\alpha,n-1}, \tag{3.91}$$

where sequence $\{t_{\alpha,n}\}$ is given by (3.83).

*Proof* We use the same notations as in Theorem 3.5.1. We follow the proof of Theorem 3.5.1 until (3.58). Then, using (3.55), (3.82) (for $G = F'$), (3.83), (3.84) and hypothesis $\alpha \geq \alpha_0(R)$, we get in turn that

$$
\begin{aligned}
\xi \, d(0, \mathrm{d}(x_k)) &\leq \xi \, \beta_{x_0}(\| x_k - x_0 \|) \, d(F(x_k), \mathcal{C}) \\
&\leq \xi \, \beta_{x_0}(\| x_k - x_0 \|) \, \| F(x_k) - F(x_{k-1}) - F'(x_{k-1})(x_k - x_{k-1}) \| \\
&\leq \xi \, \beta_{x_0}(\| x_k - x_0 \|) \int_0^1 \| (F'(x_k^\theta) - F'(x_{k-1}))(x_k - x_{k-1}) \, d\theta \| \\
&\leq \xi \, \frac{\beta_{x_0}(t_{\alpha,k})}{\alpha_0(R)} \int_0^1 (f_\alpha'(t_{\alpha,k}^\theta) - f_\alpha'(t_{\alpha,k-1}))(t_{\alpha,k} - t_{\alpha,k-1}) \, d\theta \\
&\leq \xi \, \frac{\beta_{x_0}(t_{\alpha,k})}{\alpha} \int_0^1 (f_\alpha'(t_{\alpha,k}^\theta) - f_\alpha'(t_{\alpha,k-1}))(t_{\alpha,k} - t_{\alpha,k-1}) \, d\theta \\
&\leq -f_\alpha'(t_{\alpha,k})^{-1} (f_\alpha(t_{\alpha,k}) - f_\alpha(t_{\alpha,k-1}) - f_\alpha'(t_{\alpha,k-1})(t_{\alpha,k} - t_{\alpha,k-1})) \\
&= -f_\alpha'(t_{\alpha,k}) f_\alpha(t_{\alpha,k}),
\end{aligned}
\tag{3.92}
$$

where $t_{\alpha,k}^\theta = \theta \, t_{\alpha,k} + (1 - \theta)(t_{\alpha,k} - t_{\alpha,k-1})$ for all $\theta \in [0, 1]$. The rest follows as in the proof of Theorem 3.5.1. That completes the proof of Theorem 3.6.1.

*Remark 3.6.1* In view of condition (3.82), there exists $f_{\alpha,0} : [0, r) \to (-\infty, +\infty)$ continuously differentiable such that

$$\| G(x) - G(x_0) \| \leq \alpha^{-1} (f_{\alpha,0}'(\| x - x_0 \|) - f_{\alpha,0}'(0)) \quad for \ all \quad x \in U(x_0, r), r \leq R. \tag{3.93}$$

Moreover

$$f'_{\alpha,0}(t) \le f'_\alpha(t) \quad \text{for all} \quad t \in [0, r] \tag{3.94}$$

holds in general and $\dfrac{f'_\alpha}{f'_{\alpha,0}}$ can be arbitrarily large (see, e.g., [1–19]). These observations motivate us to introduce tighter majorizing sequences $\{r_{\alpha,n}\}$, $\{s_{\alpha,n}\}$ by

$$
\begin{aligned}
&r_{\alpha,0} = 0, \quad r_{\alpha,1} = \eta = -\frac{f_\alpha(0)}{f'_\alpha(0)}, \\
&r_{\alpha,2} = r_{\alpha,1} - \frac{\alpha\,(f_{\alpha,0}(r_{\alpha,1}) - f_{\alpha,0}(r_{\alpha,0}) - f'_{\alpha,0}(r_{\alpha,0})(r_{\alpha,1} - r_{\alpha,0}))}{f'_{\alpha,0}(r_{\alpha,1})}, \\
&r_{\alpha,n+1} = r_{\alpha,n} - \frac{\displaystyle\int_0^1 (f'_\alpha(r^\theta_{\alpha,k}) - f'_\alpha(r_{\alpha,k-1}))\,(r_{\alpha,k} - r_{\alpha,k-1})\,d\theta}{f'_{\alpha,0}(r_{\alpha,n})}
\end{aligned}
\tag{3.95}
$$

for each $n = 2, 3, \ldots$

and

$$
\begin{aligned}
&s_{\alpha,0} = 0, \quad s_{\alpha,1} = r_{\alpha,1}, \\
&s_{\alpha,n+1} = s_{\alpha,n} - \frac{\displaystyle\int_0^1 (f'_\alpha(s^\theta_{\alpha,k}) - f'_\alpha(s_{\alpha,k-1}))\,(s_{\alpha,k} - s_{\alpha,k-1})\,d\theta}{f'_{\alpha,0}(s_{\alpha,n})}
\end{aligned}
\tag{3.96}
$$

for each $n = 0, 1, \ldots$

Then, in view of the proof of Theorem 3.6.1, $\{r_{\alpha,n}\}$, $\{s_{\alpha,n}\}$, $\lim_{n \to \infty} r_{\alpha,n} = r^{**}_\alpha$, $\lim_{n \to \infty} s_{\alpha,n} = s^*_\alpha$ can replace $\{t_{\alpha,n}\}$, $\{t_{\alpha,n}\}$, $r^*_{\alpha,n}$, $r^*_{\alpha,n}$, respectively in estimates (3.88) and (3.91), in the proof of Lemma 3.3.5,

$$r_{\alpha,n} \le s_{\alpha,n} \le t_{\alpha,n}$$

$$r_{\alpha,n+1} - r_{\alpha,n} \le s_{\alpha,n+1} - s_{\alpha,n} \le t_{\alpha,n+1} - t_{\alpha,n}$$

and

$$r^{**}_\alpha \le s^*_\alpha \le r^*_\alpha,$$

provided that

$$-\frac{(f'_\alpha(u + \theta(v - u)) - f'_\alpha(u))(v - u)}{f'_{\alpha,0}(v)} \le -\frac{(f'_\alpha(\overline{u} + \theta(\overline{v} - \overline{u})) - f'_\alpha(\overline{u}))(\overline{v} - \overline{u})}{f'_\alpha(\overline{v})}$$

for each $u \leq v$, $u \leq \bar{u}$, $\bar{u} \leq \bar{v}$, $v \leq \bar{v}$, $u, \bar{u}, v, \bar{v} \in [\eta, R]$ for each $\theta \in [0, 1]$. Notice that the preceding estimate holds in many interesting cases. Choose, e.g. $f_\alpha(t) = \frac{\alpha L t^2}{2} - t$ and $f_{\alpha,0}(t) = \frac{\alpha L_0}{2} t^2 - t$. Then, $f_\alpha, f_{\alpha,0}$ are the majorant and center majorant functions for $F$.

## 3.7 Applications

We shall provide in this section some numerical examples to validate the semilocal convergence results of (GNA).

First, we present an example to show that $L_0(t) < L(t)$ for each $u \in [0, R]$ and that hypotheses of Theorem 3.5.1 are satisfied. Then, according to Lemma 3.3.5, new sequence $\{s_{\alpha,n}\}$ is tighter than $\{t_{\alpha,n}\}$ given in [29] so that estimates (3.29)–(3.31) are satisfied.

*Example 3.7.1* Let function $h : \mathbb{R} \to \mathbb{R}$ be defined by

$$h(x) = \begin{cases} 0 \text{ if } x \leq 0 \\ x \text{ if } x \geq 0. \end{cases}$$

Define function $F$ by

$$F(x) = \begin{cases} \lambda - x + \dfrac{1}{18}x^3 + \dfrac{x^2}{1-x} & \text{if } x \leq \dfrac{1}{2} \\ \lambda - \dfrac{71}{144} + 2x^2 & \text{if } x \geq \dfrac{1}{2}, \end{cases} \tag{3.97}$$

where $\lambda > 0$ is a constant. Then, we have that $\mathcal{C} = (-\infty, 0]$,

$$F'(x) = \begin{cases} -2 + \dfrac{1}{(1-x)^2} + \dfrac{x^2}{6} & \text{if } x \leq \dfrac{1}{2} \\ 4x & \text{if } x \geq \dfrac{1}{2} \end{cases} \tag{3.98}$$

and

$$F''(x) = \begin{cases} \dfrac{2}{(1-x)^2} + \dfrac{x}{3} & \text{if } x \leq \dfrac{1}{2} \\ 4 & \text{if } x \geq \dfrac{1}{2}. \end{cases} \tag{3.99}$$

We shall first show that $F'$ satisfies the $L$-average Lipschitz condition on $U(0, 1)$, where

$$L(u) = \frac{2}{(1-u)^3} + \frac{1}{6} \quad \text{for each } u \in [0, 1) \tag{3.100}$$

and the $L_0$-average condition on $U(0, 1)$, where

$$L_0(u) = \frac{2}{(1-u)^3} + \frac{1}{12} \quad \text{for each } u \in [0, 1). \tag{3.101}$$

It follows from (3.99) that

$$L(u) < L(v) \quad \text{for each} \quad 0 \le u < v < 1 \tag{3.102}$$

and

$$0 < F''(u) < F''(|u|) < L(|u|) \quad \text{for each} \quad \frac{1}{2} \ne u < 1. \tag{3.103}$$

Let $x, y \in U(0, 1)$ with $|y| + |x - y| < 1$. Then, it follows from (3.102) and (3.103) that

$$\begin{aligned}
|F'(x) - F'(y)| &\le |x - y| \int_0^1 F''(y + t(x - y)) \, dt \\
&\le |x - y| \int_0^1 L(|y| + t|x - y|) \, dt.
\end{aligned} \tag{3.104}$$

Hence, $F'$ satisfies the $L$-average Lipschitz condition on $U(0, 1)$. Similarly, using (3.98) and (3.101), we deduce that $F'$ satisfies the center $L_0$-average condition on $U(0, 1)$. Notice that

$$L_0(u) < L(u) \quad \text{for each} \quad u \in [0, 1). \tag{3.105}$$

We also have $1 = \| T_{x_0}^{-1} \| = \beta_{x_0}(0)$ and $T_{x_0}$ carries $\mathbb{R}$ into $\mathbb{R}$ as $F'(x_0) = -1$. We have that $\xi = 1$, $\alpha = 1$, $F(x_0) = F(0) = \lambda$ and by (46) that

$$\eta = \| T_{x_0}^{-1} \| \, d(F(x_0), C) = \lambda.$$

Let us choose $\lambda = 0.05$. By Maple 13 and using estimates (3.11) and (3.16) (see also Fig. 3.1), we have that

$$\psi_\alpha(t) = \eta - t + \alpha \left( \frac{1}{12} t^2 - \frac{1}{t-1} - t - 1 \right),$$

$$\psi_{\alpha,0}(t) = \eta - t + \alpha \left( \frac{1}{24} t^2 - \frac{1}{t-1} - t - 1 \right),$$

$$r_\alpha^\star = 0.05322869296 \quad \text{and} \quad r_{\alpha,0}^\star = 0.05309455977.$$

Comparison Table 3.1 show that our error bounds $s_{\alpha,n+1} - s_{\alpha,n}$ are finer than $t_{\alpha,n+1} - t_{\alpha,n}$ given in [29].

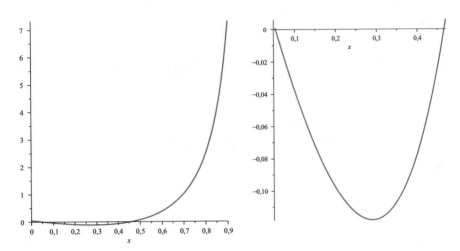

**Fig. 3.1** *Left*: function $\psi_\alpha$ on interval $[0, 0.9)$. *Right*: function $\psi_{\alpha,0}$ on interval $[0.052, 0.47)$

**Table 3.1** Comparison table of majorizing sequences for $\alpha = 1$

| $n$ | $s_{\alpha,n}$ | $s_{\alpha,n+1} - s_{\alpha,n}$ | $t_{\alpha,n}$ | $t_{\alpha,n+1} - t_{\alpha,n}$ |
|---|---|---|---|---|
| 0 | 0 | 0.05 | 0 | 0.05 |
| 1 | 0.05 | 0.00308148876 | 0.05 | 0.00321390287 |
| 2 | 0.05308148876 | 0.00001307052 | 0.05321390287 | 0.00001479064 |
| 3 | 0.05309455928 | | 0.05322869351 | |

In the rest of the examples $L_0$ and $L$ are positive constant functions. In Example 3.7.2 we show that $L/L_0$ can be arbitrarily large.

*Example 3.7.2* Let $x_0 = 0$. Define the scalar function $F$ by $F(x) = d_0 x + d_1 + d_2 \sin e^{d_3 x}$, where $d_i$, $i = 0, 1, 2, 3$ are given parameters. It can easily be seen that for $d_3$ large and $d_2$ sufficiently small, $\dfrac{L}{L_0}$ can be arbitrarily large.

Next, we present three examples in the more general setting of a Banach space. That is we consider equation $F(x) = 0$, where $F$ is a Fréchet-differentiable operator defined on an open convex subset $\Omega$ of a Banach space $\mathcal{X}$ with values in a Banach space $\mathcal{Y}$. In Example 3.7.3 we show that our condition (3.71) holds but not the Kantorovich condition [see also (3.69)]. Note that the earlier results [29, 33, 34, 39, 40] reduce to the Kantorovich condition.

*Example 3.7.3* We consider a simple example in one dimension to test conditions (3.69)–(3.71). Let $X = \mathbb{R}$, $x_0 = 1$, $\Omega = [d, 2 - d]$, $d \in [0, 0.5)$. Define function $F$ on $\Omega$ by $F(x) = x^3 - d$. We get that $\eta = (1/3)(1 - d)$ and $L = 2(2 - d)$. Kantorovich's condition [see also (3.69)] is given by $h = (2/3)(1-d)(2-d) > 0.5$ for all $d \in (0, 0.5)$. Hence, there is no guarantee that Newton's method starting at $x_0 = 1$ converges to $x^\star$. However, one can easily see that if for example $d = 0.49$,

Newton's method converges to $x^\star = \sqrt[3]{0.49}$. We can deduce the center-Lipschitz condition by

$$L_0 = 3 - d < L = 2\,(2 - d) \quad \text{for all} \quad d \in (0, 0.5). \tag{3.106}$$

We consider conditions (3.70) and (3.71). Then, we obtain that

$$h_1 = \frac{1}{12}\,(8 - 3\,d + (5\,d^2 - 24\,d + 28)^{1/2})\,(1 - d) \leq 0.5 \text{ for all } d \in [0.450339002, 0.5)$$

and

$$h_2 = \frac{1}{24}\,(1 - d)\,(12 - 4\,d + (84 - 58\,d + 10\,d^2)^{1/2} + (12 - 10\,d + 2\,d^2)^{1/2}) \leq .5$$
for all $\quad d \in [0.4271907643, 0.5)$.

In Fig. 3.2, we compare conditions (3.69)–(3.71) for $d \in (0, 0.999)$.

*Example 3.7.4* Let $\mathcal{X} = \mathcal{Y} = \mathcal{C}[0, 1]$, equipped with the max-norm. Let $\theta \in [0, 1]$ be a given parameter. Consider the "cubic" integral equation

$$u(s) = u^3(s) + \lambda\,u(s) \int_0^1 \mathcal{K}(s, t)\,u(t)\,dt + y(s) - \theta. \tag{3.107}$$

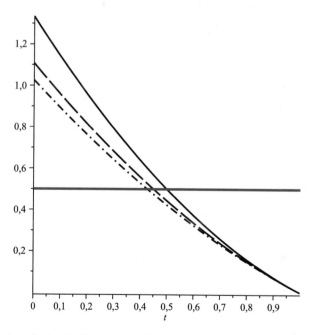

**Fig. 3.2** Functions $h$, $h_1$, $h_2$ (from *top* to *bottom*) with respect to $d$ in interval $(0, 0.999)$, respectively. The *horizontal blue line* is of equation $y = 0.5$

Nonlinear integral equations of the form (3.107) are considered Chandrasekhar-type equations (see, e.g., [18]) and they arise in the theories of radiative transfer, neutron transport and in the kinetic theory of gasses. Here, the kernel $K(s, t)$ is a continuous function of two variables $(s, t) \in [0, 1] \times [0, 1]$ satisfying

(i)  $0 < K(s, t) < 1$,
(ii) $K(s, t) + K(t, s) = 1$.

The parameter $\lambda$ is a real number called the "albedo" for scattering; $y(s)$ is a given continuous function defined on $[0, 1]$ and $x(s)$ is the unknown function sought in $C[0, 1]$. For simplicity, we choose

$$u_0(s) = y(s) = 1 \text{ and } K(s, t) = \frac{s}{s + t} \text{ for all } (s, t) \in [0, 1] \times [0, 1] \ (s + t \neq 0).$$

Let $d = U(u_0, 1 - \theta)$ and define the operator $F$ on d by

$$F(x)(s) = x^3(s) - x(s) + \lambda x(s) \int_0^1 K(s, t) x(t) \, dt + y(s) - \theta \text{ for all } s \in [0, 1].$$
$$(3.108)$$

Then every zero of $F$ satisfies Eq. (3.107). We obtain using   (3.108) and [18, Chap. 1] that

$$[F'(x) v] (s) = \lambda x(s) \int_0^1 K(s, t) v(t) \, dt + \lambda v(s) \int_0^1 K(s, t) x(t) \, dt +$$
$$3 x^2(s) v(s) - I(v(s)).$$

Therefore, the operator $F'$ satisfies the Lipschitz conditions, with

$$\eta = \frac{|\lambda| \ln 2 + 1 - \theta}{2 (1 + |\lambda| \ln 2)}, \quad L = \frac{|\lambda| \ln 2 + 3 (2 - \theta)}{1 + |\lambda| \ln 2} \text{ and } L_0 = \frac{2 |\lambda| \ln 2 + 3 (3 - \theta)}{2 (1 + |\lambda| \ln 2)}.$$

It follows from our main results that if one of conditions (3.69) or (3.70) or (3.71) holds, then problem (3.107) has a unique solution near $u_0$. This assumption is weaker than the one given before using the Newton-Kantorovich hypothesis. Note also that $L_0 < L$ for all $\theta \in [0, 1]$ (see also Fig. 3.3).

Next, we pick some values of $\lambda$ and $\theta$ such that all hypotheses are satisfied, so we can compare conditions (3.69) or (3.70) or (3.71) (see Table 3.2).

*Example 3.7.5* Let $\mathcal{X}$ and $\mathcal{Y}$ as in Example 3.7.4. Consider the following nonlinear boundary value problem (see, e.g., [18])

$$\begin{cases} u'' = -u^3 - \gamma u^2 \\ u(0) = 0, \quad u(1) = 1. \end{cases}$$

**Fig. 3.3** Functions $L_0$ and $L$ in 3d with respect to $(\lambda, \theta)$ in $(-3, 3) \times (0, 1)$. $L$ is above $L_0$

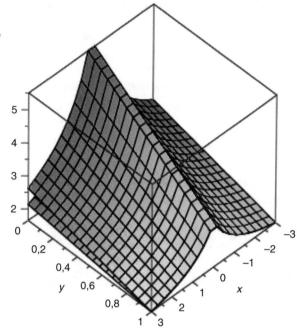

**Table 3.2** Comparison table of conditions (3.69), (3.70) and (3.71) for problem (3.108)

| $\lambda$ | $\theta$ | $h$ | $h_1$ | $h_2$ |
|---|---|---|---|---|
| 0.97548 | 0.954585 | 0.4895734208 | 0.4851994045 | 0.4837355633 |
| 0.8457858 | 0.999987 | 0.4177974405 | 0.4177963046 | 0.4177959260 |
| 0.3245894 | 0.815456854 | 0.5156159025 | 0.4967293568 | 0.4903278739 |
| 0.3569994 | 0.8198589998 | 0.5204140737 | 0.5018519741 | 0.4955632842 |
| 0.3789994 | 0.8198589998 | 0.5281518448 | 0.5093892893 | 0.5030331107 |
| 0.458785 | 0.5489756 | 1.033941504 | 0.9590659445 | 0.9332478337 |

It is well known that this problem can be formulated as the integral equation (see, e.g., [18, Chap. 1])

$$u(s) = s + \int_0^1 \mathcal{Q}(s, t) \left( u^3(t) + \gamma\, u^2(t) \right) dt \qquad (3.109)$$

where $\mathcal{Q}$ is the Green function given by

$$\mathcal{Q}(s, t) = \begin{cases} t(1 - s), & t \le s \\ s(1 - t), & s < t. \end{cases}$$

Then problem (3.109) is in the form $F(x) = 0$, where $F : \mathrm{d} \to \mathcal{Y}$ is defined as

$$[F(x)](s) = x(s) - s - \int_0^1 \mathcal{Q}(s,t) (x^3(t) + \gamma\, x^2(t))\, dt.$$

Set $u_0(s) = s$ and $\mathrm{d} = U(u_0, R_0)$. The Fréchet derivative of $F$ is given by (see, e.g., [18, Chap. 1])

$$[F'(x)v](s) = v(s) - \int_0^1 \mathcal{Q}(s,t) (3\,x^2(t) + 2\,\gamma\, x(t))\, v(t)\, dt.$$

It is easy to verify that $U(u_0, R_0) \subset U(0, R_0 + 1)$ since $\| u_0 \| = 1$. If $2\,\gamma < 5$, the operator $F'$ satisfies the Lipschitz conditions, with

$$\eta = \frac{1+\gamma}{5-2\gamma}, \quad L = \frac{\gamma + 6R_0 + 3}{4(5-2\gamma)} \quad \text{and} \quad L_0 = \frac{2\gamma + 3R_0 + 6}{8(5-2\gamma)}.$$

Note that $L_0 < L$ (see also Fig. 3.4).

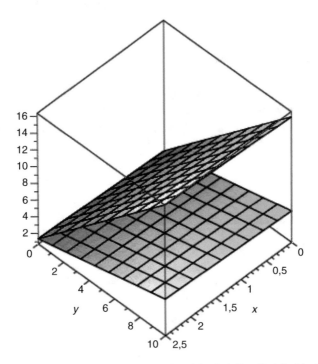

**Fig. 3.4** Functions $L_0$ and $L$ in 3d with respect to $(\gamma, R_0)$ in $(0, 2.5) \times (0, 10)$. $L$ is also above $L_0$

**Table 3.3** Comparison table of conditions (3.69), (3.70) and (3.71) for problem (3.109)

| $\gamma$ | $R_0$ | $h$ | $h_1$ | $h_2$ |
|---|---|---|---|---|
| 0.00025 | 1 | 0.4501700201 | 0.3376306412 | 0.2946446274 |
| 0.25 | 0.986587 | 0.6367723612 | 0.4826181423 | 0.4240511567 |
| 0.358979 | 0.986587 | 0.7361726023 | 0.5600481163 | 0.4932612622 |
| 0.358979 | 1.5698564 | 1.013838328 | 0.7335891949 | 0.6245310288 |
| 0.341378 | 1.7698764 | 1.084400750 | 0.7750792917 | 0.6539239239 |

Next, we pick some values of $\gamma$ and $R_0$ such that conditions (3.69)–(3.71) are satisfied, so we can compare these conditions (see Table 3.3).

*Example 3.7.6* Let also $\mathcal{X} = \mathcal{Y} = C[0, 1]$ as in Example 3.7.4 and $d = U(0, r)$ for some $r > 1$. Define $F$ on d by (see, e.g., [18, Chap. 1])

$$F(x)(s) = x(s) - y(s) - \mu \int_0^1 \mathcal{Q}(s, t) x^3(t) \, dt, \quad x \in C[0, 1], \ s \in [0, 1]. \quad (3.110)$$

$y \in C[0, 1]$ is given, $\mu$ is a real parameter and the Kernel $\mathcal{Q}$ is the Green's function defined as in Example 3.7.5. Then, the Fréchet-derivative of $F$ is defined by

$$(F'(x)(w))(s) = w(s) - 3\mu \int_0^1 \mathcal{Q}(s, t) x^2(t) y(t) \, dt, \quad w \in C[0, 1], \ s \in [0, 1].$$

Let us choose $x_0(s) = y(s) = 1$ and $|\mu| < 8/3$. Then, we have that (see, e.g., [18, Chap. 1])

$$\| \mathcal{I} - F'(x_0) \| < \frac{3}{8} |\mu|, \quad F'(x_0)^{-1} \in \mathcal{L}(\mathcal{Y}, \mathcal{X}),$$

$$\| F'(x_0)^{-1} \| \leq \frac{8}{8 - 3|\mu|}, \quad \eta = \frac{|\mu|}{8 - 3|\mu|}, \quad L_0 = \frac{12|\mu|}{8 - 3|\mu|},$$

$$L = \frac{6r|\mu|}{8 - 3|\mu|} \quad \text{and} \quad h = \frac{6r|\mu|^2}{(8 - 3|\mu|)^2}.$$

In Table 3.4, we compare as the last examples conditions (3.69)–(3.71). Hence, Table 3.4 shows that our new condition $h_2$ is always better than the Newton–Kantorovich conditions $h$ (see the first and third columns of Table 3.4).

Next, we present the following specialization of Theorem 3.5.1 under the Robinson condition. For simplicity we shall drop "$\alpha$" from the definition of the majorizing sequences.

**Table 3.4** Comparison table of conditions (3.69), (3.70) and (3.71) for problem (3.110) with $\mu = 1$

| $h$ | $h_1$ | $h_2$ |
|---|---|---|
| 1.007515200 | 0.9837576000 | 0.9757536188 |
| 1.055505600 | 1.007752800 | 0.9915015816 |
| 1.102065600 | 1.031032800 | 1.006635036 |
| 1.485824160 | 1.222912080 | 1.127023800 |

**Theorem 3.7.1** *Let* $F : U(x_0, R) \to \mathbb{R}^m$ *be continuously differentiable. Suppose: there exist positive constants* $L_0$ *and* $L$ *such that*

$$\|F'(x) - F'(x_0)\| \le L_0 \|x - x_0\| \quad \text{for each} \quad x \in U(x_0, R),$$
$$\|F'(x) - F'(y)\| \le L \|x - y\| \quad \text{for each} \quad x, y \in U(x_0, R),$$
$$\delta_0 = \alpha l_0 \xi \le \frac{1}{2}$$

*and*

$$\alpha \ge \frac{\eta b_0}{1 + (\eta - 1) L_0 b_0 \xi}$$

*where*

$$l_0 = \frac{1}{8}\left(4L + \sqrt{L_0 L + 8L^2} + \sqrt{L_0 L}\right).$$

*Then,*

*(a) scalar sequence* $\{q_k\}$ *defined by*

$$q_0 = 0, \quad q_1 = \xi, \quad q_2 = q_1 - \frac{\alpha L_0 (q_1 - q_0)^2}{2(1 - \alpha L_0 q_1)}$$
$$q_{k+1} = q_k - \frac{\alpha L(q_k - q_{k-1})^2}{2(1 - \alpha L_0 q_k)} \quad \text{for each} \quad k = 2, 3, \ldots$$

*is increasingly convergent to its unique least upper bound* $q^*$.
*(b) Sequence* $\{x_k\}$ *generated by (GNA) is well defined, remains in* $U(x_0, q^*)$ *for each* $k = 0, 1, 2, \ldots$ *and converges to a limit point* $x^* \in \overline{U}(x_0, q^*)$ *satisfying* $F(x^*) \in C$. *Moreover, the following estimates hold for each* $k = 0, 1, 2, \ldots$

$$\|x_{k+1} - x_k\| \le q_{k+1} - q_k$$

*and*

$$\|x_k - x^*\| \le q^* - q_k.$$

Notice that we can choose $f_{\alpha,0}(t) = \frac{\alpha L_0 t^2}{2} - t$ and $f_\alpha(t) = \frac{\alpha L t^2}{2} - t$.

*Remark 3.7.1*

(a) In particular, if $C = \{0\}$ and $n = m$, the Robinson condition is equivalent to the condition that $F'(x_0)^{-1}$ is non-singular. Hence, for $\eta = 1$ we obtain the semilocal convergence for Newton's method defined by

$$x_{k+1} = x_k - F'(x_k)^{-1} F(x_k) \quad \text{for each} \quad k = 0, 1, 2, \ldots$$

under the Lipschitz condition [6, 19, 28]. However, the convergence condition in [6, 19, 30, 31, 35] given by

$$\delta = \alpha L \xi \leq \frac{1}{2}.$$

Notice again that

$$l_0 \leq L$$

holds in general and $\frac{L}{l_0}$ can be arbitrarily large. Moreover, the corresponding majorizing sequence $\{t_k\}$ is defined by

$$t_0 = 0, \quad t_1 = \xi, \quad t_{k+1} = t_k - \frac{\alpha L(t_k - t_{k-1})^2}{2(1 - \alpha L t_k)} \quad \text{for each} \quad k = 1, 2, \ldots$$

Then, we have for $l_0 < L$ (i. e. for $L_0 < L$) that

$$q_k < t_k \quad \text{for each} \quad k = 2, 3, \ldots$$

$$q_{k+1} - q_k < t_{k+1} - t_k \quad \text{for each} \quad k = 1, 2, \ldots$$

and

$$q^* \leq t^*.$$

Finally, notice that

$$\delta \leq \frac{1}{2} \Rightarrow \delta_0 \leq \frac{1}{2}$$

(but not necessarily viceversa unless if $L_0 = L$) and

$$\frac{\delta_0}{\delta} \to 0 \quad \text{as} \quad \frac{L_0}{L} \to 0.$$

The preceding estimate shows by how many times at most the applicability of (GNA) is expanded under our new technique.

(b) If $n \neq m$ notice also that if $L_0 < L$ the $\alpha$ given in the preceding result is larger than the old one using $L$ instead of $L_0$. Clearly the rest of the advantages stated in (a) also hold in the setting.

Notice that the inequalities at the end of Remark 3.6.1 are satisfied for $L_0 \leq L$.

Hence, the applicability of Newton's method or (GNA) under the Robinson condition is expanded under the same computational cost, since in practice the computation of constant $L$ requires the computation of $L_0$ as a special case.

## 3.8 Conclusion

Using a combination of average and center-average type conditions, we presented a semilocal convergence analysis for (GNA) to approximate a locally solution of a convex composite optimization problem in finite dimensional spaces setting. Our analysis extends the applicability of (GNA) under the same computational cost as in earlier studies such as [29, 33, 34, 39, 40].

**Acknowledgements** This scientific work has been supported by the 'Proyecto Prometeo' of the Ministry of Higher Education Science, Technology and Innovation of the Republic of Ecuador.

## References

1. Argyros, I.K.: On the Newton-Kantorovich hypothesis for solving equations. J. Comput. Appl. Math. **169**, 315–332 (2004)
2. Argyros, I.K.: A unifying local-semilocal convergence analysis and applications for two-point Newton-like methods in Banach space. J. Math. Anal. Appl. **298**, 374–397 (2004)
3. Argyros, I.K.: On the semilocal convergence of the Gauss-Newton method. Adv. Nonlinear Var. Inequal. **8**, 93–99 (2005)
4. Argyros, I.K.: Approximating solutions of equations using Newton's method with a modified Newton's method iterate as a starting point. Rev. Anal. Numér. Théor. Approx. **36**, 123–138 (2007)
5. Argyros, I.K.: Concerning the semilocal convergence of Newton's method and convex majorants. Rend. Circ. Mat. Palermo **57**, 331–341 (2008)
6. Argyros, I.K.: Convergence and Applications of Newton-Type Iterations. Springer, New York (2009)
7. Argyros, I.K.: Concerning the convergence of Newton's method and quadratic majorants, J. Appl. Math. Comput. **29**, 391–400 (2009)
8. Argyros, I.K.: On a class of Newton-like methods for solving nonlinear equations. J. Comput. Appl. Math. **228**, 115–122 (2009)
9. Argyros, I.K.: Local convergence of Newton's method using Kantorovich's convex majorants. Revue Anal. Numér. Théor. Approx. **39**, 97–106 (2010)
10. Argyros, I.K.: A semilocal convergence analysis for directional Newton methods. Math. Comput. **80**, 327–343 (2011)
11. Argyros, I.K., Hilout, S.: Extending the Newton-Kantorovich hypothesis for solving equations. J. Comput. Appl. Math. **234**, 2993–3006 (2010)

12. Argyros, I.K., Hilout, S.: Improved generalized differentiability conditions for Newton-like methods. J. Complex. **26**, 316–333 (2010)
13. Argyros, I.K., Hilout, S.: On the solution of systems of equations with constant rank derivatives. Numer. Algorithms **57**, 235–253 (2011)
14. Argyros, I.K., Hilout, S.: Extending the applicability of the Gauss-Newton method under average Lipschitz-type conditions. Numer. Algorithms **58**, 23–52 (2011)
15. Argyros, I.K., Hilout, S.: On the semilocal convergence of Newton's method using majorants and recurrent functions. J. Nonlinear Funct. Anal. Appl. **1**, 254–267 (2016)
16. Argyros, I.K., Hilout, S.: Improved local convergence of Newton's method under weak majorant condition. J. Comput. Appl. Math. **236**, 1892–1902 (2012)
17. Argyros, I.K., Hilout, S.: Weaker conditions for the convergence of Newton's method. J. Complex. **28**, 364–387 (2012)
18. Argyros, I.K., Hilout, S.: Computational Methods in Nonlinear Analysis. Efficient Algorithms, Fixed Point Theory and Applications. World Scientific, Singapore (2013)
19. Argyros, I.K., Cho, Y.J., Hilout, S.: Numerical Methods for Equations and Its Applications. CRC Press/Taylor and Francis Group, New York (2012)
20. Burke, J.V., Ferris, M.C.: A Gauss-Newton method for convex composite optimization. Math. Program. Ser. A **71**, 179–194 (1995)
21. Chen, X., Yamamoto, T.: Convergence domains of certain iterative methods for solving nonlinear equations. Numer. Funct. Anal. Optim. **10**, 37–48 (1989)
22. Ferreira, O.P., Svaiter, B.F.: Kantorovich's majorants principle for Newton's method. Comput. Optim. Appl. **42**, 213–229 (2009)
23. Giannessi, F., Mastroeni, G., Pellegrini, L.: On the theory of vector optimization and variational inequalities. Image space analysis and separation. In: Vector Variational Inequalities and Vector Equilibria. Nonconvex Optimization and Its Applications, vol. 38, pp. 153–215. Kluwer Academic Publishers, Dordrecht (2000)
24. Giannessi, F., Moldovan, A., Pellegrini, L.: Metric regular maps and regularity for constrained extremum problems. In: Nonlinear Analysis and Optimization II. Optimization. Contemporary Mathematics, vol. 514, pp. 143–154. American Mathematical Society, Providence, RI (2010)
25. Häubler, W.M.: A Kantorovich-type convergence analysis for the Gauss-Newton method. Numer. Math. **48**, 119–125 (1986)
26. Hiriart-Urruty, J.B, Lemaréchal, C.: Convex Analysis and Minimization Algorithms (two volumes). I. Fundamentals, II. Advanced Theory and Bundle Methods, vols. 305 and 306. Springer, Berlin (1993)
27. Kantorovich, L.V.: The majorant principle and Newton's method. Doklady Akademii Nauk SSSR **76**, 17–20 (1951) [in Russian]
28. Kantorovich, L.V., Akilov, G.P.: Functional Analysis. Pergamon Press, Oxford (1982)
29. Li, C., Ng, K.F.: Majorizing functions and convergence of the Gauss-Newton method for convex composite optimization. SIAM J. Optim. **18**, 613–642 (2007)
30. Li, C., Wang, X.H.: On convergence of the Gauss-Newton method for convex composite optimization. Math. Program. Ser. A **91**, 349–356 (2002)
31. Li, C., Zhang, W.-H., Jin, X.-Q.: Convergence and uniqueness properties of Gauss-Newton's method. Comput. Math. Appl. **47**, 1057–1067 (2004)
32. Li, C., Hu, N., Wang, J.: Convergence bahavior of Gauss-Newton's method and extensions to the Smale point estimate theory. J. Complex. **26**, 268–295 (2010)
33. Moldovan, A., Pellegrini, L.: On regularity for constrained extremum problems. I. Sufficient optimality conditions. J. Optim. Theory Appl. **142**, 147–163 (2009)
34. Moldovan, A., Pellegrini, L.: On regularity for constrained extremum problems. II. Necessary optimality conditions. J. Optim. Theory Appl. **142**, 165–183 (2009)
35. Ng, K.F., Zheng, X.Y.: Characterizations of error bounds for convex multifunctions on Banach spaces. Math. Oper. Res. **29**, 45–63 (2004)
36. Robinson, S.M.: Extension of Newton's method to nonlinear functions with values in a cone. Numer. Math. **19**, 341–347 (1972)

37. Robinson, S.M.: Stability theory for systems of inequalities. I. Linear systems. SIAM J. Numer. Anal. **12**, 754–769 (1975)
38. Rockafellar, R.T.: Convex Analysis, Princeton Mathematical Series, vol. 28. Princeton University Press, Princeton, NJ (1970)
39. Wang, X.H.: Convergence of Newton's method and inverse function theorem in Banach space. Math. Comput. **68**, 169–186 (1999)
40. Wang, X.H.: Convergence of Newton's method and uniqueness of the solution of equations in Banach space. IMA J. Numer. Anal. **20**, 123–134 (2000)
41. Xu, X.B., Li, C.: Convergence of Newton's method for systems of equations with constant rank derivatives. J. Comput. Math. **25**, 705–718 (2007)
42. Xu, X.B., Li, C.: Convergence criterion of Newton's method for singular systems with constant rank derivatives. J. Math. Anal. Appl. **345**, 689–701 (2008)
43. Zabrejko, P.P., Nguen, D.F.: The majorant method in the theory of Newton–Kantorovich approximations and the Pták error estimates. Numer. Funct. Anal. Optim. **9**, 671–684 (1987)

# Chapter 4
# Inexact Newton Methods on Riemannian Manifolds

## I.K. Argyros and Á.A. Magreñán

**Abstract** In this chapter we study of the Inexact Newton Method in order to solve problems on a Riemannian Manifold. We present standard notation and previous results on Riemannian manifolds. A local convergence study is presented and some special cases are also provided.

## 4.1 Introduction

Many problems in the Applied Sciences and other disciplines including engineering, optimization, dynamic economic systems, physics, biological problems can be formulated as equation in abstract spaces using Mathematical Modelling. Moreover, other problems in Computational Science such as linear and nonlinear programming problems, linear semi-definite programming problems, interior point problems, Hamiltonian and gradient flow problems and control problems to mention a few require finding a singularity of a differentiable vector field defined on a complete Riemannian manifold [1–36]. The solution of these equations can rarely be found in closed form. Therefore the solution methods for these equations are usually iterative. Recently, attention has been paid in studying iterative procedures on manifolds, since there are many numerical problems posed on manifolds that arise in many contexts. Examples include eigenvalue problems, minimization problems with orthogonality constraints, optimization problems with equality constraints, invariant subspace computations. For these problems, one has to compute solutions of equations or to find zeros of a vector field on Riemannian manifolds. In Applied Sciences, the practice of Numerical Analysis for finding such solutions is essentially connected to Newton-type method.

I.K. Argyros
Department of Mathematics Sciences, Cameron University, Lawton, OK 73505, USA
e-mail: ioannisa@cameron.edu

Á.A. Magreñán (✉)
Escuela Superior de Ingeniería y Tecnología, Universidad Internacional de La Rioja (UNIR),
Logroño (La Rioja), Spain
e-mail: alberto.magrenan@unir.net

© Springer International Publishing Switzerland 2016
S. Amat, S. Busquier (eds.), *Advances in Iterative Methods for Nonlinear Equations*, SEMA SIMAI Springer Series 10,
DOI 10.1007/978-3-319-39228-8_4

The singularity can be found in closed form only in special cases. That is why most solution methods for finding such singularities are usually iterative. In particular, the practice of Numerical Functional Analysis for finding such singularities is essentially connected to Newton-like methods. The study about convergence matter of iterative procedures in general is usually centered on two types: semilocal and local convergence analysis. The semilocal convergence matter is, based on the information around an initial point, to give criteria ensuring the convergence of iterative procedures; while the local one is based on the information around a solution, to find estimates of the radii of the convergence balls. A plethora of sufficient conditions for the local as well the semilocal convergence of Newton-like methods as well as error analysis for such methods can be found in [1–36].

First, since the singularity-finding problem is indeed a metric-free problem, the choice of a particular Riemannian structure for implementing Newton's method is a strategy among others. The choice of an adequate distance measure is of primary importance, not only because of its dramatic consequences for obtaining good basic estimations, but also for the well-posedness of the method. Let us also observe that the sequence generated by Newton's method in Riemannian manifold may strongly depend on the metric. This contrasts with the case of $\mathbb{R}^n$ viewed as an Euclidean space, a case for which Newton's iterates never depend on the choice of an inner product.

The basic idea of Newton's method is linearization. Starting from an initial guess $x_0 \in D$, we can have the linear approximation of $F(x)$ in the neighborhood $D_0$ of $x_0 (D_0 \subset D) : F(x_0 + w) \approx F(x_0) + F'(x_0)w$ and solve the resulting linear equation $F(x_0) + F'(x_0)w = 0$, leading to the recurrent Newton's method as follows

$$x_{n+1} = x_n - F'(x_n)^{-1} F(x_n) \text{ for each } n = 0, 1, \ldots$$

Here, $F'(x)$ denotes the Fréchet-derivative of $F$ at $x \in D$. Isaac Newton in 1669, inaugurated his method through the use of numerical examples to solve equations (for polynomial only defined on the real line), but did not use the current iterative expression. Later, in 1690, Raphson proposed the usage of Newton's method for general functions F. That is why the method is often called the Newton-Raphson method. In 1818, Fourier proved that the method converges quadratically in a neighborhood of the root, while Cauchy in 1829 and 1847 provided the multidimensional extension of Newton method. Kantorovich in 1948 published an important paper extending Newton's method for functional spaces (the Newton-Kantorovich method). Ever, since thousands of papers have been written in a Banach space setting for the Newton-Kantorovich method as well as Newton-type methods and their applications. Newton's method is currently and undoubtedly the most popular one-point iterative procedure for generating a sequence approximating $x^*$. We refer the reader to [1–36] for more results and the references therein.

Recently, the local convergence analysis of Inexact Newton method with relative residual error tolerance under the majorant condition in a Riemannian manifold setting was given by Bittencourt and Ferreira in [15]. By using the majorant function, they showed that the Inexact Newton method converges $Q$-linearly.

In the present chapter, we show under the same computational cost of the majorant function and the parameters involved as in [15] that the local convergence analysis on the Inexact Newton method can be improved with advantages as already stated in the abstract on this chapter (see also Remark 3.2 (b) that follows). This is the new contribution of the chapter.

The rest of the chapter is organized as follows: the mathematical background is given in Sect. 4.2; the local convergence analysis of the Inexact Newton method in Sect. 4.3. Finally, special cases appear in the concluding Sect. 4.4.

## 4.2 Background

In order to make the chapter as self contained as possible, we briefly present some standard notation and results on the Riemannian manifolds taken from [8–11, 15, 19, 24].

Throughout the chapter, we denote $\mathcal{M}$ as a smooth manifold and $\mathbb{C}^1(\mathcal{M})$ in the class of all continuously differentiable functions on $\mathcal{M}$. Moreover, we denote the space of vector fields on $\mathcal{M}$ by $\mathcal{X}(\mathcal{M})$, the tangent space of $\mathcal{M}$ at $p$ by $T_p\mathcal{M}$ and the tangent bundle of $\mathcal{M}$ by $TM = U_{x \in \mathcal{M}} T_x \mathcal{M}$. Let $\mathcal{M}$ be equipped with a Riemannian metric $< ., . >$ with corresponding norm denoted by $\|.\|$, so that $\mathcal{M}$ is a Riemannian manifold. Let us recall that the metric can be used to define the length of a piecewise $\mathbb{C}^1$ curve $\xi : [a; b] \to \mathcal{M}$ joining $p$ to $q$, i.e., such that $\xi(a) = p$ and $\xi(b) = q$, by $l(\xi) = \int_a^b \|\xi'(t)\| dt$. Minimizing this length functional over the set of all such curves we obtain a distance $d(p, q)$, which induces the original topology on $\mathcal{M}$. We define the open and closed balls of radius $r > 0$ centered at $p$, respectively, as

$$U(p, r) := \{q \in \mathcal{M} : d(p, q) < r\}$$

and

$$\bar{U}(p, r) := \{q \in \mathcal{M} : d(p, q) \leq r\}.$$

Also the metric induces a map $f \in \mathbb{C}^1(\mathcal{M}) \mapsto gradf \in \mathcal{X}(\mathcal{M})$, which associates to each f its gradient via the rule $< gradf, \mathcal{X} > = df(\mathcal{X})$, for all $X \in \mathcal{X}(\mathcal{M})$. The chain rule generalizes to this setting in the usual way: $(f \circ \xi)'(t) = < gradf(\xi(t)), \xi'(t) >$, for all curves $\xi \in \mathbb{C}^1$. Let $\xi$ be a curve joining the points $p$ and $q$ in $\mathcal{M}$ and let $\nabla$ be a Levi-Civita connection associated to $(\mathcal{M}, <, >)$. For each $t \in [a, b]$, $\nabla$ induces an isometry, relative to $<, >$,

$$P_{\xi, a, t} : T_{\xi, (a)}\mathcal{M} \to T_{\xi(t)}\mathcal{M}$$
$$v \mapsto P_{\xi, a, t} v = V(t), \tag{4.1}$$

where $V$ is the unique vector field on $\xi$ such that the following conditions are satisfied

$$\nabla_{\xi'(t)} V(t) = 0$$

and

$$V(a) = v,$$

the so-called parallel translation along $\xi$ from $\xi(a)$ to $\xi(t)$. Note also that

$$P_{\xi,b_1,b_2} \circ P_{\xi,a,b_1} = P_{\xi,a,b_2}, \quad P_{\xi,b,a} = P_{\xi,a,b}^{-1}.$$

A vector field $V$ along $\xi$ is said to be parallel if $\nabla_{\xi'} V = 0$. If $\xi'$ itself is parallel, then we say that $\xi$ is a geodesic. The equation $\nabla_{\xi'} \xi' = 0$ is clearly a second order nonlinear ordinary differential equation, so the geodesic $\xi$ is determined by its position $p$ and velocity $v$ at $p$. It is easy to check that $\|\xi'\|$ is constant. Moreover, if $\|\xi'\| = 1$ we say that $\xi$ is normalized. If the length of a geodesic $\xi : [a, b] \to \mathcal{M}$ is equal the distance of its end points, i.e. $l(\xi) = d(\xi(a)\xi(b))$, is said minimal.

A Riemannian manifold is complete if its geodesics are defined for any values of $t$. The Hopf–Rinow's theorem asserts that if this is the case then any pair of points, say $p$ and $q$, in $\mathcal{M}$ can be joined by a (not necessarily unique) minimal geodesic segment. Moreover, $(\mathcal{M}, d)$ is a complete metric space and bounded and closed subsets are compact.

The exponential map at $p$, $exp_p : T_p\mathcal{M} \to \mathcal{M}$ is defined by $exp_p v = \xi_v(1)$, where $v$ is the geodesic defined by its position $p$ and velocity $v$ at $p$ and $\xi_v(t) = exp_p tv$ for any value of $t$. For $p \in \mathcal{M}$, let

$$r_p := sup\{r > 0 : exp_{p|U(o_p,r)} \text{ is a diffeomorphism}\},$$

where $o_p$ denotes the origin of $T_p\mathcal{M}$ and $U(o_p, r) := \{v \in T_p\mathcal{M} : \|vo_p\| < r\}$. Note that if $0 < \delta < r_p$ then $exp_p U(o_p, \delta) = U(p, \delta)$. The number $r_p$ is called the injectivity radius of $\mathcal{M}$ at $p$.

**Definition 4.2.1** Let $p \in \mathcal{M}$ and $r_p$ the radius of injectivity at $p$. Define the quantity

$$K_p := sup\{\frac{d(exp_q u, exp_q v)}{\|u - v\|} : q \in U(p, r_p), u, v \in T_q\mathcal{M}, u \neq v, \|v\| \leq r_p,$$

$$\|u - v\| \leq r_p\}.$$

Let $\mathcal{M}$ be a complete connected $m$-dimensional Riemannian manifold with the Levi-Civita connection $\nabla$ on $\mathcal{M}$. Let $p \in \mathcal{M}$, and let $T_p\mathcal{M}$ denote the tangent space at $p$ to $\mathcal{M}$. Let $< \cdot, \cdot >$ be the scalar product on $T_p\mathcal{M}$ with the associated norm $\| \cdot \|_p$, where the subscript $p$ is sometimes omitted. For any two distinct elements $p, q \in \mathcal{M}$, let $c : [0, 1] \to \mathcal{M}$ be a piecewise smooth curve connecting $p$ and $q$. Then the arc-length of $c$ is defined by $l(c) := \int_0^1 \|c'(t)\| dt$, and the Riemannian distance from $p$ to $q$ by $d(p,q) := \inf_c l(c)$, where the infimum is taken over all piecewise smooth curves $c : [0,1] \to \mathcal{M}$ connecting $p$ and $q$. Thus, by the Hopf–Rinow Theorem (see [18]), $(\mathcal{M},d)$ is a complete metric space and the exponential map at $p$, $\exp_p : T_p\mathcal{M} \to \mathcal{M}$ is well-defined on $T_p\mathcal{M}$.

Recall that a geodesic $c$ in $\mathcal{M}$ connecting $p$ and $q$ is called a minimizing geodesic if its arc-length equals its Riemannian distance between $p$ and $q$. Clearly, a curve $c : [0,1] \to \mathcal{M}$ is a minimizing geodesic connecting $p$ and $q$ if and only if there exists a vector $v \in T_p\mathcal{M}$ such that $\|v\| = d(p,q)$ and $c(t) = \exp_p(tv)$ for each $t \in [0,1]$.

Let $c : \mathbb{R} \to \mathcal{M}$ be a $C^\infty$ curve and let $P_{c,\cdot,\cdot}$ denote the parallel transport along $c$, which is defined by

$$P_{c,c(b),c(a)}(v) = V(c(b)), \ \forall a, b \in \mathbb{R} \text{ and } v \in T_{c(a)}M,$$

where $V$ is the unique $C^\infty$ vector field satisfying $\nabla_{c'(t)} V = 0$ and $V(c(a)) = v$. Then, for any $a, b \in \mathbb{R}$, $P_{c,c(b),c(a)}$ is an isometry from $T_{c(a)}\mathcal{M}$ to $T_{c(b)}\mathcal{M}$. Note that, for any $a, b, b_1, b_2 \in \mathbb{R}$,

$$P_{c,c(b_2),c(b_1)} \circ P_{c,c(b_1),c(a)} = P_{c,c(b_2),c(a)} \text{ and } P_{c,c(b),c(a)}^{-1} = P_{c,c(a),c(b)}.$$

In particular, we write $P_{q,p}$ for $P_{c,q,p}$ in the case when $c$ is a minimizing geodesic connecting both points $p$ and $q$. Let $C^1(T\mathcal{M})$ denote the set of all the $C^1$-vector fields on $\mathcal{M}$ and $C^i(\mathcal{M})$ the set of all $C^i$-functions from $\mathcal{M}$ to $\mathbb{R}$ ($i = 0, 1$, where $C^0$-mappings mean continuous mappings), respectively. Let $F : \mathcal{M} \to \mathbb{R}^l$ be a $C^1$ function such that

$$F = (F_1, F_2, \ldots, F_n)$$

with $F_i \in C^1(M)$ for each $i = 1, 2, \ldots, n$. Then, the derivative of $F$ along the vector field $X$ is defined by

$$\nabla_X F = (\nabla_X F_1, \nabla_X F_2, \ldots, \nabla_X F_n) = (X(F_1), X(F_2), \ldots, X(F_n)).$$

Thus, the derivative of $F$ is a mapping $DF : (C^1(T\mathcal{M})) \to (C^0(\mathcal{M}))^n$ defined by

$$DF(X) = \nabla_X F \text{ for each } X \in C^1(T\mathcal{M}). \tag{4.2}$$

We use $DF(p)$ to denote the derivative of $F$ at $p$. Let $v \in T_p\mathcal{M}$. Taking $X \in C^1(T\mathcal{M})$ such that $X(p) = v$, and any nontrivial smooth curve $c : (-\varepsilon, \varepsilon) \to \mathcal{M}$ with $c(0) = p$ and $c'(0) = v$, one has that

$$DF(p)v := DF(X)(p) = \nabla_X F(p) = \left(\frac{d}{dt}(F \circ c)(t)\right)_{t=0}, \tag{4.3}$$

which only depends on the tangent vector $v$.

Let $W$ be a closed convex subset of $\mathbb{R}^l$ (or $\mathbb{R}^m$). The negative polar of $W$ denoted by $W^\ominus$ is defined as

$$W^\ominus = \{z : <z, w> \le 0 \quad \text{for each} \quad w \in W\}. \tag{4.4}$$

*Remark 4.2.1* The quantity $K_p$ measures how fast the geodesics spread apart in $\mathcal{M}$: In particular, when $u = 0$ or more generally when $u$ and $v$ are on the same line through $o_q$,

$$d(exp_q u, exp_q v) = \|u - v\|.$$

So $K_p \ge 1$ for all $p \in \mathcal{M}$. When $\mathcal{M}$ has non-negative sectional curvature, the geodesics spread apart less than the rays ([19], Chap. 5) so that

$$d(exp_q u, exp_q v) \le \|u - v\|.$$

As a consequence $K_p = 1$ for all $p \in \mathcal{M}$. Finally it is worth noticing that radii less than $r_p$ could be used as well (although this would require added notation such as $K_p(p)$ for $r_p$). In this spread apart might decrease, thereby providing slightly stronger results so long as the radius was not too much less than $r_p$.

Let $\mathcal{X}$ be a $C^1$ vector field on $\mathcal{M}$. The covariant derivative of $\mathcal{X}$ determined by the Levi-Civita connection $r$ defines at each $p \in \mathcal{M}$ a linear map $\nabla \mathcal{X}(p) := T_p \mathcal{M} \to T_p \mathcal{M}$ given by

$$\nabla \mathcal{X}(p) v := \nabla_y \mathcal{X}(p), \tag{4.5}$$

where $\mathcal{Y}$ is a vector field such that $\mathcal{Y}(p) = v$.

**Definition 4.2.2** Let $\mathcal{M}$ be a complete Riemannian manifold and $\mathcal{Y}_1, \ldots, \mathcal{Y}_n$ be vector fields on $\mathcal{M}$. Then, the $n$th covariant derivative of $\mathcal{X}$ with respect to $\mathcal{Y}_1, \ldots, \mathcal{Y}_n$ is defined inductively by

$$\nabla^2_{\{\mathcal{Y}_1, \mathcal{Y}_2\}} \mathcal{X} := \nabla_{\mathcal{Y}_2} \nabla_{\mathcal{Y}_1} \mathcal{X}, \quad \nabla^n_{\{\mathcal{Y}_i\}_{i=1}^n} \mathcal{X} := \nabla_{\mathcal{Y}_n} (\nabla_{Y_{n-1}} \cdots \nabla_{Y_1} \mathcal{X}).$$

**Definition 4.2.3** Let $\mathcal{M}$ be a complete Riemannian manifold, and $p \in \mathcal{M}$. Then, the $n$th covariant derivative of $\mathcal{X}$ at $p$ is the $n$th multilinear map $\nabla^n_{\{\mathcal{Y}_i\}_{i=1}^n}(p) : T_p \mathcal{M} \times \cdots \times T_p \mathcal{M}$ defined by

$$\nabla^n \mathcal{X}(p)(v_1, \ldots, v_n) := \nabla^n_{\{\mathcal{Y}_i\}_{i=1}^n} \mathcal{X}(p)$$

where $\mathcal{Y}_1, \ldots, \mathcal{Y}_n$ are vector fields on $\mathcal{M}$ such that $\mathcal{Y}_1(p) = v_1, \ldots, \mathcal{Y}_n(p) = v_n$.

We remark that Definition 23 only depends on the $n$-tuple of vectors $(v_1, \ldots, v_n)$ since the covariant derivative is tensorial in each vector field $\mathcal{Y}_i$.

**Definition 4.2.4** Let $\mathcal{M}$ be a complete Riemannian manifold and $p \in \mathcal{M}$. The norm of an $n$th multilinear map $A.T_p\mathcal{M} \times \cdots T_p\mathcal{M} \to T_p\mathcal{M}$ is defined by

$$\|A\| = sup\{\|A(v_1, \ldots, v_n)\| : v_1, \ldots, v_n \in T_p\mathcal{M}, \|v_i\| = 1, i = 1, \ldots, n\}.$$

In particular the norm of the $n$th covariant derivative of $\mathcal{X}$ at $p$ is given by

$$\|\nabla^n \mathcal{X}(p)\| = sup\{\|\nabla^n \mathcal{X}(p)(v_1, \ldots, v_n)\| : v_1, \ldots, v_n \in T_p\mathcal{M}, \|v_i\| = 1,$$
$$i = 1, \ldots, n\}.$$

**Lemma 4.2.1** Let $D$ be an open subset of $\mathcal{M}$, $\mathcal{X}$ a $C^1$ vector field defined on $D$ and $\xi : [a, b] \to D$ a $C^\infty$ curve. Then

$$P_{\xi,t,a}\mathcal{X}(\xi(t)) = \mathcal{X}(\xi(a)) + \int_a^t P_{\xi,s,a}\nabla\mathcal{X}(\xi(s))\xi'(s)ds, \quad t \in [a, b].$$

*Proof* See [21].

**Lemma 4.2.2** Let $D$ be an open subset of $\mathcal{M}$, $\mathcal{X}$ a $C^1$ vector field defined on $D$ and $\xi : [a, b] \to D$ a $C^\infty$ curve. Then for all $\mathcal{Y} \in \mathcal{X}(\mathcal{M})$ we have that

$$P_{\xi,t,a}\mathcal{X}(\xi(t))\mathcal{Y}(\xi(t)) = \nabla\mathcal{X}(\xi(a))\mathcal{Y}(\xi(a)) + \int_a^t P_{\xi,s,a}\nabla^2\mathcal{X}(\xi(s))\mathcal{Y}(\xi(s))\xi'(s)ds,$$
$$t \in [a, b].$$

*Proof* See [25].

**Lemma 4.2.3 (Banach's Lemma [15])** Let $B$ be a linear operator and let $I_p$ be the identity operator in $T_p\mathcal{M}$. If $\|B - I_p\| < 1$ then $B$ is invertible and $\|B^{-1}\| \leq 1/(1 - \|B - I_p\|)$.

## 4.3   Local Convergence Analysis

We shall show the main local convergence result for Inexact Newton method with relative residual error tolerance under majorant and center majorant condition in a Riemannian manifold context instead of only using the majorant condition.

**Theorem 4.3.1** *Let $\mathcal{M}$ be a Riemannian manifold, $D \subseteq \mathcal{M}$ an open set and $\mathcal{X}$ : $D \to T\mathcal{M}$ a continuously differentiable vector field. Let $p_* \in D$, $R > 0$ and $\kappa :=$ $\sup\{t \in [0, R) \to U(p_*, t) \subset D\}$. Suppose that $\mathcal{X}(p_*) = 0$, $\nabla \mathcal{X}(p_*)$ is invertible and there exist $f_0, f : [0, R) \to \mathbb{R}$ continuously differentiable such that*

$$\|\nabla \mathcal{X}(p_*)^{-1}[P_{\xi,1,0}\nabla \mathcal{X}(p) - P_{\xi,0,0}\nabla \mathcal{X}(\xi(0))P_{\xi,1,0}]\| \leq f_0'(d(p_*,p)) - f_0'(0) \quad (4.6)$$

$$\|\nabla \mathcal{X}(p_*)^{-1}[P_{\xi,1,0}\nabla \mathcal{X}(p_*) - P_{\xi,\tau,0}\nabla \mathcal{X}(\xi(\tau))P_{\xi,1,\tau}]\| \leq f'(d(p_*,p)) - f'(\tau d(p_*,p)), \quad (4.7)$$

*for all $\tau[0, 1]$, $p \in B_\kappa(p_*)$, where $\xi : [0, 1] \to \mathcal{M}$ is a minimizing geodesic from $p_*$ to $p$ and*

$(h_1)$  $f_0(0) = f(0) = 0$ and $f_0'(0) = f'(0) = -1$.

$(h_2)$  $f_0', f'$ are strictly increasing, $f_0(t) \leq f(t)$ and $f_0'(t) \leq f'(t)$ for each $t \in [0, R)$.
 *Let $0 \leq \vartheta < 1K_{p_*}$, $v_0 = \sup\{t \in [0, R) : f_0'(t) < 0\}$, $\rho_0 = \sup\{\delta \in (0, v_0) :$*

$$[(1 + \vartheta)|\frac{f(t) - tf'(t)}{tf_0'(t)}| + \vartheta] < \frac{1}{K_{p_*}}, \ t \in (0, \delta)\} \ and$$

$$r_0 = min\{\kappa, \rho_0, r_{p_*}\}.$$

*Then the sequence $\{p_k\}$ generated by the Inexact Newton method for solving $\mathcal{X}(p) = 0$ with starting point $p_0 \in U(p_*, r_0)\backslash\{p_*\}$ and residual relative error tolerance $\theta$,*

$$p_{k+1} = exp_{p_k}(S_k), \quad \|X(p_k) + \nabla X(p_k)S_k\| \leq \theta\|X(p_k)\|, k = 0, 1, \ldots, \quad (4.8)$$

$$0 \leq cond(\nabla X(p_*))\vartheta = \vartheta/[2/|f_0'(d(p_*,p_0))| - 1], \quad (4.9)$$

*is well defined (for any particular choice of each $S_k \in T_{p_k}\mathcal{M}$), contained in $U(p_*, r_0)$ and converges to the unique zero $p_*$ of $X$ in $U(p_*, \sigma_0)$, where $\sigma_0 :=$ $\sup\{t \in (0, \kappa) : f_0(t) < 0\}$. Furthermore, we have that:*

$$d(p_*, p_{k+1}) \leq \beta_{k+1}^{0,0}, \quad (4.10)$$

*where*

$$\beta_{k+1}^{0,0} = $$
$$K_{p_*}\left[(1 + \vartheta)\frac{\left|\frac{d(p_*,p_k)f'(d(p_*,p_k)) - f(d(p_*,p_k))}{f_0'(d(p_*,p_k))}\right|}{d(p_*,p_k)} + \vartheta\right]d(p_*,p_k),$$
$$k = 0, 1, \ldots$$

*and $\{p_k\}$ converges linearly to $p_*$. If, in addition, the function $f$ satisfies the following condition*

($h_3$)   $f_0, f'$ are convex, then there holds

$$d(p_*, p_{k+1}) \leq \beta_{k+1}^{1,0}, \tag{4.11}$$

where

$$\beta_{k+1}^{1,0} =$$

$$K_{p_*} \left[ (1 + \vartheta) \left| \frac{\frac{d(p_*, p_0) f'(d(p_*, p_0)) - f(d(p_*, p_0))}{f_0'(d(p_*, p_0))}}{d(p_*, p_0)} \right| d(p_*, p_k) + \vartheta \right] d(p_*, p_k),$$

$$\times d(p_*, p_k), k = 0, 1, \ldots.$$

Consequently, the sequence $\{p_k\}$ converges to $p_*$ with linear rate as follows

$$d(p_*, p_{k+1}) \leq \beta_{k+1}^{2,0}, \tag{4.12}$$

where

$$\beta_{k+1}^{2,0} =$$

$$K_{p_*} \left[ (1 + \vartheta) \frac{\left| d(p_*, p_0) * \frac{f'(d(p_*, p_0))}{f_0'(d(p_*, p_0))} - \frac{f(d(p_*, p_0))}{f_0'(d(p_*, p_0))} \right|}{d(p_*, p_0)} + \vartheta \right] d(p_*, p_k),$$

$$k = 0, 1, \ldots.$$

Remark 4.3.1

(a)   We have that

$$\left[ \frac{\left| d(p_*, p_k) * \frac{f'(d(p_*, p_k))}{f_0'(d(p_*, p_k))} - \frac{f(d(p_*, p_k))}{f_0'(d(p_*, p_k))} \right|}{d(p_*, p_k)} \right.$$

$$\left. = \left| \frac{f'(d(p_*, p_k))}{f_0'(d(p_*, p_k))} - \frac{1}{f_0'(d(p_*, p_k))} \frac{f(d(p_*, p_k)) - f(0)}{d(p_*, p_k) - 0} \right| \right]$$

Since the sequence $\{p_k\}$ is contained in $U(p_*, r)$ and converges to the point $p_*$ then it is easy to see that when $k$ goes to infinity the right hand side of last equality goes to zero. Therefore in Theorem 31 if taking $\vartheta = \vartheta_k$ in each iteration and letting $\vartheta_k$ goes to zero (in this case, $\theta = \theta_k$ also goes to zero) as $k$ goes to infinity, then (3.5) implies that $\{p_k\}$ converges to $p_*$ with asymptotic superlinear rate.

Note that letting $\vartheta = 0$ in Theorem 31 which implies from (3.4) that $\theta = 0$, the linear equation in (3.3) is solved exactly. Therefore (3.6) implies that $\{p_k\}$ converges to $p_*$ with quadratic rate.

(b) If $f_0(t) = f(t)$ for each $t \in [0, R)$, then our Theorem 3.1 reduces to Theorem 31
in [15]. Otherwise, i.e. if $f_0(t) < f(t)$ and $f_0'(t) < f'(t)$ for each $t \in [0, R)$, then,
we have

$$v < v_0,$$

$$\rho < \rho_0,$$

$$r < r_0,$$

$$\sigma < \sigma_0$$

and since $|f'(t)| < |f_0'(t)|$ we also have for each $k = 0, 1, 2, \ldots$

$$\beta_{k+1}^{0,0} < \beta_{k+1}^0,$$

$$\beta_{k+1}^{0,1} < \beta_{k+1}^1$$

$$\beta_{k+1}^{0,2} < \beta_{k+1}^2,$$

where $v, \rho, r, \sigma, \beta_{k+1}^0, \beta_{k+1}^1, \beta_{k+1}^2$ are defined in [15] by setting $f_0 = f$ in the
definition of $v_0, \rho_0, r_0, \sigma_0, \beta_{k+1}^{0,0}, \beta_{k+1}^{0,1}, \beta_{k+1}^{0,2}$ respectively.

Hence, we obtain the following advantages: a larger radius of convergence
for the Inexact Newton method (i.e. leading to a wider choice of initial guesses
$p_0$); more precise error estimates on the distance $d(p_*, p_k)$ (i.e. in practice
less iterates must be computed to obtain a desired error tolerance) and the
uniqueness ball is larger. These observations are important in Computational
Mathematics. Notice also that condition (3.2) always implies (3.1) [simply set
$f_0' = f'$ in (3.2) to obtain (3.1)] but not necessarily vice versa. We also have that

$$f_0'(t) \le f'(t) \text{ for each } t \in [0, R)$$

holds in general and $\frac{f'(t)}{f_0'(t)}$ can be arbitrarily large [8–14].

Finally, notice that the new results (i.e. the new advantages) are obtained
under the same computational cost of the majorant function $f$, since in practice
the computation of function $f$ requires the computation of function $f_0$ as a special
case (see also the special cases).

We assume from now on, that the hypotheses of Theorem 3.1 hold with the
exception of $h_3$, which will be considered to hold only when explicitly stated.

We begin by proving that the constants $\kappa$, $v_0$ and $\sigma_0$ are positive.

**Proposition 4.3.1** *The constants $\kappa$, $v_0$ and $\sigma_0$ are positive and $\dfrac{tf'(t) - f(t)}{f_0'(t)} < 0$ for*

*all $t \in (0, v_0)$.*

*Proof* $D$ is open and $p_* \in D$, then, $\kappa > 0$. As $f_0'$ is continuous at $t = 0$ with
$f'(0) = -1$, there exists $\delta > 0$ such that $f_0'(t) < 0$ for all $t \in (0, \delta)$. That is $v_0 > 0$.

Since, $f(0) = 0$ and $f_0'$ is continuous in $0$ with $f_0'(0) = -1$, there exists $\delta > 0$ such that $f_0(t) < 0$ for all $t \in (0, \delta)$, which implies $\sigma_0 > 0$.

Assumption $h_2$ implies that $f$ is strictly convex, so using the strict convexity of $f$ and the first equality in assumption $h_1$ we have $f(t) - tf'(t) < f(0) = 0$ for all $t \in (0, v_0)$ then $f_0'(t) < 0$. $\qquad\square$

It follows from $h_2$ and the definition of $v_0$, we have $f_0'(t) < 0$ for all $t \in [0, v_0)$. Therefore Newton iteration map for $f$ is well defined in $[0, v_0)$. Let us call it $n_{f_0,f}$

$$n_{f_0,f} : [0, v_0) \to (-\infty, 0],$$
$$t \quad \to \quad \frac{tf'(t) - f(t)}{f_0'(t)}. \qquad (4.13)$$

Because $f_0'(t) \neq 0$ for all $t \in [0, v_0)$ the Newton iteration map $n_{f_0,f}$ is a continuous function.

**Proposition 4.3.2** $\lim_{t \to 0} |n_{f_0,f}(t)| = 0$. Then, $\lim \frac{|n_{f_0,f}|}{t} = 0 > 0$, $\rho_0 > 0$ and $(1 + \vartheta)|n_{f_0,f}(t)|/t + \vartheta < 1/K_{p_*}$ for all $t \in (0, \rho_0)$.

*Proof* Using (3.8), Proposition 31, $f(0) = 0$ and the definition of $v_0$ we get

$$\frac{|n_{f_0,f}|}{t} = \left| \left[ \frac{f(t)/f'(t) - t}{t} \right] \frac{f'(t)}{f_0'(t)} \right| = \left| \left[ \frac{1}{f'(t)} \frac{f(t) - f(0)}{t - 0} - 1 \right] \frac{f'(t)}{f_0'(t)} \right| \to 0,$$

as $t \to 0$, $t \in (0, v_0)$.

Since $\vartheta < 1/K_{p_*}$ we can ensure that there exists $\delta > 0$ such that

$$\frac{f'(t)}{f_0'(t)}(1 + \vartheta)[f(t)/f'(t) - t]/t + \vartheta < 1/K_{p_*}, \quad t \in (0, \delta).$$

$\qquad\square$

**Proposition 4.3.3** *If $f$ satisfies condition $h_3$ then, the function $[0, v_0) \ni t \to \frac{n_{f_0,f}}{t^2}$ is increasing.*

*Proof* Using definition of $n_{f_0,f}$ in (3.8), Proposition 5 and $h_1$ we obtain, after simples algebraic manipulation, that

$$\frac{|n_{f_0,f}|}{t} = \frac{1}{|f'(t)|} \int_0^1 \frac{f'(t) - f'(\tau t)}{t} d\tau \left| \frac{f'(t)}{f_0'(t)} \right|, \quad \forall t \in (0, v_0). \qquad (4.14)$$

On the other hand as $f_0', f'$ are strictly increasing the map

$$[0, v_0) \ni t \mapsto \frac{f'(t) - f'(\tau t)}{t}$$

is positive for all $\tau \in (0, 1)$. From $h_3$ $f_0', f'$ are convex, so we conclude that the last map is increasing. Hence the second term in the right hand side of (3.9) is positive and increasing. Assumption $h_2$ and definition of imply that the first term in the right hand side of (3.9) is also positive and strictly increasing. Therefore we conclude that the left hand side of (3.9) is increasing. Similarly, we show that the second function is increasing. $\qquad\square$

Next, we present connections between the majorant function f and the vector field $\mathcal{X}$.

**Lemma 4.3.1** *Let $p \in D \subseteq M$. If $d(p_*, p) < min\{\kappa, v_0\}$ then $\nabla X(p)$ is invertible and*

$$\|\nabla X(p)^{-1} P_{\xi,0,1} \nabla X(p_*)\| \le 1/|f_0'(d(p_*, p))|$$

*where $\xi : [0, 1] \to M$ is a minimizing geodesic from $p_*$ to p. In particular $\nabla X(p)$ is invertible for all $p \in U(p_*, r_0)$ where $r_0$ is as defined in Theorem 31.*

*Proof* See [10, 11] or Lemma 4.4 of [20] (simply use the needed $f_0'$ [i.e. (3.1) instead of (3.2)]). $\qquad\square$

**Lemma 4.3.2** *Let $p \in M$. If $d(p_*, p) \le d(p_*, p_0) < min\{\kappa, v_0\}$, then the following hold*

$$cond(\nabla X(p)) \le cond(\nabla X(p_*)) \left[2/|f_0'(d(p_*, p_0))| - 1\right],$$

*and $cond(\nabla X(p)) \le \vartheta$.*

*Proof* Let $I_{p_*} : T_{p_*}M \to T_{p_*}M$ the identity operator, $p \in U(p_*, \kappa)$ and $\xi : [0, 1] \to M$ a minimizing geodesic from $p_*$ to p. Since $P_{\xi,0,0} = I_{p_*}$ and $P_{\xi,0,1}$ is an isometry we obtain

$$\|\nabla X(p_*)^{-1} P_{\xi,1,0} \nabla X(p) P_{\xi,0,1} - I_{p_*}\| = \\ \|\nabla X(p_*)^{-1} \left[P_{\xi,1,0} \nabla X(p) - P_{\xi,0,0} \nabla X(p_*) P_{\xi,1,0}\right]\|.$$

As $d(p_*, p) < v_0$ we have $f_0'(d(p_*, p)) < 0$. Using the last equation, (3.1) and $h_1$ we conclude that

$$\|\nabla X(p_*)^{-1} P_{\xi,1,0} \nabla X(p) P_{\xi,0,1} - I_{p_*}\| \le f_0'(d(p_*, p)) + 1.$$

Since $P_{\xi,0,1}$ is an isometry and

$$\|\nabla X(p)\| \le \|\nabla X(p)\| \|\nabla X(p_*)^{-1} P_{\xi,1,0} \nabla X(p) P_{\xi,0,1}\|,$$

triangular inequality together with above inequality imply

$$\|\nabla X(p)\| \le \|\nabla X(p)\| \left[f_0'(d(p_*, p)) + 2\right].$$

On the other hand, it is easy to see from Lemma 31 that

$$\|\nabla X(p)^{-1}\| \leq \|\nabla X(p_*)\|/f_0'(d(p_*,p))|.$$

Therefore, combining two last inequalities and definition of condition number we obtain

$$cond(\nabla X(p)) \leq cond(\nabla X(p_*)) \left[ 2/|f_0'(d(p_*,p))| - 1 \right].$$

Since $f_0'$ is strictly increasing, $f_0' < 0$ in $[0, v_0)$ and $d(p_*,p) \leq d(p_*,p_0) < min\{\kappa, v_0\}$, the first inequality of the lemma follows from last inequality.
The last inequality of the lemma follows from (3.4) and first inequality.     □.

The linearization error of $\mathcal{X}$ at a point in $B_\kappa(p_*)$ is defined by:

$$E_X(p_*,p) := X(p_*) - P_{\alpha,0,1} \left[ X(p) + \nabla X(p)\alpha'(0) \right], \quad p \in B_\kappa(p_*),$$

where $\alpha : [0, 1] \to M$ is a minimizing geodesic from $p$ to $p_*$. We will bound this error by the error in the linearization on the majorant function $f$,

$$e_f(t, u) := f(u) - [f(t) + f'(t)(u - t)], \quad t, u \in [0, R).$$

**Lemma 4.3.3** *Let $p \in D \subset M$. If $d(p_*,p) \leq \kappa$ then $\|\nabla X(p_*)^{-1}E_X(p_*,p)\| \leq e_f(d(p_*,p), 0)$.*

*Proof* See [10, 11] or Lemma 4.5 of [20].                              □

**Lemma 4.3.4** *Let $p \in D \subset M$. If $d(p_*,p) < r_0$ then*

$$\|\nabla X(p)^{-1}X(p)\| \leq g(p_*,p), p \in U(p_*, r_0),$$

*where*

$$g(s, t) = (-d(s, t) + \frac{f(d(s, t))}{f'(d(s, t))}) \frac{|f'(d(s, t))|}{|f_0'(d(s, t))|} + d(s, t).$$

*Proof* Taking into account that $X(p_*) = 0$, the inequality is trivial for $p = p_*$. Now assume that $0 < d(p_*,p) < r_0$. Lemma 31 implies that $\nabla X(p)$ is invertible. Let $\alpha : [0, 1] \to M$ be a minimizing geodesic from $p$ to $p_*$. Because $X(p_*) = 0$, the definition of $E_X(p_*,p)$ in (1.11) and direct manipulation yields

$$\nabla X(p)^{-1}P_{\alpha,1,0}E_X(p; p_*) = \nabla X(p)^{-1}X(p) + \alpha'(0).$$

Using the above equation, Lemmas 35 and 32, it is easy to conclude that

$$\|\nabla X(p)^{-1}X(p) + \alpha'(0)\| \leq \| - \nabla X(p)^{-1}P_{\alpha,1,0}\nabla X(p_*)\|\|\nabla X(p_*)^{-1}E_F(p,p_*)\|$$

$$\leq e_f(d(p_*,p), 0)/|f_0'(d(p_*,p))|.$$

Now as $f(0) = 0$ definition of $e_f$ gives

$$e_f(d(p_*,p),0)/|f'(d(p_*,p)) = -d(p_*,p) + f(d(p_*,p))/f'(d(p_*,p)),$$

which combined with last inequality yields

$$\|\nabla X(p)^{-1}X(p) + \alpha'(0)\| \leq \left(-d(p_*,p) + \frac{f(d(p_*,p))}{f'(d(p_*,p))}\right) \frac{|f'(d_*,d)|}{|f_0'(d_*,d)|}.$$

Moreover, as $\|\alpha'(0)\| = d(p_*,p)$, using simple algebraic manipulation, it is easy to see that

$$\|\nabla X(p)^{-1}X(p)\| \leq \|\nabla X(p)^{-1}X(p) + \alpha'(0)\| + d(p_*,p),$$

which combined with preceding inequality completes the proof.                    □.

The outcome of an Inexact Newton iteration is any point satisfying some error tolerance. Hence, instead of a mapping for Newton iteration, we shall deal with a family of mappings describing all possible inexact iterations.

**Definition 4.3.1** For $0 \leq \theta$, $N_\theta$ is the family of maps $N_\theta : U(p_*, r_0) \to X$ such that

$$\|X(p) + \nabla X(p)exp_p^{-1}N_\theta(p)\| \leq \theta\|X(p)\|, \quad p \in U(p_*, r_0).$$

If $p \in U(p_*, r_0)$ then $\nabla X(p)$ is non-singular. Therefore for $\theta = 0$ the family $N_0$ has a single element, namely, the exact Newton iteration map

$$N_0 : U(p_*, r_0) \to M$$
$$p \qquad \mapsto exp_p(-\nabla X(p)^{-1}X(p)).$$

Trivially, if $0 <\leq \theta \leq \theta'$ then $N_0 \subset N_\theta \subset N_{\theta'}$. Hence $N_\theta$ is non-empty for all $\theta \geq 0$.

*Remark 4.3.2* For any $\theta \in (0,1)$ and $N_\theta \in N_\theta$

$$N_\theta(p) = p \Leftrightarrow X(p) = 0; p \in U(p_*, r_0).$$

This means that the fixed points of the Inexact Newton iteration $N_\theta$ are the same fixed points of the exact Newton iteration, namely, the zeros of $X$.

**Lemma 4.3.5** *Let $\theta$ be such that $0 \leq cond(\nabla X(p_*)) \leq \vartheta/[-1+2/|f'(d(p_*,p_0))|]$ and $p \in D \subset M$. If $d(p_*,p) \leq d(p_*,p_0) < r_0$ and $N_\theta \in N_\theta$, then*

$$(d(p_*, N_\theta(p)) \leq K_{p_*}\left[(1+\vartheta)\frac{|n_{f_0,f}d(p_*,p)|}{d(p_*,p)} + \theta\right]d(p_*,p), \quad p \in U(p_*, r_0).$$

*As a consequence, $N_\theta(U(p_*, r))) \subset U(p_*, r)$.*

*Proof* As $X(p_*) = 0$, the inequality is trivial for $p = p_*$. Now, assume that $0 < d(p_*, p) \leq r$. Let $\alpha : [0; 1] \rightarrow M$ be a minimizing geodesic from $p$ to $p_*$. Using simple algebraic manipulations, triangular inequality and definition of the linearization error we obtain

$$\|exp_p^{-1} N_\theta(p) - \alpha'(0)\| \leq$$
$$\nabla X(p)^{-1} [\nabla X(p)^{-1} exp_p^{-1} N_\theta(p) + X(p)]\| + \|\nabla X(p)^{-1}\| E_X(p_*, p)\|. \tag{4.15}$$

Using Definition (3.4) it is obvious that

$$\|\nabla X(p)^{-1} [\nabla X(p) exp_p^{-1} N_\theta(p) + X(p)]\| \leq \|\nabla X(p)^{-1}\| \theta \|X(p)\|.$$

Now, since $\|X(p)\| \leq \|\nabla X(p)\| \|\nabla X(p)^{-1} X(p)\|$ we obtain from Lemma 34 that

$$\|X(p)\| \leq \|\nabla X(p)\| g(p_*, p).$$

Definition of condition number and two above inequalities imply

$$\|\nabla X(p)^{-1} [\nabla X(p) exp_p^{-1} N_\theta(p) + X(p)\| \leq \theta cond(\nabla X(p)) g(p_*, p)\|. \tag{4.16}$$

Now, combining Lemmas 31 and 33 the second term in (3.10) is bounded by

$$\|\nabla X(p)^{-1} E_X(p_*, p)\| \leq \frac{1}{|f_0'(d(p_*, p))|} e_f(d(p_*, p), 0).$$

Therefore, (3.10), (3.11) and last inequality give us

$$\|exp_p^{-1} N_\theta(p) - \alpha'(0)\| \leq \theta cond(\nabla X(p)) g(p_*, p)\| + \frac{1}{f_0'(d(p_*, p))} e_f(d(p_*, p), 0).$$

Since Lemma 32 implies $\theta cond(\nabla X(p)) \leq \vartheta$, after simple algebraic manipulation and considering definitions of $e_f$ and $n_{f_0, f}$ the above inequality becomes

$$\|exp_p^{-1} N_\theta(p) - \alpha'(0)\| \leq \left[ (1 + \vartheta) \frac{|n_{f_0, f} d(p_*, p)|}{d(p_*, p)} + \vartheta \right] d(p_*, p).$$

Note that, as $d(p_*, p) \leq r_0 < \rho$, second part of Proposition 32 implies that the term in brackets of last inequality is less than $1/K_{p_*} \leq 1$. So left hand side of last inequality is less than $r \leq r_{p_*}$. Therefore letting $p = p_*$, $q = p$, $v = \alpha'(0)$, $u = exp_p^{-1} N\theta(p)$ in Definition 21 we conclude that

$$d(p_*, N_\theta(p)) \leq K_{p_*} \|exp_p^{-1} N_\theta(p) - \alpha'(0)\|.$$

Finally combining two above inequalities the inequality of the lemma follows.

Take $p \in U(p_*, r_0)$. Since $d(p_*, p) < r_0$ and $r_0 \leq \rho_0$, the first part of the lemma and the second part of Proposition 32 imply that $d(p_*, N_X(p)) < d(p_*, p)$ and the result follows.

We can now prove Theorem 31. Let $0 \leq \theta$ satisfying (3.4) and $N_\theta \in \mathbb{N}_\theta$ , where $N_\theta$ is defined in Definition 31. Therefore (3.3) together with Definition 31 implies that the sequence $\{p_k\}$ satisfies

$$p_{k+1} = N_\theta(p_k), \quad k = 0, 1, \ldots, \tag{4.17}$$

which is indeed an equivalent definition of this sequence.

*Proof of Theorem 31* Since $p_0 \in U(p_*, r_0)$, $r \leq v$ and $0 < \theta cond(\nabla X(p_*)) \leq \vartheta/[2/|f_0'(d(p_*, p_0))| - 1]$, combining (3.12), the inclusion $N_\theta(U(p_*, r_0)) \subset U(p_*, r_0)$ in Lemmas 31 and 35, it is easy to conclude that by an induction argument the sequence $\{p_k\}$ is well defined and remains in $U(p_*, r_0)$. Next, we are going to prove that $\{p_k\}$ converges towards $p_*$. Since $d(p_*, p_k) < r_0$, for $k = 0, 1, \ldots,$ , we obtain from (3.12) and Lemma 31 that

$$d(p_*, p_{k+1}) \leq K_{p_*} \left[ (1 + \vartheta) \frac{|n_{f_0, f} d(p_*, p_k))|}{d(p_*, p_k)} + \vartheta \right] d(p_*, p_k). \tag{4.18}$$

As $d(p_*, p_k) < r_0 \leq \rho$ , for $k = 0, 1, \ldots$, using second statement in Proposition 32 and last inequality we conclude that $0\ d(p_*, p_{k+1}) < d(p_*, p_k)$, for $k = 0, 1, \ldots$. So $\{d(p_*, p_k)\}$ is strictly decreasing and bounded below which implies that it converges. Let $l_* := \lim_{k \to \infty} d(p_*, p_k)$. Because $\{d(p_*, p_k)\}$ rests in $(0, \rho_0)$ and is strictly decreasing we have $0 \leq l_* < \rho_0$. We are going to show that $l_* = 0$. If $0 < l_*$ then letting $k$ goes to infinity in (3.13), the continuity of $n_{f_0, f}$ in $[0, \rho)$ and Proposition 32 imply that

$$l_* \leq K_{p_*} \left[ (1 + \vartheta) \frac{|n_{f_0, f}(l_*)|}{(l_*)} + \vartheta \right] l_* < l_* \tag{4.19}$$

which is a contradiction. Hence we must have $l_* = 0$. Therefore the convergence of $\{p_k\}$ to $p_*$ is proved. The uniqueness of $p_*$ in $B_{\sigma_0}(p_*)$ was proved in Lemma 5.1 of [20].

To prove (3.5) it is sufficient to use Eq. (3.12) and definition of $n_{f_0, f}$ in (3.8). As $d(p_*, p_k) < r_0 \leq \rho_0$ , for $k = 0, 1, \ldots$, $\lim_{k \to \infty} d(p_*, p_k) = 0$ and by hypothesis $\vartheta < 1/K_{p_*}$ thus using definition of $n_{f_0, f}$ and first statement in Proposition 32, we conclude

$$\lim_{k \to \infty} K_{p_*} \left[ (1 + \vartheta) \frac{|d(p_*, p_k) \frac{f(d(p_*, p_k))}{f_0'(d(p_*, p_k))} - \frac{f(d(p_*, p_k))|}{f_0'(d(p_*, p_k))}}{d(p_*, p_k)} + \vartheta \right] = K_{p_*} \vartheta < 1 \tag{4.20}$$

which implies the linear convergence of $\{p_k\}$ to $p_*$ in (3.5).

We must show the inequality in (3.6): If $f$ satisfies $h_3$ then using definition of $n_{f_0,f}$ and Proposition 33 we conclude

$$(1+\vartheta)\frac{|d(p_*,p_k)\frac{f(d(p_*,p_k))}{f_0'(d(p_*,p_k))} - \frac{f(d(p_*,p_k))}{f_0'(d(p_*,p_k))}|d(p_*,p_k)}{d^2(p_*,p_k)} + \vartheta$$

$$\leq (1+\vartheta)\frac{|d(p_*,p_0)\frac{f(d(p_*,p_0))}{f_0'(d(p_*,p_0))} - \frac{f(d(p_*,p_0))}{f_0'(d(p_*,p_0))}|d(p_*,p_k)}{d^2(p_*,p_0)} + \vartheta.$$

As the quantity of the left hand side of the last inequality is equal to quantity in the brackets of (3.5), the inequality in (3.6) follows from (3.5) and last inequality. Since $\{d(p_*,p_k)\}$ is strictly decreasing, the inequality in (3.7) follows from (3.6). $\square$

## 4.4   Special Cases

We present two special cases as examples in this section.

First, for null error tolerance we present the next result on Inexact Newton method under a Hölder-like condition.

**Theorem 4.4.1** *Let $M$ be a Riemannian manifold, $D \subset M$ an open set and $X : D \to TM$ a continuously differentiable vector field. Take $p_* \in D$, $R > 0$ and let $\kappa := \sup\{t \in [0,R) : U(p_*,t)\}$. Suppose that $X(p_*) = 0$, $\nabla X(p_*)$ is invertible and there exist constants $L_0, L > 0$ and $0 \leq \eta < 1$ such that*

$$\|\nabla X(p_*)^{-1}[P_{\xi,1,0}\nabla X(p) - P_{\xi,0,0}\nabla X(\xi(\tau))P_{\xi,1,0}]\| \leq L_0 d(p_*,p)^\eta, \qquad (4.21)$$

$$\|\nabla X(p_*)^{-1}[P_{\xi,1,0}\nabla X(p) - P_{\xi,\tau,0}\nabla X(\xi(\tau))P_{\xi,1,\tau}]\| \leq L(1-\tau^\eta)d(p_*,p)^\eta, \qquad (4.22)$$

*for all $\tau \in [0,1]$ and $p \in U(p_*,\kappa)$, where $\xi : [0;1] \to M$ is a minimizing geodesic from $p_*$ to $p$. Let $r_{p_*}$ be the injectivity radius of $M$ in $p_*$, $K_{p_*}$ as in Definition 1, $0 \leq \vartheta < 1/K_{p_*}$ and*

$$r_0 := \min\left\{\kappa, \left[\frac{LK_{p_*}(1+\theta)\eta}{1-K_{p_*}\theta} + L_0(\eta+1)\right]^{\frac{1}{\eta}}, r_{p_*}\right\}.$$

*Then the sequence generated by the Inexact Newton method for solving $X(p) = 0$ with starting point $p_0 \in U(p_*,r_0) \setminus \{p_*\}$ and residual relative error tolerance $\theta$,*

$$p_{k+1} = exp_{p_k}(Sk), \quad \|X(p_k) + \nabla X(p_k)S_k\| \leq \theta\|X(p_k)\|, k = 0,1,\dots, \qquad (4.23)$$

$$0 \leq cond(\nabla X(p_*))\vartheta \frac{1 + L_0 d(p_*,p_0)^\eta}{1 - L_0 d(p_*,p_0)^\eta}, \qquad (4.24)$$

*is well defined (for any particular choice of each $S_k \in T_{p_k}M$), contained in $U(p_*, r_0)$*
*and converges to the unique zero $p_*$ of $X$ in $U(p_*, (\frac{\eta+1}{L_0})^{\frac{1}{\eta}})$ and we have that:*

$$d(p_*, p_{k+1}) \le K_{p_*}\left[(1 + \vartheta\frac{\eta Ld(p_*, p_k)^{\eta}}{(\eta + 1)[1 - L_0 d(p_*, p_k)^{\eta}]} + \vartheta\right]d(p_*, p_k),$$

$$k = 0, 1, \ldots.$$

*Moreover, the sequence $\{p_k\}$ converges linearly to $p_*$. If, in addition, $\eta = 1$ then*
*there holds*

$$d(p_*, p_{k+1}) \le K_{p_*}\left[(1 + \vartheta)\frac{L}{2[1 - L_0 d(p_*, p_0)]}d(p_*, p_k) + \vartheta\right]d(p_*, p_k),$$

$$k = 0, 1, \ldots.$$

*Consequently, the sequence $\{p_k\}$ converges to $p_*$ with linear rate as follows*

$$d(p_*, p_{k+1}) \le K_{p_*}\left[(1 + \vartheta)\frac{Ld(p_*, p_0)}{2[1 - L_0 d(p_*, p_0)]}) + \vartheta\right]d(p_*, p_k), \quad k = 0, 1, \ldots.$$

*Proof* We can prove that $X$, $p_*$, $f_0$ and $f_0, f : [0, +\infty) \to \mathbb{R}$, defined $f_0(t) = L_0 t^{\eta+1}/(\eta + 1) - t$ and by $f(t) = Lt^{\eta+1}/(\eta + 1) - t$, satisfy (3.1), (3.2), respectively and the conditions $h_1$ and $h_2$ in Theorem 4. Moreover, if $\eta = 1$ then $f$ satisfies condition $h_3$. It is easy to see that $\rho_0$, $v_0$ and $\sigma_0$, as defined in Theorem 31, satisfy

$$\rho_0 = \left[\frac{LK_{p_*}(1 + \theta)\eta}{1 - K_{p_*}\theta} + L_0(\eta + 1)\right]^{\frac{1}{\eta}} \le v_0 = \frac{1}{L^{1/\eta}}, \quad \sigma_0 = [(\eta + 1)/L_0]^{1/\eta}$$

Therefore, the result follows by invoking Theorem 31.                                            □.

*Remark 4.4.1*

(a) Note that if the vector field $X$ is Lipschitz with constant $L$ and center Lipschitz with constant $L_0$ conditions (4.1) and (4.2) are satisfied with $\eta = 1$.

(b) We remark that letting $\vartheta = 0$ in Theorem 41 which implies from (4.3) that $\vartheta = 0$, the linear equation (4.2) is solved exactly. Therefore (4.4) implies that if $\eta = 1$ then $\{p_k\}$ converges to $p_*$ with quadratic rate.

(c) If $L_0 = L$ Theorem 41 reduces to Theorem 1.3 in [15]. however, if $L_0 < L$, then the advantages of our approach as stated in Remark 31 (b) hold.

(d) Another special result can easily be given using the Smale-Wang conditions [30, 33, 34]. The choices for functions $f_0$ and $f$ have already be given in [13, 14]. Then, again the same advantages are obtained over the corresponding work in [15]. However, we leave the tails to the motivated reader. Other choice of function $f$ (and $f_0$) can be found in [5–9].

(e) In order for us to explain the difference between the affine majorant condition and the generalized Lipschitz condition in [35], let $M = \mathbb{R}^n$ and suppose that there exists a positive integrable function $L : [0, R) \to \mathbb{R}$ such that,

$$\|F'(x^*)^{-1}[F'(x) - F'(x_* + \tau(y - x_*))]\| \leq \int_{\tau\|x-x^*\|}^{x-x_*} L(u)\,du \qquad (4.25)$$

for each $t \in [0, 1]$ and each $x \in \bar{U}(x_*, \kappa)$. Define function $f : [0, R) \to \mathbb{R}$ by

$$f'(t) = \int_0^t L(u)\,du = 1.$$

Suppose that $L$ is nondecreasing. Then (3.2), function $f'$ is strictly increasing and convex. In this case (4.25) and (3.2) are equivalent. However, if $f'$ is strictly increasing and not necessarily convex, then (4.25) and (3.2) are not equivalent, since there exist functions strictly increasing, continuous, with derivative zero almost everywhere (see [28, 32]). Moreover, these functions are not absolutely continuous, so they cannot be represented by an integral.

Secondly, for null error tolerance, we present the next theorem on Inexact Newton's method under Wang's condition.

**Theorem 4.4.2** *Let $\mathcal{M}$ be a Riemannian manifold, $D \subset \mathcal{M}$ an open set and $X : D \to \mathcal{M}$ an analytic vector field. Take $p_* \in D$ and let $\kappa := \sup\{t \in [0, R) : U(p_*, t) \subset D\}$. Suppose that $X(p_*) = 0$, $\nabla X(p_*)$ is invertible and there exist $\gamma_0 > 0$ and $\gamma > 0$ such that*

$$\left\| \nabla X(p_*)^{-1} \left[ P_{\xi,1,0}\nabla X(p) - P_{\xi,0,0}\nabla X(\xi(0))P_{\xi,1,0} \right] \right\| \leq$$

$$\frac{1}{(1 - \gamma_0 d(p_*, p))^2} - 1,$$

*and*

$$\left\| \nabla X(p_*)^{-1} \left[ P_{\xi,1,0}\nabla X(p) - P_{\xi,\tau,0}\nabla X(\xi(\tau))P_{\xi,1,\tau} \right] \right\| \leq$$

$$\frac{1}{(1 - \gamma d(p_*, p))^2} - \frac{1}{(1 - \tau\gamma d(p_*, p))^2}$$

*for each $\tau \in [0, 1]$, $p \in U(p_*, 1/\gamma)$, where $\xi : [0, 1] \to \mathcal{M}$ is a minimizing geodesic from $p_*$ to $p$.*

*Let $r_{p_*}$ be the injectivity radius of $\mathcal{M}$ in $p_*$, $K_{p_*}$ as in Definition 21, $0 \leq \vartheta < 1/K_{p_*}$.*

*Then, the sequence $\{p_k\}$ generated by the Inexact Newton method for solving $X(p) = 0$ with starting point $p_0 \in U(p_*, r) \setminus \{p_*\}$ and residual relative error tolerance $\theta$,*

$$p_{k+1} = exp_{p_k}(S_k), \quad \|X(p_k) + \nabla X(p_k)S_k\| \leq \theta \|X(p_k)\|, \quad k = 0, 1, \ldots$$

$$0 \leq cond(\nabla X(p_*))\theta \leq \vartheta[2[1 - \gamma_0 d(p_*, p_0)]^2 - 1],$$

*is well defined (for any particular choice of each $S_k \in T_{p_k}\mathcal{M}$), is contained in $U(p_*, r)$ and converges to the point the unique zero $p_*$ of $X$ in $U(p_*, \frac{1}{2\gamma_0})$ and we have that*

$$d(p_*, p_{k+1}) \leq K_{p_*} \left[ (1 + \vartheta) \frac{\gamma}{2[1 - \gamma_0 d(p_*, p_0)]^2 - 1} d(p_*, p_k) + \vartheta \right] d(p_*, p_k),$$

$$k = 0, 1, \ldots.$$

*Consequently, the sequence $\{p_k\}$ converges to $p_*$ with linear rate as follows*

$$d(p_*, p_{k+1}) \leq K_{p_*} \left[ (1 + \vartheta) \frac{\gamma d(p_*, p_0)}{2[1 - \gamma_0 d(p_*, p_0)]^2 - 1} + \vartheta \right] d(p_*, p_k),$$

$$k = 0, 1, \ldots$$

*Proof* Simply, choose

$$f_0(t) = \frac{t}{1 - \gamma_0 t} - 2t \quad \text{and} \quad f(t) = \frac{t}{1 - \gamma t} - 2t.$$

Then all hypotheses of Theorem 31 are satisfied.                                    □

*Remark 4.4.2* If $\gamma_0 = \gamma$ Theorem 51 reduces to Theorem 14 in [15]. Otherwise, i.e. if $\gamma_0 < \gamma$, then our Theorem 51 is an improvement with advantages as already stated in Remark 31. Notice that

$$\gamma_0 < \gamma$$

holds in general and $\frac{\gamma}{\gamma_0}$ can be arbitrarily large [8–14]. Examples where $\gamma_0 < \gamma$ can be found in [13, 15]. Notice that Theorem 14 in [15] extended earlier results in the literature [16, 20, 25, 26]. Moreover, it is worth noticing that the computation of constant $\gamma$ requires the computation of constant $\gamma_0$ as a special case. Hence, the advantages of our approach over the earlier ones are obtained under the same computational cost on the constants involved. Finally, notice that if $\mathcal{M}$ is an analytic Riemannian manifold and

$$\gamma_0 = \gamma = \sup_{n>1} \left\| \frac{\nabla X(p_*)^{-1} \nabla^n X(p_*)}{n!} \right\|^{\frac{1}{n-1}} < +\infty,$$

then our results merge to Smale's [30].

**Acknowledgements** The research has been partially funded by UNIR Research (http://research. unir.net), Universidad Internacional de La Rioja (UNIR, http://www.unir.net), under the Research Support Strategy 3 [2015–2017], Research Group: MOdelación Matemática Aplicada a la INgeniería (MOMAIN), by the Grant SENECA 19374/PI/14 and by the project MTM2014-52016-C2-1-P of the Spanish Ministry of Economy and Competitiveness.

# References

1. Amat, S., Argyros, I.K., Busquier, S., Castro, R., Hilout, S., Plaza, S.: Traub-type high order iterative procedures on Riemannian manifolds. SeMA J. Boletin de la Sociedad Española de Matematica Aplicada **63**, 27–52 (2014)
2. Amat, S., Argyros, I.K., Busquier, S., Castro, R., Hilout, S., Plaza, S.: Newton-type methods on Riemannian manifolds under Kantorovich-type conditions. Appl. Math. Comput. **227**, 762–787 (2014)
3. Amat, S., Busquier, S., Castro, R., Plaza, S.: Third-order methods on Riemannian manifolds under Kantorovich conditions. J. Comput. Appl. Math. **255**, 106–121 (2014)
4. Amat, S., Argyros, I.K., Busquier, S., Castro, R., Hilout, S., Plaza, S.: On a bilinear operator free third order method on Riemannian manifolds. Appl. Math. Comput. **219**(14), 7429–7444 (2013)
5. Apell, J., De Pascale, E., Lysenko, J.V., Zabrejko, P.P.: New results on Newton-Kantorovich approximations with applications to nonlinear integral equations. Numer. Funct. Anal. Optim. **18**(1 and 2), 1–17 (1997)
6. Apell, J., De Pascale, E., Zabrejko, P.P.: On the application of the Newton-Kantorovich method to nonlinear integral equations of Uryson type. Numer. Funct. Anal. Optim. **12**(3), 271–283 (1991)
7. Apell, J., Zabrejko, P.P.: Nonlinear Superposition Operators. Cambridge University Press, Cambridge (1990)
8. Argyros, I.K.: Convergence and Applications of Newton–Type Iterations. Springer, New York (2008)
9. Argyros, I.K., Hilout, S.: Computational Methods in Nonlinear Analysis. Efficient Algorithms, Fixed Point Theory and Applications. World Scientific, Singapore (2013)
10. Argyros, I.K., Hilout, S.: Newton's method for approximating zeros of vector fields on Riemannian manifolds. J. Appl. Math. Comput. **29**, 417–427 (2009)
11. Argyros, I.K.: An improved unifying convergence analysis of Newton's method in Riemannian manifolds. J. Appl. Math. Comput. **25**, 345–351 (2007)
12. Argyros, I.K.: A semilocal convergence analysis for directional Newton methods. Math. Comput. **80**, 327–343 (2011)
13. Argyros, I.K., Hilout, S.: Expanding the applicability of Newton's method using the Smale Alpha theory. J. Comput. Appl. Math. **261**, 183–200 (2014)
14. Argyros, I.K., Hilout, S.: Expanding the applicability of inexact Newton's method using the Smale Alpha theory. Appl. Math. Comput. **224**, 224–237 (2014)
15. Bittencourt, T., Ferreira, O.P.: Local convergence analysis of Inexact Newton method with relative residual error tolerance under majorant condition in Riemannian manifolds. http://orizon.mat.ufg.br/p/3371-publications (2015)
16. Blum, L., Cucker, F., Shub, M., Smale, S.: Complexity and Real Computation. Springer, New York (1998). With a foreword by Richard M. Karp
17. Chen, J., Li, W.: Convergence behaviour of inexact Newton methods under weak Lipschitz condition. J. Comput. Appl. Math. **191**(1), 143–164 (2006)
18. Dembo, R.S., Eisenstat, S.C., Steihaug, T.: Inexact Newton methods. SIAM J. Numer. Anal. **19**(2), 400–408 (1982)
19. Do Carmo, M.P.: Riemannian Geometry. Birkhauser, Basel (1992)

20. Ferreira, O.P., Silva, R.C.M.: Local convergence of Newton's method under a majorant condition in Riemannian manifolds. IMA J. Numer. Anal. **32**(4), 1696–1713 (2012)
21. Ferreira, O.P., Svaiter, B.F.: Kantorovich's theorem on Newton's method in Riemannian manifolds. J. Complex. **18**(1), 304–329 (2002)
22. Gondzio, J.: Convergence analysis of an inexact feasible interior point method for convex quadratic programming. SIAM J. Optim. **23**(3), 1810–1527 (2013)
23. Huang, Z.: The convergence ball of Newton's method and the uniqueness ball of equations under Hölder-type continuous derivatives. Comput. Math. Appl. **47**, 247–251 (2004)
24. Lang, S.: Differential and Riemannian Manifolds. Springer, Berlin (1995)
25. Li, C., Wang, J.: Newton's method on Riemannian manifolds: Smale's point estimate theory under the $\gamma$-condition. IMA J. Numer. Anal. **26**(2), 228–251 (2006)
26. Li, C., Wang, J.: Newton's method for sections on Riemannian manifolds: generalized covariant $\alpha$-theory. J. Complex. **24**(3), 423–451 (2008)
27. Nurekenov, T.K.: Necessary and sufficient conditions for Uryson operators to satisfy a Lipschitz condition (Russian). Izv. Akad. Nauk. Kaz. SSR **3**, 79–82 (1983)
28. Okamoto, H., Wunsch, M.: A geometric construction of continuous, strictly increasing singular functions. Proc. Jpn. Acad. Ser. A Math. Sci. **83**(7), 114–118 (2007)
29. Potra, F.A.: The Kantorovich theorem and interior point methods. Math. Program. **102**(1, Ser. A), 47–70 (2005)
30. Smale, S.: Newton's method estimates from data at one point. In: The Merging of Disciplines: New Directions in Pure, Applied, and Computational Mathematics (Laramie, Wyo., 1985), pp. 185–196. Springer, New York (1986)
31. Smith, S.T.: Optimization techniques on Riemannian manifolds. In: Hamiltonian and Gradient Ows, Algorithms and Control. Fields Institute Communications, vol. 3, pp. 113–136. American Mathematical Society, Providence, RI (1994)
32. Takács, L.: An increasing continuous singular function. Am. Math. Mon. **85**(1), 35–37 (1978)
33. Wang, J.H.: Convergence of Newton's method for sections on Riemannian manifolds. J. Optim. Theory Appl. **148**(1), 125–145 (2011)
34. Wang, J.-H., Yao, J.-C., Li, C.: Gauss-Newton method for convex composite optimizations on Riemannian manifolds. J. Glob. Optim. **53**(1), 5–28 (2012)
35. Wang, X.: Convergence of Newton's method and inverse function theorem in Banach space. Math. Comput. **68**(225), 169–186 (1999)
36. Wayne, C.E.: An introduction to KAM theory. In: Dynamical Systems and Probabilistic Methods in Partial Differential Equations (Berkeley, CA, 1994). Lectures in Applied Mathematics, vol. 31, pp. 3–29. American Mathematical Society, Providence, RI (1996)

# Chapter 5
# On the Design of Optimal Iterative Methods for Solving Nonlinear Equations

Alicia Cordero and Juan R. Torregrosa

**Abstract** A survey on the existing techniques used to design optimal iterative schemes for solving nonlinear equations is presented. The attention is focused on such procedures that use some evaluations of the derivative of the nonlinear function. After introducing some elementary concepts, the methods are classified depending on the optimal order reached and also some general families of arbitrary order are presented. Later on, some techniques of complex dynamics are introduced, as this is a resource recently used for many authors in order to classify and compare iterative methods of the same order of convergence. Finally, some numerical test are made to show the performance of several mentioned procedures and some conclusions are stated.

## 5.1 Introduction

Many real problems in different fields of Science and Technology require finding the solution of an equation or a system of nonlinear equations. In particular, the numerical solutions of equations and systems are needed in the study of dynamic models of chemical reactors, on radioactive transfer, interpolation problems in Astronomy, in climatological simulations, in problems of turbulence, in the resolution of integral equations or partial differential equations, in the preliminary determination of satellite orbits, in global positioning systems, etc.

The solution of equations and systems of nonlinear equations has been, and remains today, one of the most active topics of Numerical Analysis and has produced a great amount of publications, which is confirmed in the texts of, among others, Ostrowski [59], Traub [69], Ortega and Rheinbolt [58] and Petković et al. [61].

A. Cordero • J.R. Torregrosa (✉)

Instituto de Matemáticas Multidisciplinar, Universidad Politécnica de Valencia, Camino de Vera, s/n 46022 Valencia, Spain

e-mail: acordero@mat.upv.es; jrtorre@mat.upv.es

© Springer International Publishing Switzerland 2016

S. Amat, S. Busquier (eds.), *Advances in Iterative Methods for Nonlinear Equations*, SEMA SIMAI Springer Series 10, DOI 10.1007/978-3-319-39228-8_5

In this chapter, we will focus on the fixed point iterative methods that, under certain conditions, approximate a simple root $\alpha$ of a nonlinear equation $f(x) = 0$, where $f : I \subseteq \mathbb{R} \rightarrow \mathbb{R}$ is a real function defined in an open interval $I$. This class of methods can be classified in one-point methods and multipoint schemes. In addition, each one of these classes can be divided in methods with or without memory, but in this chapter we will only take care of iterative schemes without memory.

The one-point iterative schemes are those in which the $(k + 1)$th-iterate is obtained by using functional evaluations only of $k$th-iterate, that is,

$$x_{k+1} = \Phi(x_k), \quad k = 0, 1, 2, \ldots$$

One of the most known and commonly used iterative method of this type is Newton's method, given by

$$x_{k+1} = x_k - \frac{f(x_k)}{f'(x_k)}, \quad k = 0, 1, 2, \ldots,$$

and the family of Chebyshev-Halley methods, whose iterative expression is

$$x_{k+1} = x_k - \left(1 + \frac{1}{2} \frac{L_f(x_k)}{1 - \beta L_f(x_k)}\right) \frac{f(x_k)}{f'(x_k)}, \quad k = 0, 1, 2, \ldots,$$

where $\beta$ is a parameter and $L_f(x_k) = \dfrac{f(x_k)f''(x_k)}{f'(x_k)^2}$ is called the degree of logarithmic convexity. Different values of parameter $\beta$ give rise to classical methods as Chebyshev's scheme, for $\beta = 0$, Halley's procedure, for $\beta = 1/2$ or super-Halley scheme for $\beta = 1$.

It is known that if $x_{k+1} = \Phi(x_k)$ is a one-point iterative method which use $d$ functional evaluations per step, then its order of convergence is at most $p = d$. On the other hand, Traub [69] proved that for designing a one-point method of order p, the iterative expression must contain derivatives at least of order $p - 1$. So, one-point methods are not a good idea in order to increase the order of convergence and the computational efficiency.

These restrictions of one-point methods are an explanation of the increasing interest that the researchers are showing in the last years by the multipoint iterative methods. In this type of schemes, also called predictor-corrector schemes, $(k+1)$th-iterate is obtained by using functional evaluations of the $k$th-iterate and also other intermediate points. For example, a multipoint scheme of two steps will have the expression

$$\begin{aligned} y_k &= \Psi(x_k), \\ x_{k+1} &= \Phi(x_k, y_k), \quad k = 0, 1, 2, \ldots \end{aligned}$$

The main objective and motivation for designing new iterative schemes is to increase the order of convergence without adding many functional evaluations. For the sake of clarity and to make this chapter self-content, we are going to recall some concepts related to iterative methods.

**Definition 5.1.1** Let $\{x_k\}_{k\geq 0}$ be a sequence generated by an iterative method, that converges to $\alpha$. If there exist a real number $p$ and a positive constant $C$ ($C < 1$ if $p = 1$) such that

$$\lim_{k\to+\infty} \frac{|x_{k+1} - \alpha|}{|x_k - \alpha|^p} = C,$$

then $p$ is called the *order of convergence* and $C$ is the asymptotic error constant.

Let $e_k = x_k - \alpha$ be the error of the approximation in the $k$th-iteration. Then, an iterative method of order $p$ satisfies the equation

$$e_{k+1} = e_k^p + O(e_k^{p+1}),$$

which is called the *error equation* of the method.

Some examples show that this definition is rather restrictive, which motivated Ortega and Rheinboldt to introduce in [58] the concepts of $Q$- and $R$-order of convergence. Nevertheless, they proved that the $Q$-, $R$- and Traub's $C$-order coincide when $0 < C < +\infty$ exists for some $p \in [1, +\infty[$. Since the asymptotic error constant $C$ always satisfies this condition for all methods considered in this chapter, we will work with the previous definition of order of convergence.

There are in the literature other measures for comparing different iterative procedures. Traub in [69] defined the *informational efficiency* of an iterative method M as

$$I(M) = \frac{p}{d},$$

where $p$ is the order of convergence and $d$ the number of functional evaluations per iteration. On the other hand, Ostrowski in [59] introduced the *efficiency index*, given by

$$EI(M) = p^{1/d}.$$

On the other hand, by using the tools of complex dynamics is possible to compare different algorithms in terms of their basins of attraction, the dynamical behavior of the rational function associated to the iterative method on polynomials of low degree, etc. Varona [70], Amat et al. [1], Neta et al. [57], Cordero et al. [27] and Magreñán [52], among others, have analyzed many schemes and parametric families of methods under this point of view, obtaining interesting results about their stability and reliability. The Cayley's test introduced by Babajee et al. [9] is also a good tool to classify iterative schemes for solving nonlinear equations.

When testing new methods, either to check the order of convergence or to estimate how much it differs from the theoretical order in practical implementation, it is of interest to use the *computational order of convergence* (COC), introduced by Weerakoon and Fernando in [72] as

$$p \approx COC = \frac{\ln |(x_{k+1} - \alpha)/(x_k - \alpha)|}{\ln |(x_k - \alpha)/(x_{k-1} - \alpha)|}, \quad k = 1, 2, \ldots$$

where $x_{k+1}$, $x_k$ and $x_{k-1}$ are the last three successive approximations of $\alpha$ obtained in the iterative process. The value of zero $\alpha$ is unknown in practice. So, the *approximated computational order of convergence*, defined by Cordero and Torregrosa in [22], is often used as

$$p \approx ACOC = \frac{\ln |(x_{k+1} - x_k)/(x_k - x_{k-1})|}{\ln |(x_k - x_{k-1})/(x_{k-1} - x_{k-2})|}, \quad k = 2, 3, \ldots \tag{5.1}$$

ACOC is a vector which gives us interesting information only if its components are stable.

The major goal in designing new multipoint iterative methods is closely related to the Kung-Traub conjecture [50]. Kung and Traub conjectured that the order of convergence of an iterative method without memory, which uses $d$ functional evaluations per iteration, is at most $2^{d-1}$. When this bound is reached the method is called *optimal*. Kung-Traub conjecture is supported by many families of multipoint methods of arbitrary order $p$, and an important number of particular schemes developed after 1980. In this chapter we will describe different techniques for designing optimal methods and we will present a review of some optimal methods and parametric families of methods existing in the literature.

A standard way to increase the order of convergence of an iterative scheme is by using the so-called *technique of composition*. It can be proved [69] the following result.

**Theorem 5.1.1** *Let $\Phi_1(x)$ and $\Phi_2(x)$ be iterative functions of orders $p_1$ and $p_2$, respectively. Then the iterative scheme resulting by the composition of the previous ones, $x_{k+1} = \Phi_1(\Phi_2(x_k))$, $k = 0, 1, 2, \ldots$, has order of convergence $p_1 p_2$.*

However, this composition always increase the number of functional evaluations. So, in order to preserve the optimality it is necessary to use some tools that allow us to reduce the amount of functional evaluations, such as approximating the last evaluations by means of interpolation polynomials, Padé approximants, inverse interpolation, Adomian polynomials, ... where we use the value of the function and the derivatives at points already known.

Weight function procedure also allows us to increase the order of convergence without new functional evaluations. It is possible to use weight functions with one or several variables as well as different functions added, multiplied, etc.

The rest of this chapter is organized as follows. Firstly, in Sect. 5.2 a review of the techniques used to design optimal fourth-order methods with derivatives is made, showing some of the most representative resulting schemes or families; Sect. 5.3 is devoted to the following steps of the improvement of efficiency index: the design of optimal eighth- and sixteenth-order schemes. Some general families of iterative schemes using derivatives are presented in Sect. 5.4. The classification and comparison between this schemes is focused on complex dynamics techniques in Sect. 5.5, using for this aim Cayley Test and also the behavior of procedure on third-degree polynomials. Some numerical performances are presented in Sect. 5.6 and, finally, some conclusions are shown in Sect. 5.7.

## 5.2   Optimal Fourth-Order Methods

As far as we know, the first researches on multipoint methods were presented by Traub in [69]. Although, in general, those methods are not optimal, the employed techniques have had a big influence in subsequent studies on multipoint methods. We consider the scheme resulting by composing Newton's method with itself

$$
\begin{aligned}
y_k &= x_k - \frac{f(x_k)}{f'(x_k)}, \\
x_{k+1} &= y_k - \frac{f(y_k)}{f'(y_k)}, \quad k = 0, 1, 2, \dots
\end{aligned}
\tag{5.2}
$$

This scheme is simple and its order of convergence is four, which is a consequence of the fact that Newton's method is of second order and Theorem 5.1.1. Nevertheless, it is not optimal in the sense of Kung-Traub conjecture, since it needs four functional evaluations per iteration. One way to reduce the number of functional evaluations is to "frozen" the derivative, which gives Traub's scheme [69] (also known as Potra-Pták method)

$$
\begin{aligned}
y_k &= x_k - \frac{f(x_k)}{f'(x_k)}, \\
x_{k+1} &= y_k - \frac{f(y_k)}{f'(x_k)}, \quad k = 0, 1, 2, \dots
\end{aligned}
\tag{5.3}
$$

but, this method is still not optimal because it has order three. To reduce the number of functional evaluations without decreasing the order of convergence it can be used the idea of Chun [18] to approximate $f'(y)$ by

$$
f'(y_k) \approx \frac{f'(x_k)}{G(t_k)}, \quad t_k = \frac{f(y_k)}{f(x_k)},
$$

assuming that real function $G$ is sufficiently differentiable in a neighborhood of 0. Then, the two-step scheme (5.3) becomes

$$
\begin{aligned}
y_k &= x_k - \frac{f(x_k)}{f'(x_k)}, \\
x_{k+1} &= y_k - G(t_k)\frac{f(y_k)}{f'(x_k)}, \quad t_k = \frac{f(y_k)}{f(x_k)}, \quad k = 0, 1, 2, \ldots,
\end{aligned}
\tag{5.4}
$$

whose convergence is established in the following result (its proof can be found in [18]).

**Theorem 5.2.1** *Let $f : I \subseteq \mathbb{R} \to \mathbb{R}$ be a real sufficiently differentiable function and $\alpha \in I$ a simple root of $f(x) = 0$. Let $G(t)$ be a real function satisfying $G(0) = 1$, $G'(0) = 2$ and $|G''(0)| < +\infty$. If $x_0$ is close enough to $\alpha$, then the order of convergence of family (5.4) is four and its error equation*

$$
e_{k+1} = \left[ c_2^3 \left( 5 - \frac{G''(0)}{2} \right) - c_2 c_3 \right] e_k^4 + O(e_k^5),
$$

*where $c_k = \dfrac{1}{k!}\dfrac{f^{(k)}(\alpha)}{f'(\alpha)}$, $k = 2, 3, \ldots$ and $e_k = x_k - \alpha$.*

A similar result was obtained by Artidiello [5] by using a weight function of two variables $G(t_k, u_k)$, $t_k = \dfrac{f(y_k)}{f(x_k)}$, $u_k = \dfrac{f(x_k)}{f'(x_k)}$ and assuming that function $G$ is sufficiently differentiable in a neighborhood of $(0, 0)$. He also generalized these results in the following way. Let us consider the iterative scheme

$$
\begin{aligned}
y_k &= x_k - \gamma \frac{f(x_k)}{f'(x_k)}, \\
x_{k+1} &= y_k - G(\mu_k)\frac{f(y_k)}{f'(x_k)}, \quad k = 0, 1, 2, \ldots,
\end{aligned}
\tag{5.5}
$$

where $\gamma$ is a real parameter and $G(\mu(x))$ is a real function of variable $\mu(x) = \dfrac{a_1 f(x) + a_2 f(y)}{b_1 f(x) + b_2 f(y)}$, $a_1, a_2, b_1, b_2 \in \mathbb{R}$, sufficiently differentiable in a neighborhood of $c = a_1/b_1$. The following result analyzes the convergence of this family.

**Theorem 5.2.2** *Let $f : I \subseteq \mathbb{R} \to \mathbb{R}$ be a real function sufficiently differentiable and $\alpha \in I$ a simple root of $f(x) = 0$. Let $G(\mu(x))$ be a real function satisfying $G(c) = 1$, $G'(c) = \dfrac{2b_1^2}{a_2 b_1 - a_1 b_2}$ and $|G''(c)| < +\infty$. If $\gamma = 1$, $a_1 \neq 0$, $b_1 \neq 0$ and $x_0$ is close enough to $\alpha$, then the order of convergence of family (5.5) is four and its error* *equation*

$$e_{k+1} = \left[ \frac{10b_1^4 + 4b_1^3b_2 + (2a_1a_2b_1b_2 - a_2^2b_1^2 - a_1^2b_2^2)G''(c)}{2b_1^4} c_2^3 - c_2c_3 \right] e_k^4 + O(e_k^5),$$

*where* $c_k = \dfrac{1}{k!}\dfrac{f^{(k)}(\alpha)}{f'(\alpha)}$, $k = 2, 3, \ldots$ *and* $e_k = x_k - \alpha$.

*Proof* By using Taylor series around $\alpha$, we have

$$f(x_k) = f'(\alpha)\left[ e_k + c_2e_k^2 + c_3e_k^3 + c_4e_k^4 \right] + O(e_k^5) \tag{5.6}$$

and

$$f'(x_k) = f'(\alpha)\left[ 1 + 2c_2e_k + 3c_3e_k^2 + 4c_4e_k^3 \right] + O(e_k^4). \tag{5.7}$$

From these expressions, we get

$$y_k - \alpha = (1 - \gamma)e_k + \gamma c_2e_k^2 - 2\gamma(c_2^2 - c_3)e_k^3 + \gamma(4c_2^3 - 7c_2c_3 + 3c_4)e_k^4 + O(e_k^5).$$

Furthermore, we have

$$
\begin{aligned}
f(y_k) = f'(\alpha)\Big[ & (1 - \gamma)e_k + (1 - \gamma + \gamma^2)c_2e_k^2 \\
& +(-2\gamma^2c_2^2 - (-1 + \gamma - 3\gamma^2 + \gamma^3)c_3)e_k^3 + (5\gamma^2c_2^3 + \gamma^2(-10 + 3\gamma)c_2c_3 \\
& +(1 - \gamma + 6\gamma^2 - 4\gamma^3 + \gamma^4)c_4)e_k^4 \Big] + O(e_k^5). \tag{5.8}
\end{aligned}
$$

Now, Taylor series for $\mu(x_k)$ gives,

$$
\begin{aligned}
\frac{a_1f(x_k) + a_2f(y_k)}{b_1f(x_k) + b_2f(y_k)} = \; & \frac{1}{B}(a_1 + a_2 - a_2\gamma) + \frac{A}{B^2}\gamma^2c_2e_k \\
& -\frac{1}{B^3}\left( A\gamma^2((3b_1 + b_2(3 - 3\gamma + \gamma^2))c_2^2 + (-3 + \gamma)Bc_3 \right)e_k^2 \\
& +\frac{1}{B^4}A\gamma^2\Big( (8b_1^2 + b_2^2(-2 + \gamma)^2(2 - 2\gamma + \gamma^2) \\
& +2b_1b_2(8 - 8\gamma + 3\gamma^2))c_2^3 + 2(b_1^2(-7 + 2\gamma) \\
& +b_1b_2(-14 + 18\gamma - 7\gamma^2 + \gamma^3) \\
& -b_2^2(7 - 16\gamma + 14\gamma^2 - 6\gamma^3 + \gamma^4))c_2c_3 \\
& +B^2(6 - 4\gamma + \gamma^2)c_4 \Big)e_k^3 + O(e_k^4),
\end{aligned}
$$

where $A = a_2b_1 - a_1b_2$ and $B = b_1 + b_2 - b_2\gamma$.

Let us represent function $G$ by its Taylor's polynomial of the second order at the point $c$,

$$G(\mu(x_k)) \approx G(c) + G'(c)(\mu(x_k) - c) + \frac{G''(c)}{2}(\mu(x_k) - c)^2. \qquad (5.9)$$

Now, using (5.7)–(5.9), we obtain an expression for $e_{k+1} = y_k - \alpha - G(\mu(x_k))\frac{f(y_k)}{f'(x_k)}$ in terms of powers of $e_k$. In order to cancel the coefficients of $e_k$, $e_k^2$ and $e_k^3$ it is necessary to assume $\gamma = 1$, $G(c) = 1$, $G'(c) = \frac{2b_1^2}{a_2b_1 - a_1b_2}$, and $|G''(c)| < \infty$. In this case, the error equation is

$$e_{k+1} = \left[ \frac{10b_1^4 + 4b_1^3b_2 + (2a_1a_2b_1b_2 - a_2^2b_1^2 - a_1^2b_2^2)G''(c)}{2b_1^4} c_2^3 - c_2c_3 \right] e_k^4 + O(e_k^5),$$

and the proof is finished. □

The first fourth-order optimal method was constructed by Ostrowski [59], who derived his method using interpolation techniques. This method can also be designed starting from double Newton's scheme and replacing $f'(y_k)$ (the derivative of the second step) by a suitable approximation which does not require new information. The iterative expression is

$$\begin{aligned} y_k &= x_k - \frac{f(x_k)}{f'(x_k)}, \\ x_{k+1} &= y_k - \frac{f(x_k)}{f(x_k) - 2f(y_k)}\frac{f(y_k)}{f'(x_k)}, \quad k = 0, 1, 2, \ldots \end{aligned} \qquad (5.10)$$

This method can be obtained as a particular case of many parametric families of fourth-order methods developed in later researches, see for example, the Chun-Ham family [19], the family derived by Kou et al. [48], etc. Ostrowski's scheme has good convergence properties (see [73]) and an interesting stability behavior [9, 26]. This fact has propitiated that many researchers construct high order optimal schemes using Ostrowski's method as a predictor.

A generalization of Ostrowski's method was proposed by King [47] using the approximation of $f'(y_k)$ in (5.2)

$$f'(y_k) \approx f'(x_k)\frac{f(x_k) + \gamma f(y_k)}{f(x_k) + \beta f(y_k)},$$

where $\beta$ and $\gamma$ are parameters. He obtained the parametric family of fourth-order iterative methods

$$\begin{aligned} y_k &= x_k - \frac{f(x_k)}{f'(x_k)}, \\ x_{k+1} &= y_k - \frac{f(x_k) + \beta f(y_k)}{f(x_k) + (\beta - 2)f(y_k)}\frac{f(y_k)}{f'(x_k)}, \quad k = 0, 1, 2, \ldots \end{aligned} \qquad (5.11)$$

It is easy to observe that Ostrowski's method is a particular case of (5.11) for $\beta=0$. Many published works after [47] have King's family as a particular case, even though the authors obtained their methods with different procedures. Kou et al. [48] designed an iterative scheme by using a linear combination of Traub' and Newton-Steffensen methods, which turned out to be a particular case of King's family. Also, Kou et al. [49] obtained King's family by composing Chebyshev's method, free of second derivative, and Newton's scheme.

Different selections of function $G(t)$ in (5.4), satisfying the convergence conditions, allow us to obtain many other optimal fourth-order methods. For example, we consider the following special cases for function $G(t)$, $t \in \mathbb{R}$

1. $G(t) = \dfrac{1 + \beta t}{1 + (\beta - 2)t}$, $\beta \in \mathbb{R}$.

2. $G(t) = 1 + 2t$.

3. $G(t) = \dfrac{4}{4 - 2t - t^2}$.

4. $G(t) = \dfrac{t^2 + (\gamma - 2)t - 1}{\gamma t - 1}$, $\gamma \in \mathbb{R}$.

Such choices in (5.4) produce new methods or particular cases of known ones. For example, from function (1) we obtain King's family of fourth-order methods. From (2) the iterative expression is

$$
\begin{aligned}
y_k &= x_k - \frac{f(x_k)}{f'(x_k)}, \\
x_{k+1} &= y_k - \frac{f(x_k) + 2f(y_k)}{f(x_k)} \frac{f(y_k)}{f'(x_k)}, \quad k = 0, 1, 2, \ldots
\end{aligned}
$$

which is a particular case of King's family.

From (3) we get the new method

$$
\begin{aligned}
y_k &= x_k - \frac{f(x_k)}{f'(x_k)}, \\
x_{k+1} &= y_k - \frac{(f(x_k) + f(y_k))^2}{f(x_k)^2 - 5f(y_k)^2} \frac{f(y_k)}{f'(x_k)}, \quad k = 0, 1, 2, \ldots
\end{aligned}
$$

Finally, choosing $G(t)$ of (4) we obtain the parametric family

$$
\begin{aligned}
y_k &= x_k - \frac{f(x_k)}{f'(x_k)}, \\
x_{k+1} &= y_k - \left[1 + \frac{f(y_k)(f(y_k) - 2f(x_k))}{f(x_k)(\gamma f(y_k) - f(x_k))}\right] \frac{f(y_k)}{f'(x_k)}, \quad k = 0, 1, 2, \ldots
\end{aligned}
$$

Taking $\gamma = 1$ we obtain Maheshvari's scheme [53] as an special case.

Other procedures for designing iterative methods, different from weight functions are possible. By composing Traub's method and Newton's scheme with "frozen" derivative, a non-optimal fourth-order method results

$$
\begin{aligned}
y_k &= x_k - \frac{f(x_k)}{f'(x_k)}, \\
z_k &= x_k - \frac{f(x_k) + f(y_k)}{f'(x_k)}, \\
x_{k+1} &= z_k - \frac{f(z_k)}{f'(x_k)}, \quad k = 0, 1, 2, \ldots
\end{aligned}
$$

By using the second-order Taylor expansion of $f(z_k)$ and $f(y_k)$ around $x_k$, we obtain

$$
f(z_k) \approx f(x_k) + f'(x_k)(z_k - x_k) + \frac{1}{2} f''(x_k)(z_k - x_k)^2 \tag{5.12}
$$

and

$$
f(y_k) \approx f(x_k) + f'(x_k)(y_k - x_k) + \frac{1}{2} f''(x_k)(y_k - x_k)^2,
$$

getting the following approximation

$$
\frac{1}{2} f''(x_k) \approx \frac{f(y_k) - f(x_k) - f'(x_k)(y_k - x_k)}{(y_k - x_k)^2} = \frac{f(y_k) f'(x_k)^2}{f(x_k)^2},
$$

which only depends on already evaluated functional values $f(x_k), f(y_k)$ and $f'(x_k)$. By substituting this expression in (5.12) we have

$$
f(z_k) \approx \frac{f(y_k)^2 (2f(x_k) + f(y_k))}{f(x_k)^2}.
$$

So, the new method is expressed as

$$
\begin{aligned}
y_k &= x_k - \frac{f(x_k)}{f'(x_k)}, \\
z_k &= x_k - \frac{f(x_k) + f(y_k)}{f'(x_k)}, \\
x_{k+1} &= z_k - \frac{f(y_k)^2 (2f(x_k) + f(y_k))}{f(x_k)^2 f'(x_k)}, \quad k = 0, 1, 2, \ldots
\end{aligned} \tag{5.13}
$$

The following result, whose proof can be found in [23], establishes the convergence of (5.13).

**Theorem 5.2.3** *Let $f : I \subseteq \mathbb{R} \to \mathbb{R}$ be a real function sufficiently differentiable and $\alpha \in I$ a simple root of $f(x) = 0$. If $x_0$ is close enough to $\alpha$, then the method defined by (5.13) has optimal convergence order four and its error equation is*

$$e_{k+1} = (4c_2^3 - c_2c_3)e_k^4 + O(e_k^5),$$

*where $c_k = \dfrac{f^{(k)}(\alpha)}{k!f'(\alpha)}$, $k = 2, 3, \ldots$ and $e_k = x_k - \alpha$.*

The multipoint methods previously presented use two evaluations of function $f$ at different points and one evaluation of $f'$. In 1966, Jarratt [43] designed an iterative scheme that uses one function and two derivative evaluations per iteration. So, optimal two-point schemes with this type of evaluations are called Jarratt-type methods.

Jarratt in [43] analyzed a class of iterative methods of the form

$$x_{k+1} = x_k - \phi_1(x_k) - \phi_2(x_k),$$

where

$$\phi_1(x) = a_1 w_1(x) + a_2 w_2(x), \quad \phi_2(x) = \frac{f(x)}{b_1 f'(x) + b_2 f'(x + \gamma w_1(x))},$$

$$w_1(x) = \frac{f(x)}{f'(x)} \quad \text{and} \quad w_2(x) = \frac{f(x)}{f'(x + \gamma w_1(x))}.$$

By using Taylor expansion around a simple root of $f(x) = 0$, $\alpha$, Jarratt obtained the values of parameters in order to construct a fourth-order scheme. Specifically, Jarratt's method has the iterative expression

$$
\begin{aligned}
y_k &= x_k - \frac{2}{3}\frac{f(x_k)}{f'(x_k)}, \\
x_{k+1} &= x_k - \frac{1}{2}\left(\frac{3f'(y_k) + f'(x_k)}{3f'(y_k) - f'(x_k)}\right)\frac{f(x_k)}{f'(x_k)}, \quad k = 0, 1, 2, \ldots
\end{aligned}
\tag{5.14}
$$

Other families of two-point methods of fourth-order, which also uses one evaluation of $f$ and two of its derivative, have been proposed. Chun and Ham in [20] design

$$x_{k+1} = x_k - \left(1 + \frac{1}{2}\frac{J_f(x_k)}{1 - J_f(x_k)}\right)\frac{f(x_k)}{f'(x_k)}, k = 0, 1, 2, \ldots,$$

where $J_f(x_k) = \dfrac{f(x_k)}{f'(x_k)^2}\dfrac{f'(y_k) - f'(x_k)}{y_k - x_k}$ and $y_k = x_k - h(x_k)\dfrac{f(x_k)}{f'(x_k)}$, being $h$ any function satisfying $h(\alpha) = 2/3$. Also Kou et al. in [49] construct the scheme

$$
\begin{aligned}
y_k &= x_k - \frac{2}{3}\frac{f(x_k)}{f'(x_k)}, \\
x_{k+1} &= x_k - \left(1 - \frac{3}{4}\frac{(t_k - 1)(\gamma t_k + 1 - \gamma)}{(\eta t_k + 1 - \eta)(\beta t_k + 1 - \beta)}\right)\frac{f(x_k)}{f'(x_k)}, \quad k = 0, 1, 2, \ldots,
\end{aligned}
$$

where $t_k = \dfrac{f'(y_k)}{f'(x_k)}$ and $\gamma = \eta + \beta - \dfrac{3}{2}$, $\eta, \beta \in \mathbb{R}$.

## 5.3   High Order Optimal Multipoint Methods

In this section, we consider some classes of optimal three-point methods with order eight. As we will see, once an optimal two-point methods of order four is stated, it is easy to construct three-point methods of order eight that require four functional evaluations, that is, optimal schemes. It can be constructed by derivative estimation, inverse interpolation, weight function procedure, etc.

By composing a general optimal fourth-order scheme with Newton's method we obtain a general method of order 8

$$
\begin{aligned}
y_k &= x_k - \frac{f(x_k)}{f'(x_k)}, \\
z_k &= \varphi_4(x_k, y_k), \\
x_{k+1} &= z_k - \frac{f(z_k)}{f'(z_k)},
\end{aligned}
\tag{5.15}
$$

which is not optimal because it uses five functional evaluations per step. To reduce the number of functional evaluations, we can use different procedures.

1. **Polynomial interpolation.**

We approximate $f'(z_k)$ by using the available data $f(x_k), f'(x_k), f(y_k)$ and $f(z_k)$. The polynomial interpolation of third order is

$$
\begin{aligned}
p_3(t) = {} & f(z_k) + (t - z_k)f[z_k, y_k] + (t - z_k)(t - y_k)f[z_k, y_k, x_k] \\
& + (t - z_k)(t - y_k)(t - x_k)f[z_k, y_k, x_k, x_k],
\end{aligned}
$$

where $f[\,\cdot\,]$ denotes the divided differences of several orders. Hence,

$$
\begin{aligned}
p_3'(z_k) &= f[z_k, y_k] + (z_k - y_k)f[z_k, y_k, x_k] + (z_k - y_k)(z_k - x_k)f[z_k, y_k, x_k, x_k] \\
&= f[z_k, y_k] + 2(z_k - y_k)f[z_k, y_k, x_k] - (z_k - y_k)f[y_k, x_k, x_k]
\end{aligned}
$$

and we use the approximation $f'(z_k) \approx p'_3(z_k)$. From a free from second derivative variant of Chebyshev-Halley method and this idea, the authors in [30] designed an eighth-order optimal family, whose iterative expression is

$$
\begin{aligned}
y_k &= x_k - \frac{f(x_k)}{f'(x_k)}, \\
z_k &= x_k - \left(1 + \frac{f(y_k)}{f(x_k) - 2\beta f(y_k)}\right) \frac{f(x_k)}{f'(x_k)}, \\
x_{k+1} &= z_k - \frac{f(z_k)}{f[z_k, y_k] + 2(z_k - y_k)f[z_k, y_k, x_k] - (z_k - y_k)f[y_k, x_k, x_k]},
\end{aligned}
\tag{5.16}
$$

for $k = 0, 1, 2, \ldots$, where $\beta$ is a real parameter.

## 2. Hermite's interpolating polynomial.

This polynomial of third order for the given data has the form

$$
h_3(t) = a_0 + a_1(t - x_k) + a_2(t - x_k)^2 + a_3(t - x_k)^3.
$$

The unknown coefficients are determined from the conditions

$$
h_3(x_k) = f(x_k), \quad h_3(y_k) = f(y_k), \quad h_3(z_k) = f(z_k), \quad h'_3(x_k) = f'(x_k).
$$

Then, the values of the coefficients are

$$
\begin{aligned}
a_0 &= f(x_k), \\
a_1 &= f'(x_k), \\
a_2 &= \frac{(z_k - x_k)f[y_k, z_k]}{(z_k - y_k)(y_k - x_k)} - \frac{(y_k - x_k)f[z_k, x_k]}{(z_k - y_k)(z_k - x_k)} - f'(x_k)\left(\frac{1}{z_k - x_k} + \frac{1}{y_k - x_k}\right), \\
a_3 &= \frac{f[z_k, x_k]}{(z_k - y_k)(z_k - x_k)} - \frac{f[y_k, x_k]}{(z_k - y_k)(y_k - x_k)} + \frac{f'(x_k)}{(z_k - x_k)(y_k - x_k)}.
\end{aligned}
$$

Replacing the obtained coefficients, we get the expression of $h'_3(z_k)$,

$$
h'_3(z_k) = f[z_k, x_k]\left(2 + \frac{z_k - x_k}{z_k - y_k}\right) - \frac{(z_k - x_k)^2}{(y_k - x_k)(z_k - y_k)}f[y_k, x_k] + f'(x_k)\frac{z_k - y_k}{y_k - x_k}
$$

and by using the approximation $f'(z_k) \approx h'_3(z_k)$ in the third step of (5.15) the following optimal eighth-order scheme is constructed

$$
\begin{aligned}
y_k &= x_k - \frac{f(x_k)}{f'(x_k)}, \\
z_k &= \varphi_4(x_k, y_k), \\
x_{k+1} &= z_k - \frac{f(z_k)}{h'_3(z_k)}, \quad k = 0, 1, 2, \ldots
\end{aligned}
$$

This idea was employed, among other references, in [60, 71].

## 3. **Weigh function procedure.**

It is possible to deal with weight functions of one, two or more variables, or combine two or more weight functions with one o more arguments. These weight functions and their variables must use only available information to keep the number of functional evaluations not greater than four. Many optimal three-point methods have been constructed in this way (see for example [7, 36, 39] and the references therein).

In [7], the authors proposed the following three-step method, which is a generalization of the one proposed in [36], with weight functions in the second and third step,

$$
\begin{aligned}
y_k &= x_k - \gamma \frac{f(x_k)}{f'(x_k)}, \\
z_k &= y_k - H(u_k)\frac{f(y_k)}{f'(x_k)}, \\
x_{k+1} &= z_k - G(u_k, v_k)\frac{f(z_k)}{f'(x_k)},
\end{aligned}
\tag{5.17}
$$

where $u_k = \dfrac{f(y_k)}{b_1 f(x_k) + b_2 f(y_k)}$, being $b_1$ and $b_2$ arbitrary real parameters and $v_k = \dfrac{f(z_k)}{f(y_k)}$. Let us observe that we use the composition of Newton's method with "frozen" derivative and weight functions of one and two variables. The following result was proved in [7]. In it, $G_u(0,0)$ and $G_{uu}(0,0)$ denote the first and second partial derivative of $G$, respect to $u$, evaluated at $(0,0)$.

**Theorem 5.3.1** *Let $\alpha \in I$ be a simple zero of a sufficiently differentiable function $f : I \subset \mathbb{R} \to \mathbb{R}$ on an open interval $I$. Let $H$ and $G$ be sufficiently differentiable real functions and $x_0$ be an initial approximation close enough to $\alpha$. If $\gamma = 1$ and $H$ and $G$ satisfy $H(0) = 1$, $H'(0) = 2b_1$, $H''(0) = 2b_1(2b_1 + b_2)$, $G(0,0) = G_v(0,0) = 1$, $G_u(0,0) = 2b_1$, $G_{uv}(0,0) = 4b_1$, $G_{uu}(0,0) = 2b_1(3b_1 + b_2)$ and $|G_{vv}(0,0)| < +\infty$, then the methods of the family described by (5.17) have order of convergence eight for any value of parameters $b_1$ and $b_2$, $b_1 \neq 0$. The error equation is*

$$
e_{k+1} = \frac{1}{2b_1^3}[(3b_1 + b_2)c_2^2 - b_1 c_3][r_3 c_3^2 + r_4 c_2^4 + r_5 c_2^2 c_3 - 2b_1^2 c_2 c_4]e_k^8 + O(e_k^9),
$$

*where $c_j = \dfrac{f^{(j)}(\alpha)}{j! f'(\alpha)}, j = 2, 3, \ldots, r_3 = b_1^2(G_{vv}(0,0) - 2), r_4 = 9b_1^2(G_{vv}(0,0) - 6) + b_2^2(G_{vv}(0,0) - 2) + 2b_1 b_2(3G_{vv}(0,0) - 13)$ and $r_5 = b_1^2(34 - G_{vv}) - 2b_1 b_2(G_{vv} - 5)$.*

Other procedures used by the researchers in order to reduce the number of functional evaluations in (5.15) are the rational interpolation and the inverse interpolation (see, for example [62]). However, let us note that weight function procedure contains rational functions and Hermite's interpolating polynomial as special cases.

The end of this section is devoted to present in short different families of optimal eighth-order methods. Bi et al. [12] used King's family as optimal fourth-order method, the approximation of $f'(z_k)$,

$$f'(z_k) \approx f[z_k, y_k] + f[z_k, x_k, x_k](z_k - y_k)$$

and a weight function in the third step for designing the following family

$$
\begin{aligned}
y_k &= x_k - \frac{f(x_k)}{f'(x_k)}, \\
z_k &= y_k - \frac{f(x_k) + \beta f(y_k)}{f(x_k) + (\beta - 2)f(y_k)} \frac{f(y_k)}{f'(x_k)}, \\
x_{k+1} &= z_k - p(s_k) \frac{f(z_k)}{f[z_k, y_k] + f[z_k, x_k, x_k](z_k - y_k)},
\end{aligned}
\tag{5.18}
$$

where $s_k = f(z_k)/f(x_k)$ and $p(s)$ is a real function satisfying the conditions showed in the following result, whose proof can be found in [12].

**Theorem 5.3.2** *Let $\alpha$ be a simple zero of a sufficiently differentiable function $f$. If $x_0$ is close enough to $\alpha$, then the sequence $\{x_k\}$ generated by (5.18) converges to $\alpha$ with order eight if $\beta = -\frac{1}{2}$ and function $p$ satisfies the properties*

$$p(0) = 1, \quad p'(0) = 2, \quad |p''(0)| < +\infty.$$

Other three-step optimal schemes that use King's family or Ostrowski's method for the two first steps was constructed by Cordero et al. in [24, 25]. On the other hand, from the Kung-Traub's optimal method of order four [50] and by using different combinations of weight functions, the authors in [29] presented two three-step families of optimal eighth-order schemes.

In a similar way, as we have passed through optimal fourth-order to get optimal eighth-order, by using a combination of the described procedures it is possible to design optimal methods of order 16, 32,... and so on. If we add a new step to family described by (5.17) with a similar structure to the previous one, we can obtain a four-step family of optimal sixteenth-order methods. Let us consider the four-step family

$$
\begin{aligned}
y_k &= x_k - \gamma \frac{f(x_k)}{f'(x_k)}, \\
z_k &= y_k - H(u_k) \frac{f(y_k)}{f'(x_k)}, \\
s_k &= z_k - G(u_k, v_k) \frac{f(z_k)}{f'(x_k)}, \\
x_{k+1} &= s_k - T(u_k, v_k, w_k) \frac{f(s_k)}{f'(x_k)},
\end{aligned}
\tag{5.19}
$$

where $u_k = f(y_k)/f(x_k)$, $v_k = f(z_k)/f(y_k)$ and $w_k = f(s_k)/f(z_k)$. The authors proved in [6] that if $\gamma = 1$ and functions $H$, $G$ and $T$ satisfy several conditions, the iterative methods described in (5.19) reach sixteenth-order of convergence. These conditions are:

$$
\begin{array}{llll}
H(0) = 1, & H'(0) = 2, & H''(0) = 0, & H'''(0) = 24, \\
H^{(iv)}(0) = -72, & G(0,0) = 1, & G_u(0,0) = 2, & G_v(0,0) = 1, \\
G_{uu}(0,0) = 2, & G_{uv}(0,0) = 4, & G_{uuu}(0,0) = 0, & G_{uuuv}(0,0) = 24, \\
G_{uuvv}(0,0) = -16, & G_{uuv}(0,0) = 6, & G_{uuuu}(0,0) = 0, & T_u(0,0,0) = 2, \\
T_v(0,0,0) = 1, & T_w(0,0,0) = 1, & T_{uu}(0,0,0) = 2, & T_{uv}(0,0,0) = 4, \\
T_{vw}(0,0,0) = 2, & T_{uuv}(0,0,0) = 8, & T_{uuu}(0,0,0) = 0, & T_{uvw}(0,0,0) = 8, \\
T_{uw}(0,0,0) = 2, & T_{uuw}(0,0,0) = 2,
\end{array}
$$

and the conditions involving more than a weight function

$$
\begin{aligned}
T_{vv}(0,0,0) &= G_{vv}(0,0), \\
T_{uvv}(0,0,0) &= 4 + G_{uvv}(0,0), \\
G_{uvv}(0,0) &= 8 - (1/3)(G_{uvvv}(0,0) + 6G_{vv}(0,0)), \\
T_{vvv}(0,0,0) &= -6 + 3G_{vv}(0,0).
\end{aligned}
$$

We can find some other optimal sixteenth-order methods existing in the literature. For example, Sharma et al. in [67] designed a scheme of this type from Ostrowski's method. Khattri and Argyros constructed a sixteenth-order method which may converge even if the derivative vanishes during the iteration process [45]. Also Babajee and Thukral in [8], from King's family and by using weight functions with two and three variables, designed a method with this order.

## 5.4 General Classes of Optimal Multipoint Methods

There exist several optimal $n$-point families with arbitrary number of steps that support the Kung-Traub conjecture. In particular, those authors presented in [50] two families, one of them with derivative-free methods and the other one with the following structure: for any $n$, they defined iterative function $q_j(f)$, as follows:

$$
\begin{aligned}
q_1(f)(x) &= x, \\
q_2(f)(x) &= x - f(x)/f'(x), \\
&\vdots \\
q_{j+1}(f)(x) &= S_j(0),
\end{aligned}
$$

for $j = 2, 3, \ldots, n-1$, where $S_j(y)$ is the inverse interpolatory polynomial of degree at most $j$ such that

$$S_j(f(x)) = x, \quad S_j'(f(x)) = 1/f'(x), \quad S_j(f(q_r(f)(x))) = q_r(x), r = 2, 3, \ldots, j.$$

The iterative method is defined by

$$x_{k+1} = q_n(f)(x_k), \tag{5.20}$$

starting with an initial guess $x_0$. In [50] the following result is proved.

**Theorem 5.4.1** *Let $q_n(f)(x)$ be defined by (5.20) and let $x_0$ be close enough to a simple zero $\alpha$ of function $f$. Then the order of convergence of the iterative method (5.20) is $2^n$.*

On the other hand, Petković in [60] presented a class of optimal $n$-point methods for any $n \geq 3$ from an arbitrary optimal fourth-order scheme in the first two steps and by using Hermite interpolating polynomial. Given $x_0$, for $k \geq 0$ calculate the following $n$ steps:

(1)    $\phi_1(x_k) = N(x_k) = x_k - \dfrac{f(x_k)}{f'(x_k)}.$

(2)    $\phi_2(x_k) = \psi_4(x_k, \phi_1(x_k)).$

(3)    $\phi_3(x_k) = \bar{N}(\phi_2(x_k)) = \phi_2(x_k) - \dfrac{f(\phi_2(x_k))}{h_3'(\phi_2(x_k))},$

$$\vdots \tag{5.21}$$

$(n-1)$  $\phi_{n-1}(x_k) = \bar{N}(\phi_{n-2}(x_k)) = \phi_{n-2}(x_k) - \dfrac{f(\phi_{n-2}(x_k))}{h_{n-1}'(\phi_{n-2}(x_k))},$

$(n)$   $\phi_n(x_k) = \bar{N}(\phi_{n-1}(x_k)) = \phi_{n-1}(x_k) - \dfrac{f(\phi_{n-1}(x_k))}{h_n'(\phi_{n-1}(x_k))},$

$$x_{k+1} = \phi_n(x_k).$$

In this procedure

$$\bar{N}(\phi_m) = \phi_m - \frac{f(\phi_m)}{\bar{f}'(\phi_m)},$$

where $\bar{f}'(\phi_m)$ is the approximation of $f'(\phi_m)$ by using the derivative of the Hermite interpolating polynomial

$$h_{m+1}(t) = a_0 + a_1(t - \phi_0) + a_2(t - \phi_0)^2 + \cdots + a_{m+1}(t - \phi_0)^{m+1}$$

of degree $m + 1$ at the nodes $\phi_0, \phi_1, \ldots, \phi_m$ constructed using the conditions

$$f'(\phi_0) = h'_{m+1}(\phi_0), \quad f(\phi_j) = h_{m+1}(\phi_j), j = 0, 1, \ldots, m \leq n - 1.$$

In [61] the following result is presented.

**Theorem 5.4.2** *If $x_0$ is sufficiently close to a simple zero $\alpha$ of function $f$, then the family of n-point methods described in (5.21) has order $2^n$.*

There exist in the literature more general families of arbitrary optimal order of convergence, but their members are derivative-free iterative schemes which are out of the subject of this chapter.

## 5.5    Dynamical Behavior of Optimal Methods

In the literature, optimal methods are analyzed under different points of view. A research area that is getting strength nowadays consists of applying discrete dynamics techniques to the associated fixed point operator of iterative methods. The dynamical behavior of such operators when applied on the simplest function (a low degree polynomial) gives us relevant information about its stability and performance. This study is focused on the asymptotic behavior of fixed points, as well as in its associated basins of attraction. Indeed, in case of families of iterative schemes, the analysis of critical points (where the derivative of the rational function is null), different from the roots of the polynomial, not only allows to select those members of the class with better properties of stability, but also to classify optimal methods of the same order in terms of their dynamics.

The application of iterative methods for solving nonlinear equations $f(z) = 0$, with $f : \mathbb{C} \to \mathbb{C}$, gives rise to rational functions whose dynamics are not well-known. There is an extensive literature on the study of iteration of rational mappings of a complex variable (see, for example, [32, 34]). The simplest model is obtained when $f(z)$ is a quadratic polynomial and the iterative process is Newton's method. The dynamics of this iterative scheme has been widely studied (see, for instance, [14, 32, 37]).

The analysis of the dynamics of Newton's method has been extended to other one-point iterative methods, used for solving nonlinear equations with convergence order higher than two (see, for example, [3, 21, 40, 64, 70]).

Most of the iterative methods analyzed from the dynamical point of view are schemes with derivatives in their iterative expressions. Unlike Newton's method, the derivative-free scheme of Steffensen has been less studied under this point of view. We can find some dynamical comments on this method and its variants in [1, 16, 28, 70].

In order to analyze the dynamical behavior of an iterative method when is applied on a polynomial $p(z)$, it is necessary to recall some basic dynamical concepts. For a more extensive and comprehensive review of these concepts, see [10, 13, 33].

Let $R : \hat{\mathbb{C}} \to \hat{\mathbb{C}}$ be a rational function, where $\hat{\mathbb{C}}$ is the Riemann sphere. The orbit of a point $z_0 \in \hat{\mathbb{C}}$ is defined as

$$\{z_0, R(z_0), \ldots, R^2 (z_0), \ldots, R^n(z_0), \ldots\}.$$

The dynamical behavior of the orbit of a point on the complex plane can be classified depending on its asymptotic behavior on the initial estimation $z_0$ used. To get this aim, the fixed and periodic point of rational operator $R$ must be classified. In this terms, a point $z_0 \in \hat{\mathbb{C}}$ is a *fixed point of R* if $R(z_0) = z_0$. If a fixed point is not a root of $p(z)$, it is called *strange fixed point*. Moreover, a *periodic point* $z_0$ of period $p > 1$ is a point satisfying $R^p (z_0) = z_0$ and $R^k (z_0) \neq z_0, k < p$. A *pre-periodic point* $z_0$ is not a periodic point but there exists $k > 0$ such that $R^k (z_0)$ is a periodic point.

So, the behavior of a fixed point depends of the value of the stability function $|R'(z)|$ on it, that is called *multiplier of the fixed point*. Then, a fixed point is called

- *attracting* if its multiplier is lower than one, $|R'(z_0)| < 1$,
- *superattracting* when it is null, $|R'(z_0)| = 0$,
- *repelling* if the multiplier is greater than one, $|R'(z_0)| > 1$ or
- *neutral* when it is equal to one, $|R'(z_0)| = 1$.

Let us remark that the roots of $p(z)$ will be always fixed points of the rational operator $R$. Moreover, they will be superattracting if the order of convergence of the iterative method is, at least, two.

Also, the stability of a periodic orbit is defined by the magnitude of its multiplier, $|R'(z_1) \cdot \ldots \cdot R'(z_p)|$, being $\{z_1, \ldots, z_p\}$ those points defining the periodic orbit of period $p$.

On the other hand, a critical point $z_0$ validates $R' (z_0) = 0$ and, if it does not coincide with a root of $p(z)$ (that is, if it is not a superattracting fixed point), it is called *free critical point*.

Let $z^*$ be an attracting fixed point of rational function $R$. The *basin of attraction* of $z^*$, $\mathcal{A}(z^*)$, is defined as the set of pre-images of any order such that

$$\mathcal{A}(z^*) = \left\{z_0 \in \hat{\mathbb{C}} : R^n(z_0) \to z^*, n \to \infty\right\}.$$

The set of points $z \in \hat{\mathbb{C}}$ such that their families $\{R^n (z)\}_{n \in \mathbb{N}}$ are normal in a neighborhood $U(z)$ is the *Fatou set*, $\mathcal{F}(R)$, that is, Fatou set is composed by those points whose orbits tend to an attractor (fixed point, periodic orbit or infinity). The complementary set, the *Julia set*, $\mathcal{J}(R)$, is the closure of the set consisting of its repelling fixed points or orbits and their preimages, and establishes the borders between the basins of attraction.

Mayer and Schleicher defined in [54] the *immediate basin of attraction* of an attracting fixed point $z^*$ (considered as a periodic point of period 1), as the connected component of the basin containing $z^*$. This concept is directly related with the existence of critical points.

**Theorem 5.5.1 (Fatou [38], Julia [44])** *Let R be a rational function. The imme-diate basin of attraction of any attracting periodic point holds, at least, a critical point.*

Newton's scheme has been profusely studied by using tools of complex dynamics. See, for example, the texts [14, 37, 63]. The rational function associated to the fixed point operator of Newton's scheme, acting on the generic quadratic polynomial $p(z) = z^2 + c$ is:

$$N_p(z) = \frac{z^2 - c}{2z}. \tag{5.22}$$

There exist only two different fixed points of $N_p(z)$, the roots of $p(z)$, that are superattracting. Moreover, the only critical points of operator $N_p(z)$ are the roots of $p(z)$, so there exist only two basins of attraction.

Schröder and Cayley demonstrated at the end of nineteenth century (see [15, 65], for instance) that, in case of quadratic polynomials, the basins of attraction are both semiplanes, separated by the Julia set. Later on, in 1918, Fatou y Julia were devoted to the systematic analysis of the iterated analytical functions, setting the basis of the actual complex dynamics (see [38, 44]).

In Fig. 5.1a, we observe these basins of attraction for $p(z) = z^2 - 1$, where $z = \pm 1$ are marked in their respective immediate basins as white stars. These dynamical planes have been represented by using the software described in [17]. A mesh with eight hundred points per axis is drawn; each point of the mesh is a different initial estimation which is introduced in the iterative procedure. When the method reaches the solution (under an estimation of the error of $10^{-3}$) in less than eighty iterations, this point is drawn in orange if it converges to the first root

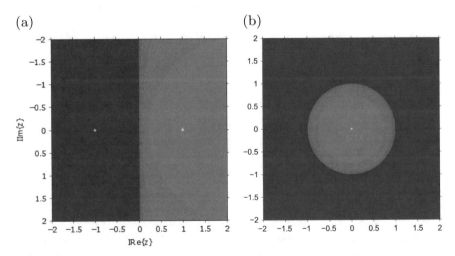

**Fig. 5.1** Dynamical planes corresponding to Newton's method. (**a**) $N_p(z) = \frac{z^2+1}{2z}$. (**b**) $N_q(z) = z^2$

and in blue if it converges to the second one. When other attracting fixed points appear, different colors are used. The color will be more intense when the number of iterations is lower. Otherwise, if the method arrives at the maximum of iterations, the point will be drawn in black.

### 5.5.1   Conjugacy Classes

It is possible to analyze the behavior of an iterative method acting on any quadratic polynomial. For this, we define the following concept.

**Definition 5.5.1** Let $f$ and $g$ be defined as $f, g : \hat{\mathbb{C}} \to \hat{\mathbb{C}}$. An analytic conjugation between $f$ and $g$ is a diffeomorphism $h : \hat{\mathbb{C}} \to \hat{\mathbb{C}}$ such that $h \circ f = g \circ h$.

In the next result from Curry, Garnet and Sullivan, it is shown that the rational functions resulting from the iteration of Newton's method on affine-conjugated analytic functions are also conjugated and so, their dynamics are equivalent.

**Theorem 5.5.2 (Scaling Theorem for Newton's Method, [32])** *Let $g(z)$ be an analytic function, and let $A(z) = \alpha z + \beta$, with $\alpha \neq 0$, be an affine map. Let $h(z) = \lambda (g \circ A)(z)$, with $\lambda \neq 0$. Let $O_p(z)$ be the fixed point operator of Newton's method. Then, $\left( A \circ O_h \circ A^{-1} \right)(z) = O_g(z)$, i.e., $O_g$ and $O_h$ are affine conjugated by $A$.*

Moreover, it is possible to allow up the knowledge of a family of polynomials with just the analysis of a few cases, from a suitable scaling, as it is stated in the following known result for quadratic polynomials.

**Theorem 5.5.3** *Let $p(z) = a_1 z^2 + a_2 z + a_3$, $a_1 \neq 0$, be a generic quadratic polynomial with simple roots. Then $p(z)$ can be reduced to $q(z) = z^2 + c$, where $c = 4a_1 a_3 - a_2^2$, by means of an affine map. This affine transformation induces a conjugation between $N_q$ and $N_p$.*

In fact, a classical result from Cayley [15] and Schröder [65] shows that Julia and Fatou sets of two conjugated rational functions (by means of a Möbius map) are also conjugated.

**Theorem 5.5.4** *Let $R_1$ and $R_2$ be two rational functions and let $\psi$ be a Möbius map satisfying, $R_2 = \psi \circ R_1 \circ \psi^{-1}$. Then, $\mathcal{J}(R_2) = \psi(\mathcal{J}(R_1))$ and $\mathcal{F}(R_2) = \psi(\mathcal{F}(R_1))$.*

The following result, from Cayley and Schröder, made the difference in the analysis of Newton's method. They analyzed the dynamical behavior of Newton's scheme on quadratic polynomials in a simpler way, in the complex plane. This motivated the later works in the area by Fatou and Julia.

**Theorem 5.5.5 (Cayley [15], Schröder [65])** *Let*

$$N_p(z) = \frac{z^2 - ab}{2z - (b+a)}$$

*be the rational operator associated to Newton's method on the quadratic polynomial* $p(z) = (z - a)(z - b)$, *being* $a \neq b$. *Then* $N_p$ *is conjugated to* $z \to z^2$ *by Möbius map* $h(z) = \dfrac{z-a}{z-b}$, *being its Julia set* $\mathcal{J}(N_p)$, *in the complex plane, the straight line equidistant between* $a$ *and* $b$, *or, equivalently, the unit circle after Möbius transformation.*

Then, by applying Möbius map,

$$h(z) = \frac{z-a}{z-b} \tag{5.23}$$

satisfying the following properties: $h(\infty) = 1$, $h(a) = 0$, $h(b) = \infty$, points 0 and $\infty$ play the role of the roots of the polynomial $a$ and $b$ in $N_q$ and also the rational operator associated to Newton's scheme is free of parameters $a$ and $b$ vanish from the rational function,

$$N_q(z) = \left(h \circ N_p \circ h^{-1}\right)(z) = z^2.$$

The dynamical plane associated to $N_q$ can be observed in Fig. 5.1b.

### 5.5.2 Optimal Methods on Quadratic Polynomials

As Newton's scheme shows global convergence, a first classification of optimal iterative methods is to behave "as well as Newton's method" on quadratic polynomials. This is the simplest dynamics and, if an iterative scheme has as associated rational function $z^k$, $k \geq 1$, it satisfies the so-called *Cayley Test*. In this case, there exist no critical points different from 0 and $\infty$ (corresponding to the original roots of polynomial $p(z)$), and its corresponding dynamical plane will have the same appearance as Fig. 5.1b.

In [9], a large collection of optimal iterative methods are analyzed and classified in terms of satisfying Cayley Test: if we consider families of iterative methods, such as Chebyshev–Halley-like family due to Nedzhibov et al. [56], the only elements of the class satisfying Cayley Test are those that are optimal, with $\lambda = 1$; the dynamical planes of some elements of these family are showed in Fig. 5.2.

Other element of this family is Ostrowski's method ([59], appearing for $\lambda = \beta = 1$, see Fig. 5.2b), as well as in King's family [47] whose dynamics was analyzed in [26]. The optimal class of iterative methods constructed by Kim in [46] has order four, independently from the values of parameters $\lambda$, $\beta$ and $\mu$. However, some

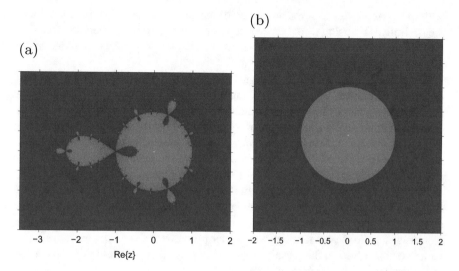

**Fig. 5.2** Dynamical planes of members of Chebyshev–Halley-like class. (**a**) $\lambda = 0$. (**b**) $\lambda = 1$

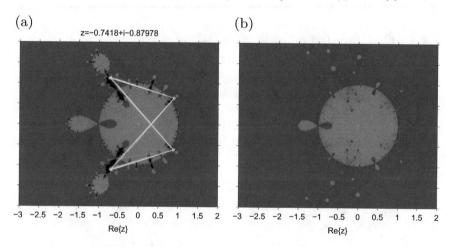

**Fig. 5.3** Dynamical plane corresponding to elements of Kim's family. (**a**) $\lambda = 1, \beta = 0, \mu = 2$. (**b**) $\lambda = 1, \beta = 0, \mu = 5$

conditions must be imposed to pass Cayley Test: the associated rational function is $z^4$ if $\lambda = 1$ and $\mu = 2\beta$, and it is conjugated to $z^5$ for the particular procedure $\lambda = 0$, $\beta = -1$ and $\mu = 1$. A couple of examples of members of this class showing unstable (an attracting 4-periodic orbit in Fig. 5.3a) and stable behavior (see Fig. 5.3b) are presented in Fig. 5.3.

On the other hand, Neta et al. in [57] have analyzed the conjugacy classes associated to several optimal eight-order methods, generated by composition of an optimal fourth-order scheme (Jarratt, Kung-Traub, Ostrowski) or King's family with a step of Newton and replacing one of the new evaluations (of the function or the derivative) by using direct or inverse interpolation. Some of them are found to satisfy Cayley Test and for the rest of them, the strange fixed points are numerically calculated and their stability studied.

Some other optimal fourth- or eight-order families have been deeply analyzed in terms of complex dynamics, as King's family in [26], where the authors get the conjugacy class of this family and calculate both fixed and critical points. This allowed to analytically study the stability of strange fixed points and, by using the explicit expression of free critical points, to get the parameter plane and completely analyze the stability of the methods depending on the behavior of the parameter. Also the class designed by Behl in [11] which was studied by Argyros et al. in [4] or the optimal eighth-order family from [57] have been dynamically studied in [31]. Similar analysis were previously made on non-optimal schemes or families, as for example [1, 2, 16, 27, 41], among others. This kind of study allows to select those elements of the different families with better properties of stability, as the value of parameters involved in the iterative expression plays a crucial role in the behavior of the corresponding iterative scheme.

### 5.5.3  Optimal Methods on Cubic Polynomials

Let us consider now those optimal iterative schemes satisfying Cayley Test for quadratic polynomials. Although all these optimal schemes show a completely stable behavior on quadratic polynomials, when they are applied on cubic ones, the result is uneven, as it was showed in [9]. In Fig. 5.4 it seems clear: the behavior of Ostrowski's scheme on $z^3 - 1$ showed in Fig. 5.4a is like the one of Newton's but

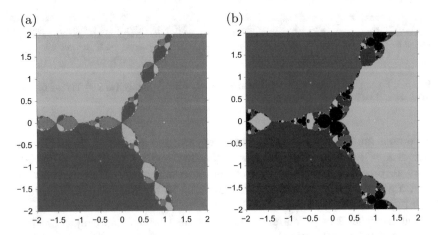

**Fig. 5.4** Some optimal 4th-order methods on $z^3 - 1$. (**a**) Ostrowski's method. (**b**) Kim's method $\lambda = 0, \beta = -1$ and $\mu = 1$

with optimal fourth-order of convergence, however other optimal scheme, which also satisfies Cayley Test for quadratic polynomials, posses black areas of no convergence to the roots in the dynamical plane showed in Fig. 5.4b. This shows the necessity of establishing two categories inside optimal methods: depending of the satisfaction of Cayley Test, and among those that satisfy it, depending of their behavior on cubic polynomials.

Indeed, McMullen in [55] showed that there exists one third-order iterative method with global convergence for cubic polynomials $q(z) = z^3 + az + b$ (or, in general, any cubic polynomial), obtained by applying Newton's method on $r(z) = \dfrac{q(z)}{3az^2 + 9bz - a^2}$; the resulting rational function is

$$N_q(z) = z - \frac{(z^3 + az + b)(3az^2 + 9bz - a^2)}{3az^4 + 18bz^3 - 6a^2z^2 - 6abz - 9b^2 - a^3},$$

which coincides with Halley's method applied on $z^3 - 1$. We can see the associated dynamical plane in Fig. 5.5a, whereas in Fig. 5.5b we observe the behavior of an element of King's family ($\beta = 1$) on $z^3 - 1$.

Also Hawkins in [42], showed that it is not possible to design globally convergent methods from rational functions of any degree. In fact, he showed that, for $q(z) = z^3 - 1$,

$$R_7(z) = \frac{z(14 + 35z^3 + 5z^6)}{5 + 35z^3 + 14z^6}$$

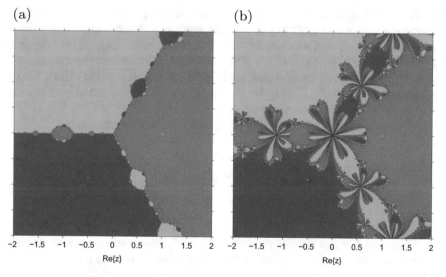

**Fig. 5.5** Some dynamical planes on $z^3 - 1$. (**a**) $R_4(z)$. (**b**) King's method ($\beta = 1$)

is the only algorithm of degree 7 with order of convergence higher than three (in fact it has fifth-order convergence) and

$$R_{10}(z) = \frac{z(7 + 42z^3 + 30z^6 + 2z^9)}{2 + 30z^3 + 42z^6 + 7z^9}$$

is the only algorithm of degree 10 with seventh-order convergence. McMullen third-order scheme is, in this case,

$$R_4(z) = \frac{z(2 + z^3)}{1 + 2z^3}.$$

As Hawkins also showed that there cannot be any generally convergent methods which are of order precisely 4, optimal fourth-order schemes are not able to converge to a root for almost every initial guess and for almost every polynomial.

Does it means that it is not possible to classify optimal methods in terms of their behavior on cubic polynomials? Not necessarily, as we can analyze the number of critical points of the rational operator associated to an iterative scheme on cubic polynomials. The lower this number is, the more stable will be the process (as in the immediate basin of attraction of any attracting periodic point there is at least one critical point).

## 5.6 Numerical Performances

In the following, we apply some of the described iterative schemes with order of convergence four and eight to estimate the solution of some particular nonlinear equations. In particular, we will compare Newton's scheme (NM) and the following fourth-order methods: Ostrowski's procedure (OM, [59]), whose iterative expression is (5.10); Jarratt's method (JM, [43]), with iterative expression (5.14); King's scheme (expression (5.11), [47], for the special case $\beta = 1$, denoted by KM), the variant of Potra-Pták that will be denoted by PTM, designed in [23], [Eq. (5.13)] and Chun's scheme (CM, selected in [18]), which iterative expression is

$$y_k = x_k - \frac{f(x_k)}{f'(x_k)},$$

$$x_{k+1} = y_k - \left(1 + 2\frac{f(y_k)}{f(x_k)}\right) \frac{f(y_k)}{f'(x_k)}.$$

Numerical computations have been carried out using variable precision arithmetics, with 2000 digits of mantissa, in Mathematica R2014a, in a computer with processor Intel(R) Xeon(R) CPU E5-2420 v2@2.2 GHz, 64 GB RAM. The stopping criterion used has been $|x_{k+1} - x_k| < 10^{-500}$ or $|f(x_{k+1})| < 10^{-500}$. For every method, we count the number of iterations needed to reach the wished precision, we calculate the approximated computational order of convergence $ACOC$ by using expression (5.1) and the error estimation made with the last values of $|x_{k+1} - x_k|$ and $|f(x_{k+1})|$.

The nonlinear functions $f(x)$, the desired zeros $\alpha$ and the initial estimations $x_0$ used for the tests, joint with the numerical results obtained are described in the following.

- $f_1(x) = \sin x - x^2 + 1$, $\alpha \approx 1.404492$, $x_0 = 1$.
- $f_2(x) = x^2 - e^x - 3x + 2$, $\alpha \approx 0.257530$, $x_0 = 1$.
- $f_3(x) = \cos x - x$, $\alpha \approx 0.739085$, $x_0 = 0$.
- $f_4(x) = \cos x - xe^x + x^2$, $\alpha \approx 0.639154$, $x_0 = 2$.
- $f_5(x) = e^x - 1.5 - \arctan x$, $\alpha \approx 0.767653$, $x_0 = 5$.

We show in Tables 5.1 and 5.2 the performance of second- (Newton) fourth- and eighth-order methods by means of the following items: the number of iterations $iter$, the error estimations of the last iteration, $|x_{k+1} - x_k|$ and $|f(x_{k+1})|$, the approximated order of convergence ACOC and the mean elapsed time (e-time) in seconds after 500 executions of the methods.

Let us remark that, as it was expected, the duplication of the order of convergence respect to Newton's method yields to a reduction (in a half, approximately) of the number of iterations. Is is reflected also in the mean execution time. Indeed, when a method needs one more iteration than the rest of partners (with the same order), then the precision of its results is the highest and the elapsed time is not as much high.

Respect the comparison among optimal eighth-order methods, the following schemes will be used: scheme (5.18), by Bi et al., where $p(t) = 1 + \dfrac{2t}{1+t}$, that will be denoted by BRW.

On the other hand, it will be also checked the method with the iterative expression

$$y_k = x_k - \frac{f(x_k)}{f'(x_k)},$$

$$z_k = x_k - \frac{f(x_k) - f(y_k)}{f(x_k) - 2f(y_k)} \frac{f(x_k)}{f'(x_k)},$$

$$x_{k+1} = u_k - 3\frac{f(z_k)}{f'(x_k)} \frac{u_k - z_k}{y_k - x_k},$$

where $u_k = z_k - \dfrac{f(z_k)}{f'(x_k)} \left( \dfrac{f(x_k) - f(y_k)}{f(x_k) - 2f(y_k)} + \dfrac{1}{2} \dfrac{f(z_k)}{f(y_k) - 2f(z_k)} \right)^2$ ; it corresponds to an element of the optimal eighth-order family of iterative schemes designed in [25] and will be denoted by CTV.

**Table 5.1** Numerical results for fourth-order methods

|          | Method | Iter | $|x_{k+1} - x_k|$ | $|f(x_{k+1})|$ | ACOC | e-time (s) |
|----------|--------|------|-------------------|----------------|--------|------------|
| $f_1$    | NM     | 10   | 1.867e−273        | 5.205e−546     | 2.0000 | 0.6253     |
| $x_0 = 1$| OM     | 5    | 7.315e−139        | 1.307e−553     | 4.0000 | 0.3802     |
|          | JM     | 5    | 4.093e−139        | 1.268e−554     | 4.0000 | 0.4095     |
|          | KM     | 6    | 1.982e−315        | 2.156e−1259    | 4.0000 | 0.4635     |
|          | PTM    | 6    | 5.611e−272        | 1.853e−1085    | 4.0000 | 0.4789     |
|          | CM     | 6    | 3.774e−196        | 4.751e−782     | 4.0000 | 0.4559     |
| $f_2$    | NM     | 9    | 1.869e−380        | 1.234e−760     | 2.0000 | 0.6187     |
| $x_0 = 1$| OM     | 5    | 1.756e−258        | 1.622e−1033    | 4.0000 | 0.4346     |
|          | JM     | 5    | 3.475e−286        | 3.362e−1144    | 4.0000 | 0.4375     |
|          | KM     | 5    | 6.895e−263        | 2.463e−1051    | 4.0000 | 0.4392     |
|          | PTM    | 5    | 4.41e−266         | 2.955e−1064    | 4.0000 | 0.4426     |
|          | CM     | 5    | 7.97e−271         | 1.909e−1083    | 4.0000 | 0.4522     |
| $f_3$    | NM     | 10   | 1.119e−333        | 4.625e−667     | 2.0000 | 0.6099     |
| $x_0 = 0$| OM     | 5    | 3.282e−141        | 4.964e−564     | 4.0000 | 0.3682     |
|          | JM     | 5    | 1.957e−143        | 5.78e−573      | 4.0000 | 0.3872     |
|          | KM     | 6    | 4.255e−386        | 2.585e−1543    | 4.0000 | 0.4365     |
|          | PTM    | 6    | 1.688e−344        | 7.861e−1377    | 4.0000 | 0.4522     |
|          | CM     | 6    | 5.826e−283        | 1.324e−1130    | 4.0000 | 0.4306     |
| $f_4$    | NM     | 12   | 3.372e−326        | 2.162e−651     | 2.0000 | 0.7858     |
| $x_0 = 2$| OM     | 6    | 1.44e−239         | 1.489e−956     | 4.0000 | 0.4943     |
|          | JM     | 6    | 1.23e−232         | 8.785e−929     | 4.0000 | 0.5246     |
|          | KM     | 7    | 3.786e−479        | 5.522e−1914    | 4.0000 | 0.5889     |
|          | PTM    | 7    | 2.079e−406        | 7.214e−1623    | 4.0000 | 0.5991     |
|          | CM     | 7    | 2.241e−368        | 1.267e−1470    | 4.0000 | 0.5728     |
| $f_5$    | NM     | 15   | 9.913e−375        | 1.357e−748     | 2.0000 | 0.9587     |
| $x_0 = 5$| OM     | 7    | 3.631e−161        | 1.504e−642     | 4.0000 | 0.5475     |
|          | JM     | 7    | 6.98e−154         | 2.064e−613     | 4.0000 | 0.5826     |
|          | KM     | 8    | 1.167e−216        | 5.809e−864     | 4.0000 | 0.6511     |
|          | PTM    | 8    | 1.757e−163        | 4.062e−651     | 3.9999 | 0.6805     |
|          | CM     | 8    | 1.839e−139        | 6.176e−555     | 3.9998 | 0.6652     |

We will also use the following scheme by Džunić and Petković in [35] (denoted by DP), whose two first steps are also Ostrowski's method, being the third one

$$x_{k+1} = z_k - \frac{(1 + w_k)(1 + 2v_k)}{1 - 2u_k - u_k^2} \frac{f(z_k)}{f'(x_k)},$$

where $u_k = \dfrac{f(y_k)}{f(x_k)}$, $v_k = \dfrac{f(z_k)}{f(x_k)}$ and $w_k = \dfrac{f(z_k)}{f(y_k)}$.

Another eighth-order method, denoted by LW, was designed by Liu and Wang in [51]. It is initialized also with Ostroswki's scheme and the third step of its iterative

**Table 5.2** Numerical results for optimal eighth-order methods

|  | Method | Iter | $|x_{k+1} - x_k|$ | $|f(x_{k+1})|$ | ACOC | e-time (s) |
|---|---|---|---|---|---|---|
| $f_1$ | BRW | 4 | 8.708e−239 | 1.558e−1907 | 7.9997 | 0.4127 |
| $x_0 = 1$ | CTV | 4 | 1.223e−226 | 5.861e−1809 | 8.0000 | 0.3854 |
|  | DP | 4 | 2.799e−266 | 2.697e−2008 | 8.0000 | 0.4033 |
|  | LW | 4 | 1.356e−212 | 4.021e−1696 | 7.9999 | 0.4005 |
|  | SS | 4 | 2.413e−239 | 1.454e−1910 | 8.0000 | 0.4107 |
|  | SM | 4 | 4.368e−168 | 3.572e−1340 | 8.0006 | 0.4334 |
| $f_2$ | BRW | 4 | 3.503e−484 | 0.0 | 8.0000 | 0.4686 |
| $x_0 = 1$ | CTV | 4 | 7.796e−418 | 2.697e−2008 | 8.0000 | 0.4300 |
|  | DP | 4 | 7.301e−424 | 2.697e−2008 | 8.0000 | 0.4406 |
|  | LW | 4 | 5.317e−384 | 0.0 | 8.0000 | 0.4447 |
|  | SS | 4 | 1.415e−487 | 0.0 | 8.0000 | 0.4352 |
|  | SM | 4 | 7.005e−416 | 2.697e−2008 | 8.0000 | 0.4652 |
| $f_3$ | BRW | 4 | 1.883e−306 | 1.349e−2008 | 8.0000 | 0.3767 |
| $x_0 = 0$ | CTV | 4 | 5.538e−247 | 1.903e−1974 | 7.9999 | 0.3599 |
|  | DP | 4 | 7.007e−253 | 1.349e−2008 | 8.0000 | 0.3693 |
|  | LW | 4 | 2.535e−289 | 1.349e−2008 | 8.0000 | 0.3598 |
|  | SS | 4 | 1.826e−251 | 1.349e−2008 | 8.0000 | 0.3695 |
|  | SM | 4 | 1.51e−198 | 6.153e−1587 | 8.0009 | 0.4026 |
| $f_4$ | BRW | 5 | 1.12e−381 | 1.349e−2008 | 8.0000 | 0.5537 |
| $x_0 = 2$ | CTV | 4 | 2.193e−96 | 4.608e−767 | 7.9999 | 0.4413 |
|  | DP | 4 | 5.974e−107 | 9.433e−852 | 7.9926 | 0.4475 |
|  | LW | 4 | 3.647e−106 | 1.144e−844 | 7.9988 | 0.4374 |
|  | SS | 4 | 2.958e−99 | 3.614e−790 | 8.0012 | 0.4405 |
|  | SM | 5 | 4.483e−368 | 1.349e−2008 | 8.0000 | 0.5828 |
| $f_5$ | BRW | 4 | 2.007e−70 | 1.63e−558 | 7.9326 | 0.4395 |
| $x_0 = 5$ | CTV | 5 | 8.568e−109 | 3.589e−865 | 7.9951 | 0.5219 |
|  | DP | 5 | 6.356e−128 | 2.593e−1018 | 7.9985 | 0.5403 |
|  | LW | 5 | 4.414e−89 | 5.559e−707 | 7.9872 | 0.5072 |
|  | SS | 5 | 8.742e−112 | 4.146e−889 | 7.9960 | 0.5149 |
|  | SM | – | – | – | – | – |

expression is:

$$x_{k+1} = z_k - \left[ \left( \frac{f(x_k) - f(y_k)}{f(x_k) - 2f(y_k)} \right)^2 + \frac{f(z_k)}{f(y_k) - 5f(z_k)} + \frac{4t_k}{1 - 7t_k} \right] \frac{f(z_k)}{f'(x_k)},$$

where $t_k = \dfrac{f(z_k)}{f(x_k)}$.

Again Sharma and Sharma in [66] initialized with Ostrowski's procedure, being the iterative expression of the last step:

$$x_{k+1} = z_k - (1 + t_k + t_k^2)\frac{f[x_k, y_k]f(z_k)}{f[x_k, z_k]f[y_k, z_k]},$$

where $t_k = \dfrac{f(z_k)}{f(x_k)}$. We will denote this method by SS.

Finally, we recall the procedure designed by Soleymani et al. in [68], denoted by SM, where $y_k$ is Newton's step, $z_k$ is (second) Ostrowski's step and the third one is

$$x_{k+1} = z_k - \frac{f(z_k)}{2f[y_k, x_k] - f'(x_k) + f[z_k, x_k, x_k](z_k - y_k)}A_k$$

where $A_k = \left(1 + w_k + 2v_k - 2u_k^3 + \dfrac{2}{5}\dfrac{f(z_k)}{f'(x_k)}\right)$, being $u_k = \dfrac{f(y_k)}{f(x_k)}$, $v_k = \dfrac{f(z_k)}{f(x_k)}$ and $w_k = \dfrac{f(z_k)}{f(y_k)}$.

The numerical results obtained by using these schemes are shown in Table 5.2. All schemes perform adequately, confirming theoretical order of convergence. The differences among them are not very significant and, for each problem, all of them achieve an estimation of the solution with 500 significant digits in 4 or 5 iterations. Symbol '−' means that this method diverges with the initial estimation used.

## 5.7  Conclusion

In this chapter, a survey on optimal iterative methods using derivatives for solving nonlinear equations is presented, centering the attention on an specific aspect: the techniques employed to generate them. It does not pretend to include all the designed optimal methods of any order, but to reflect the state of the art and some of the families and methods constructed by using this techniques.

It has been also pretended to show how it is necessary to compare and classify these schemes, as even optimal procedures can behave in different ways on the same functions. To get this aim, complex discrete dynamics techniques play an important role, and many researchers have employed these tools to better understanding the processes and classifying them in terms of stability and reliability.

The numerical tests made reflect the theoretical results and show that, when variable precision arithmetics is used, small differences in precision, mean execution time and number of iterations can be found when some academic test functions are solved.

**Acknowledgements** This scientific work has been supported by Ministerio de Economía y Competitividad MTM2014-52016-C02-2-P.

# References

1. Amat, S., Busquier, S., Plaza, S.: Review of some iterative root-finding methods from a dynamical point of view. Sci. Ser. A: Math. Scientia **10**, 3–35 (2004)
2. Amat, S., Busquier, S., Plaza, S.: A construction of attracting periodic orbits for some classical third-order iterative methods. J. Comput. Appl. Math. **189**, 22–33 (2006)
3. Amat, S., Busquier, S., Plaza, S.: Chaotic dynamics of a third-order Newton-type method. J. Math. Anal. Appl. **366**, 24–32 (2010)
4. Argyros, I.K., Magreñán, Á.A.: On the convergence of an optimal fourth-order family of methods and its dynamics. Appl. Math. Comput. **252**, 336–346 (2015)
5. Artidiello, S.: Diseño, implementación y convergencia de métodos iterativos para resolver ecuaciones y sistemas no lineales utilizando funciones peso. Ph.D. thesis, Universitat Politèc-nica de València (2014)
6. Artidiello, S., Cordero, A., Torregrosa, J.R., Vassileva, M.P.: Optimal high-order methods for solving nonlinear equations. J. Appl. Math. **2014**, 9 pp. (2014). ID 591638
7. Artidiello, S., Cordero, A., Torregrosa, J.R., Vassileva, M.P.: Two weighted eight-order classes of iterative root-finding methods. Int. J. Comput. Math. **92**(9), 1790–1805 (2015)
8. Babajee, D.K.R., Thukral, R.: On a 4-Point sixteenth-order King family of iterative methods for solving nonlinear equations. Int. J. Math. Math. Sci. **2012**, 13 pp. (2012). ID 979245
9. Babajee, D.K.R., Cordero, A., Torregrosa, J.R.: Study of iterative methods through the Cayley Quadratic Test. J. Comput. Appl. Math. **291**, 358–369 (2016)
10. Beardon, A.F.: Iteration of Rational Functions. Graduate Texts in Mathematics. Springer, New York (1991)
11. Behl, R.: Development and analysis of some new iterative methods for numerical solutions of nonlinear equations. Ph.D. thesis, Punjab University (2013)
12. Bi, W., Ren, H., Wu, Q.: Three-step iterative methods with eighth-order convergence for solving nonlinear equations. J. Comput. Appl. Math. **225**, 105–112 (2009)
13. Blanchard, P.: Complex analytic dynamics on the Riemann sphere. Bull. Am. Math. Soc. **11**(1), 85–141 (1984)
14. Blanchard, P.: The dynamics of Newton's method. Proc. Symp. Appl. Math. **49**, 139–154 (1994)
15. Cayley, A.: Applications of the Newton-Fourier Method to an imaginary root of an equation. Q. J. Pure Appl. Math. **16**, 179–185 (1879)
16. Chicharro, F., Cordero, A., Gutiérrez, J.M., Torregrosa, J.R.: Complex dynamics of derivative-free methods for nonlinear equations. Appl. Math. Comput. **219**, 7023–7035 (2013)
17. Chicharro, F.I., Cordero, A., Torregrosa, J.R.: Drawing dynamical and parameters planes of iterative families and methods. Sci. World J. **2013**, 11 pp. (2013). Article ID 780153. http://dx.doi.org/10.1155/2013/780153
18. Chun, C.: Some fourth-order iterative methods for solving nonlinear equations. Appl. Math. Comput. **195**, 454–459 (2008)
19. Chun, C., Ham, Y.: A one-parametric fourth-order family of iterative methods for nonlinear equations. Appl. Math. Comput. **189**, 610–614 (2007)
20. Chun, C., Ham, Y.: Some second-derivative-free variants of super Halley method with fourth-order convergence. Appl. Math. Comput. **195**, 537–541 (2008)
21. Chun, C., Lee, M.Y., Neta, B., Džunić, J.: On optimal fourth-order iterative methods free from second derivative and their dynamics. Appl. Math. Comput. **218**, 6427–6438 (2012)
22. Cordero, A., Torregrosa, J.R.: Variants of Newton's method using fifth-order quadrature formulas. Appl. Math. Comput. **190**, 686–698 (2007)
23. Cordero, A., Hueso, J.L., Martínez, E., Torregrosa, J.R.: New modifications of Potra-Pták's method with optimal fourth and eighth orders of convergence. J. Comput. Appl. Math. **234**(10), 2969–2976 (2010)
24. Cordero, A., Torregrosa, J.R., Vassileva, M.P.: A family of modified Ostrowski's methods with optimal eighth order of convergence. Appl. Math. Lett. **24**, 2082–2086 (2011)

25. Cordero, A., Torregrosa, J.R., Vassileva, M.P.: Three-step iterative methods with optimal eighth-order convergence. J. Comput. Appl. Math. **235**, 3189–3194 (2011)
26. Cordero, A., García-Maimó, J., Torregrosa, J.R., Vassileva, M.P., Vindel, P.: Chaos in King's iterative family. Appl. Math. Lett. **26**, 842–848 (2013)
27. Cordero, A., Torregrosa, J.R., Vindel, P.: Dynamics of a family of Chebyshev-Halley type methods. Appl. Math. Comput. **219**, 8568–8583 (2013)
28. Cordero, A., Soleymani, F., Torregrosa, J.R., Shateyi, S.: Basins of attraction for various Steffensen-type methods. J. Appl. Math. **2014**, 17 pp. (2014). Article ID 539707. http://dx. doi.org/10.1155/2014/539707
29. Cordero, A., Lotfi, T., Mahdiani, K., Torregrosa, J.R.: Two optimal general classes of iterative methods with eighth-order. Acta Appl. Math. **134**(1), 61–74 (2014)
30. Cordero, A., Lotfi, T., Mahdiani, K., Torregrosa, J.R.: A stable family with high order of convergence for solving nonlinear equations. Appl. Math. Comput. **254**, 240–251 (2015)
31. Cordero, A., Magreñán, Á.A., Quemada, C., Torregrosa, J.R.: Stability study of eighth-order iterative methods for solving nonlinear equations. J. Comput. Appl. Math. **291**, 348–357 (2016)
32. Curry, J., Garnet, L., Sullivan, D.: On the iteration of a rational function: computer experiments with Newton's method. Commun. Math. Phys. **91**, 267–277 (1983)
33. Devaney, R.L.: An Introduction to Chaotic Dynamical Systems. Addison-Wesley, Redwood City, CA (1989)
34. Douady, A., Hubbard, J.H.: On the dynamics of polynomials-like mappings. Ann. Sci. Ec. Norm. Sup. **18**, 287–343 (1985)
35. Džunić, J., Petković, M.: A family of three-point methods of Ostrowski's type for solving nonlinear equations. J. Appl. Math. 9 pp. (2012). ID 425867. doi:10.1155/2012/425867
36. Džunić, J., Petković, M.S., Petković, L.D.: A family of optimal three-point methods for solving nonlinear equations using two parametric functions. Appl. Math. Comput. **217**, 7612–7619 (2011)
37. Fagella, N.: Invariants in dinàmica complexa. Butlletí de la Soc. Cat. de Matemàtiques **23**(1), 29–51 (2008)
38. Fatou, P.: Sur les équations fonctionelles. Bull. Soc. Mat. Fr. **47**, 161–271 (1919); **48**, 33–94, 208–314 (1920)
39. Geum, Y.H., Kim, Y.I.: A uniparametric family of three-step eighth-order multipoint iterative methods for simple roots. Appl. Math. Lett. **24**, 929–935 (2011)
40. Gutiérrez, J.M., Hernández, M.A., Romero, N.: Dynamics of a new family of iterative processes for quadratic polynomials. J. Comput. Appl. Math. **233**, 2688–2695 (2010)
41. Gutiérrez, J.M., Plaza, S., Romero, N.: Dynamics of a fifth-order iterative method. Int. J. Comput. Math. **89**(6), 822–835 (2012)
42. Hawkins, J.M.: McMullen's root-finding algorithm for cubic polynomials. Proc. Am. Math. Soc. **130**, 2583–2592 (2002)
43. Jarratt, P.: Some fourth order multipoint iterative methods for solving equations. Math. Comput. **20**, 434–437 (1966)
44. Julia, G.: Mémoire sur l'iteration des fonctions rationnelles. J. Mat. Pur. Appl. **8**, 47–245 (1918)
45. Khattri, S.K., Argyros, I.K.: Sixteenth order iterative methods without restraint on derivatives. Appl. Math. Sci. **6**(130), 6477–6486 (2012)
46. Kim, Y.I.: A triparametric family of three-step optimal eighth-order methods for solving nonlinear equations. Int. J. Comput. Math. **89**, 1051–1059 (2012)
47. King, R.F.: A family of fourth-order methods for solving nonlinear equations. SIAM J. Numer. Anal. **10**, 876–879 (1973)
48. Kou, J., Li, Y., Wang, X.: A composite fourth-order iterative method for solving nonlinear equations. Appl. Math. Comput. **184**, 471–475 (2007)
49. Kou, J., Li, Y., Wang, X.: Fourth-order iterative methods free from second derivative. Appl. Math. Comput. **184**, 880–885 (2007)
50. Kung, H.T., Traub, J.F.: Optimal order of one-point and multipoint iteration. J. ACM **21**, 643–651 (1974)

51. Liu, L., Wang, X.: Eighth-order methods with high efficiency index for solving nonlinear equations. Appl. Math. Comput. **215**(9), 3449–3454 (2010)
52. Magreñán, Á.A.: Estudio de la dinámica del método de Newton amortiguado. Ph.D. thesis. Servicio de Publicaciones, Universidad de La Rioja (2013)
53. Maheshwari, A.K.: A fourth order iterative methods for solving nonlinear equations. Appl. Math. Comput. **211**, 383–391 (2009)
54. Mayer, S., Schleicher, D.: Immediate and virtual basins of Newton's method for entire functions. Ann. Inst. Fourier **56**(2), 325–336 (2006)
55. McMullen, C.: Families of rational maps and iterative root-finding algorithms. Ann. Math. **125**, 467–493 (1987)
56. Nedzhibov, G.H., Hasanov, V.I., Petkov, M.G.: On some families of multi-point iterative methods for solving nonlinear equations. Numer. Algorithms **42**(2), 127–136 (2006)
57. Neta, B., Chun, C., Scott, M.: Basins of attraction for optimal eighth order methods to find simple roots of nonlinear equations. Appl. Math. Comput. **227**, 567–592 (2014)
58. Ortega, J.M., Rheinboldt, W.C.: Iterative Solution of Nonlinear Equations in Several Variables. Academic, New York (1970)
59. Ostrowski, A.M.: Solution of Equations and Systems of Equations. Academic, New York (1960)
60. Petković, M.S.: On a general class of multipoint root-finding methods of high computational efficiency. SIAM J. Numer. Anal. **47**, 4402–4414 (2010)
61. Petković, M.S., Neta, B., Petković, L.D., Džunić, J.: Multipoint Methods for Solving Nonlinear Equations. Academic, Elsevier, Amsterdam (2013)
62. Petković, M.S., Neta, B., Petković, L.D., Džunić, J.: Multipoint methods for solving nonlinear equations: a survey. Appl. Math. Comput. **226**, 635–660 (2014)
63. Plaza, S., Gutiérrez, J.M.: Dinámica del método de Newton. Material Didáctico de Matemáticas 9, Universidad de La Rioja, Servicio de Publicaciones (2013)
64. Plaza, S., Romero, N.: Attracting cycles for the relaxed Newton's method. J. Comput. Appl. Math. **235**, 3238–3244 (2011)
65. Schröder, E.: Ueber iterite Functionen. Math. Ann. **3**, 296–322 (1871)
66. Sharma, J.R., Sharma, R.: A new family of modified Ostrowski's methods with accelerated eighth order convergence. Numer. Algorithms **54**(4), 445–458 (2010)
67. Sharma, J.R., Guha, R.K., Gupta, P.: Improved King's methods with optimal order of convergence based on rational approximations. Appl. Math. Lett. **26**(4), 473–480 (2013)
68. Soleymani, F., Sharifi, M., Mousavi, B.S.: An improvement of Ostrowski's and King's techniques with optimal convergence order eight. J. Optim. Theory Appl. **153**, 225–236 (2012)
69. Traub, J.F.: Iterative Methods for the Solution of Equations. Prentice-Hall, Englewood Cliffs, NJ (1964)
70. Varona, J.: Graphic and numerical comparison between iterative methods. Math. Intell. **24**(1), 37–46 (2002)
71. Wang, X., Liu, L.: Modified Ostrowski's method with eighth-order convergence and high efficiency index. Appl. Math. Lett. **23**, 549–554 (2010)
72. Weerakoon, S., Fernando, T.G.I.: A variant of Newton's method with accelerated third-order convergence. Appl. Math. Lett. **13**, 87–93 (2000)
73. Yun, B.I., Petković, M.S.: Iterative methods based on the signum function approach for solving nonlinear equations. Numer. Algorithms **52**, 649–662 (2009)

# Chapter 6
# The Theory of Kantorovich for Newton's Method: Conditions on the Second Derivative

**J.A. Ezquerro and M.A. Hernández-Verón**

**Abstract** We present, from the Kantorovich theory for Newton's method, two variants of the classic Newton-Kantorovich study that allow guaranteeing the semilocal convergence of the method for solving more nonlinear equations.

## 6.1  Introduction

One of the most studied problems in numerical mathematics is the solution of nonlinear equations. To approximate a solution of a nonlinear equation, we usually look for numerical approximations of the solutions, since finding exact solutions is usually difficult. To approximate a solution of a nonlinear equation we normally use iterative methods and Newton's method is one of the most used because of its simplicity, easy implementation and efficiency.

To give sufficient generality to the problem of approximating a solution of a nonlinear equation by Newton's method, we consider equations of the form $F(x) = 0$, where $F$ is a nonlinear operator, $F : \Omega \subseteq X \to Y$, defined on a nonempty open convex domain $\Omega$ of a Banach space $X$ with values in a Banach space $Y$. Newton's method is usually known as the Newton-Kantorovich method and the algorithm is

$$\begin{cases} x_0 \text{ given in } \Omega, \\ x_{n+1} = N_F(x_n) = x_n - [F'(x_n)]^{-1} F(x_n), \quad n = 0, 1, 2, \dots \end{cases} \tag{6.1}$$

The study about convergence matter of Newton's method is usually centered on two types: semilocal and local convergence analysis. The semilocal convergence matter is, based on the information around an initial point, to give criteria ensuring the convergence of the method; while the local one is, based on the information around a solution, to find estimates of the radii of convergence balls. In this work, we

J.A. Ezquerro (✉) • M.A. Hernández-Verón
Department of Mathematics and Computation, University of La Rioja, 26004 Logroño, Spain
e-mail: jezquer@unirioja.es; mahernan@unirioja.es

© Springer International Publishing Switzerland 2016
S. Amat, S. Busquier (eds.), *Advances in Iterative Methods for Nonlinear Equations*, SEMA SIMAI Springer Series 10, DOI 10.1007/978-3-319-39228-8_6

113

are interested in the semilocal convergence of Newton's method. The first semilocal convergence result for Newton's method in Banach spaces is due to Kantorovich, which is usually known as the Newton-Kantorovich theorem [13] and is proved under the following conditions for the operator $F$ and the starting point $x_0$:

(A1)   There exist $\Gamma_0 = [F'(x_0)]^{-1} \in \mathcal{L}(Y, X)$, for some $x_0 \in \Omega$, with $\|\Gamma_0\| \leq \beta$ and $\|\Gamma_0 F(x_0)\| \leq \eta$, where $\mathcal{L}(Y, X)$ is the set of bounded linear operators from $Y$ to $X$.

(A2)   $\|F''(x)\| \leq L$, for all $x \in \Omega$.

(A3)   $L\beta\eta \leq \dfrac{1}{2}$.

Since then many papers have appeared that study the semilocal convergence of the method. Most of them are modifications of the Newton-Kantorovich theorem in order to relax conditions (A1)–(A2)–(A3), specially condition (A2). But if the condition required to the operator $F$ is milder than (A2), as we can see in [3, 4], then condition (A3) is usually replaced by other condition more restrictive, which necessarily leads to a reduction in the domain of valid starting points for Newton's method.

The first aim of this work is not to require milder conditions to the operator $F$, but stronger, which pursue a modification, not a restriction, of the valid starting points for Newton's method, so that the method can start at points from which the Newton-Kantorovich theorem cannot guarantee its semilocal convergence, as well as improving the domains of existence and uniqueness of solution and the a priori error estimates. For this, we consider the semilocal convergence result given by Huang [12], where $F''$ satisfies a Lipschitz condition, and, from this result, centered conditions are required to the operator $F''$ [9, 10].

The second aim of this work is to generalize the semilocal convergence conditions given by Kantorovich for Newton's method, so that condition (A3) is relaxed in order to Newton's method can be applied to solve more equations. For this, we introduce a modification of condition (A3), so that $F''$ is $\omega$-bounded [6].

In this work, we follow a variation of Kantorovich's technique. In particular, we construct majorizing sequences ad hoc, so that these are adapted for particular problems, since the proposed modifications of condition (A3) give more information about the operator $F$, not just that $F''$ is bounded, as (A3) does. Our approaches go through to obtain general semilocal convergence results by using the well-known majorant principle of Kantorovich, that he developed for Newton's method [13]. From them, we see other results as particular cases.

Throughout the paper we denote $\overline{B}(x, \varrho) = \{y \in X; \|y - x\| \leq \varrho\}$ and $B(x, \varrho) = \{y \in X; \|y - x\| < \varrho\}$.

## 6.2   The Theory of Kantorovich

The famous Newton-Kantorovich theorem guarantees the semilocal convergence of Newton's method in Banach spaces and gives a priori error estimates and information about the existence and uniqueness of solution. Kantorovich proves the theorem by using two different techniques, although the most prominent one is the majorant principle, which is based on the concept of majorizing sequence. This technique has been usually used later by other authors to analyse the semilocal convergence of several iterative methods. We begin by introducing the concept of majorizing sequence and remembering how it is used to prove the convergence of sequences in Banach spaces.

**Definition 6.2.1** If $\{x_n\}$ is a sequence in a Banach space $X$ and $\{t_n\}$ is a scalar sequence, then $\{t_n\}$ is a *majorizing sequence* of $\{x_n\}$ if $\|x_n - x_{n-1}\| \le t_n - t_{n-1}$, for all $n \in \mathbb{N}$.

From the last inequality, it follows the sequence $\{t_n\}$ is nondecreasing. The interest of the majorizing sequence is that the convergence of the sequence $\{x_n\}$ in the Banach space $X$ is deduced from the convergence of the scalar sequence $\{t_n\}$, as we can see in the following result [13]:

**Lemma 6.2.1** *Let $\{x_n\}$ be a sequence in a Banach space $X$ and $\{t_n\}$ a majorizing sequence of $\{x_n\}$. Then, if $\{t_n\}$ converges to $t^* < \infty$, there exists $x^* \in X$ such that $x^* = \lim_n x_n$ and $\|x^* - x_n\| \le t^* - t_n$, for $n = 0, 1, 2, \dots$*

Once the concept of majorizing sequence is introduced, we can already establish Kantorovich's theory for Newton's method. In [13], Kantorovich proves the semilocal convergence of Newton's method under conditions (A1)–(A2)–(A3). For this, Kantorovich first considers that $F \in C^{(2)}(\Omega_0)$, with $\Omega_0 = \overline{B(x_0, r_0)} \subseteq X$, requires the existence of a real auxiliary function $f \in C^{(2)}([s_0, s'])$, with $s' - s_0 \le r_0$, and proves the following general semilocal convergence for Newton's method under the following conditions:

(K1)    There exists $\Gamma_0 = [F'(x_0)]^{-1} \in \mathcal{L}(Y, X)$ for some $x_0 \in \Omega$, $\|\Gamma_0\| \le -\dfrac{1}{f'(s_0)}$ and

$$\|\Gamma_0 F(x_0)\| \le -\frac{f(s_0)}{f'(s_0)}.$$

(K2)    $\|F''(x)\| \le f''(s)$ if $\|x - x_0\| \le s - s_0 \le r_0$.

(K3)    the equation $f(s) = 0$ has a solution in $[s_0, s']$.

**Theorem 6.2.1 (The General Semilocal Convergence Theorem)** *Let $F : \Omega_0 \subseteq X \to Y$ be a twice continuously differentiable operator defined on a nonempty open convex domain $\Omega_0 = \overline{B(x_0, r_0)}$ of a Banach space $X$ with values in a Banach space $Y$. Suppose that there exists $f \in C^{(2)}([s_0, s'])$, with $s_0, s' \in \mathbb{R}$, such that (K1)–(K2)–(K3) are satisfied. Then, Newton's sequence, given by (6.1), converges to a solution $x^*$ of the equation $F(x) = 0$, starting at $x_0$. Moreover,*

$$\|x^* - x_n\| \le s^* - s_n, \quad n = 0, 1, 2, \dots,$$

*where $s^*$ is the smallest solution of the equation*

$$f(s) = 0 \tag{6.2}$$

*in* $[s_0, s']$. *Furthermore, if* $f(s') \leq 0$ *and (6.2) has a unique solution in* $[s_0, s']$, *then* $x^*$ *is the unique solution of* $F(x) = 0$ *in* $\Omega_0$.

According to the convergence conditions required, the last result is known as general semilocal convergence theorem of Newton's method for operators with second derivative bounded in norm.

In practice, the application of the last theorem is complicated, since the scalar function $f$ is unknown. So, the following result is of particular interest, since if the operator $F$ satisfies (A2), taking into account the relationship given in (K2) and second degree polynomials are the simplest elementary functions that satisfies (A2), from (K1) and (K2) and solving the corresponding problem of interpolation fitting, we can consider the polynomial

$$p(s) = \frac{L}{2}(s - s_0)^2 - \frac{s - s_0}{\beta} + \frac{\eta}{\beta} \tag{6.3}$$

as the scalar function $f$ for Theorem 6.2.1. In this case, we obtain the classic Newton-Kantorovich theorem, that establishes the semilocal convergence of Newton's method under conditions (A1)–(A2)–(A3), that we call classic conditions of Kantorovich (not to be confused with the general conditions of Kantorovich, which are conditions (K1)–(K2)–(K3) of Theorem 6.2.1).

**Theorem 6.2.2 (The Newton-Kantorovich Theorem)** *Let* $F : \Omega \subseteq X \to Y$ *be a twice continuously differentiable operator defined on a nonempty open convex domain* $\Omega$ *of a Banach space* $X$ *with values in a Banach space* $Y$. *Suppose that conditions (A1)–(A2)–(A3) are satisfied and* $B(x_0, s^* - s_0) \subset \Omega$, *where* $s^* = s_0 + \frac{1-\sqrt{1-2L\beta\eta}}{L\beta}$, *then Newton's sequence, given by (6.1) and starting at* $x_0$, *converges to a solution* $x^*$ *of* $F(x) = 0$. *Moreover,* $x_n, x^* \in \overline{B(x_0, s^* - s_0)}$, *for all* $n \in \mathbb{N}$, *and* $x^*$ *is unique in* $B(x_0, s^{**} - s_0) \cap \Omega$, *where* $s^{**} = s_0 + \frac{1+\sqrt{1-2L\beta\eta}}{L\beta}$, *if* $L\beta\eta < \frac{1}{2}$ *or in* $\overline{B(x_0, s^* - s_0)}$ *if* $L\beta\eta = \frac{1}{2}$. *Furthermore,*

$$\|x^* - x_n\| \leq s^* - s_n, \quad n = 0, 1, 2, \ldots,$$

*where* $s_n = s_{n-1} - \frac{p(s_{n-1})}{p'(t_{n-1})}$, *with* $n \in \mathbb{N}$ *and* $p(s)$ *is given in (6.3).*

On the other hand, note that polynomial (6.3) can be obtained otherwise, without interpolation fitting, by solving the following initial value problem:

$$\begin{cases} p''(s) = L, \\ p(s_0) = \dfrac{\eta}{\beta}, \quad p'(s_0) = -\dfrac{1}{\beta}. \end{cases}$$

This new way of getting polynomial (6.3) has the advantage of being able to be generalized to other conditions, so that we can then obtain the scalar function $f$ for Theorem 6.2.1 under more general conditions, as we see later.

*Remark 6.2.1*  Observe that the Kantorovich polynomial given in (6.3) is such that

$$p(s + s_0) = \hat{p}(s), \quad \text{where} \quad \hat{p}(s) = \frac{L}{2}s^2 - \frac{s}{\beta} + \frac{\eta}{\beta}.$$

Therefore, the scalar sequences given by Newton's method with $p$ and $\hat{p}$ can be obtained, one from the other, by translation. As a consequence, Theorem 6.2.2 is independent of the value $s_0$. For this reason, Kantorovich chooses $s_0 = 0$, which simplifies considerably the expressions used.

## 6.3  $F''$ is a Lipschitz-Type Operator

Under conditions (A1)–(A2)–(A3), we obtain error estimates, domains of existence and uniqueness of solution and know whether $x_0$ is an initial point from which Newton's method converges, as we can see in the Newton-Kantorovich theorem. But, sometimes, conditions (A1)–(A2)–(A3) fail and we cannot guarantee, from the Newton-Kantorovich theorem, the convergence of Newton's method starting at $x_0$, as we see in the following example (see [12]).

*Example 6.3.1*  Let $X = Y = [-1, 1]$, $F(x) = \dfrac{x^3}{6} + \dfrac{x^2}{6} - \dfrac{5}{6}x + \dfrac{1}{3}$ and $x_0 = 0$. Then, $\beta = \dfrac{6}{5}, \eta = \dfrac{2}{5}, L = \dfrac{4}{3}$ and $L\beta\eta \le \dfrac{16}{25} > \dfrac{1}{2}$.

Next, we propose a modification, not a restriction, of the valid starting points for Newton's method, so that this method can start at points from which the Newton-Kantorovich theorem cannot guarantee its semilocal convergence. For this, different conditions are required to the operator $F''$. In [12], Huang gives new conditions,

(B1)   There exist $\Gamma_0 = [F'(x_0)]^{-1} \in \mathcal{L}(Y, X)$, for some $x_0 \in \Omega$, with $\|\Gamma_0\| \le \beta$ and $\|\Gamma_0 F(x_0)\| \le \eta$.

(B2)   $\|F''(x_0)\| \le M$.

(B3)   $\|F''(x) - F''(y)\| \le K\|x - y\|$, for all $x, y \in \Omega$.

(B4)   $\phi(\alpha) \le 0$, where $\phi$ is the function

$$\phi(t) = \frac{K}{6}t^3 + \frac{M}{2}t^2 - \frac{t}{\beta} + \frac{\eta}{\beta} \tag{6.4}$$

and $\alpha$ is the unique positive solution of $\phi'(t) = 0$,

and a new semilocal convergence result (Theorem 6.3.1, see below), under which, Newton's method converges starting at $x_0$ in Example 6.3.1.

Notice that the semilocal convergence conditions required to Newton's method in Huang's paper [12] is not exactly the same that we consider below, but it is equivalent and nothing change the semilocal convergence result established by Huang. The reason for this change is the uniformity sought throughout the present work. As a consequence, we have adapted the semilocal convergence result given in [12] to the new notation used here. Huang states in [12] that $\phi(\alpha) \leq 0$ holds provided that one of the two following conditions is satisfied:

$$6M^3\beta^3\eta + 9K^2\beta^2\eta^2 + 18KM\beta^2\eta - 3M^2\beta^2 - 8K\beta \leq 0,$$

$$3KM\beta^2 + M^3\beta^3 + 3K^2\beta^2\eta \leq \left(M^2\beta^2 + 2K\beta\right)^{3/2}.$$

Moreover, in [16], we can also see the following equivalent condition to $\phi(\alpha) \leq 0$:

$$\eta \leq \frac{4K\beta + M^2\beta^2 - M\beta\sqrt{M^2\beta^2 + 2K\beta}}{3K\beta\left(M\beta\sqrt{M^2\beta^2 + 2K\beta}\right)}.$$

*Remark 6.3.1* As $\alpha$ is the unique positive solution of $\phi'(t) = 0$, then $\alpha$ is a minimum of $\phi$ such that $\phi(\alpha) \leq 0$, so that (B4) is a necessary and sufficient condition for the existence of two positive solutions $t^*$ and $t^{**}$ of $\phi(t) = 0$ such that $0 < t^* \leq t^{**}$. Moreover, $\phi$ is a nonincreasing convex function in $[0, \alpha]$ and $\phi(\alpha) \leq 0 < \phi(0)$. Furthermore, from the Fourier conditions [14], these conditions are sufficient to guarantee the semilocal convergence of Newton's increasing real sequence,

$$t_0 = 0, \quad t_{n+1} = t_n - \frac{\phi(t_n)}{\phi'(t_n)}, \quad n = 0, 1, 2, \ldots, \tag{6.5}$$

to $t^*$.

Note that following an analogous procedure to Kantorovich, by solving the problem of interpolation fitting given by conditions (B1)–(B2)–(B3), we obtain function (6.4).

**Theorem 6.3.1 (See [12])** *Let $F : \Omega \subseteq X \to Y$ be a twice continuously differentiable operator defined on a nonempty open convex domain $\Omega$ of a Banach space X with values in a Banach space Y. Suppose that conditions (B1)–(B2)–(B3)–(B4) are satisfied and $B(x_0, t^*) \subset \Omega$. Then, Newton's sequence, given by (6.1) and starting at $x_0$, converges to a solution $x^*$ of $F(x) = 0$. Moreover, $x_n, x^* \in \overline{B(x_0, t^*)}$, for all $n \in \mathbb{N}$, and $x^*$ is unique in $B(x_0, t^{**}) \cap \Omega$ if $t^* < t^{**}$ or in $\overline{B(x_0, t^*)}$ if $t^* = t^{**}$. Furthermore,*

$$\|x^* - x_n\| \leq t^* - t_n, \quad n = 0, 1, 2, \ldots,$$

*where $\{t_n\}$ is defined in (6.5).*

From the above conditions, we can see that conditions (A1)–(A2)–(A3) and (B1)–(B2)–(B3)–(B4) are the same when $K = 0$.

Besides, from Theorem 6.3.1, the convergence of Newton's method is guaranteed, starting at $x_0 = 0$, in Example 6.3.1, since $\phi(\alpha) = -\frac{1}{6} \leq 0$, where

$\phi(t) = \frac{t^3}{6} + \frac{t^2}{6} - \frac{5}{6}t + \frac{1}{3}$ and $\alpha = 1$. In addition, the fact that Newton's method starting at $x_0$ in Example 6.3.1 converges under conditions (B1)–(B2)–(B3)–(B4) implies that higher derivatives at initial points are useful, though they may not be used in numerical processes, for the convergence of an iterative method. Therefore, the new condition has theoretical and practical value.

## 6.3.1  $F''$ is a Center Lipschitz Operator

In line with the above, if we pay attention to the proof of Theorem 6.3.1 given in [12], we see that condition (B3) is not necessary, since it is enough to be fulfilled the condition

$$\|F''(x) - F''(x_0)\| \le K\|x - x_0\|, \text{ for all } x \in \Omega, \qquad (6.6)$$

instead of condition (B3). As a consequence, we present the following corollary.

**Corollary 6.3.1** *Let* $F : \Omega \subseteq X \to Y$ *be a twice continuously differentiable operator defined on a nonempty open convex domain* $\Omega$ *of a Banach space* $X$ *with values in a Banach space* $Y$. *Suppose that conditions (B1)–(B2)–(6.6)–(B4) are satisfied and* $B(x_0, t^*) \subset \Omega$. *Then, Newton's sequence, given by (6.1) and starting at* $x_0$, *converges to a solution* $x^*$ *of* $F(x) = 0$. *Moreover,* $x_n, x^* \in \overline{B(x_0, t^*)}$, *for all* $n \in \mathbb{N}$, *and* $x^*$ *is unique in* $B(x_0, t^{**}) \cap \Omega$ *if* $t^* < t^{**}$ *or in* $\overline{B(x_0, t^*)}$ *if* $t^* = t^{**}$. *Furthermore,*

$$\|x^* - x_n\| \le t^* - t_n, \quad n = 0, 1, 2, \ldots,$$

*where* $\{t_n\}$ *is defined in (6.5).*

On the other hand, the Newton-Kantorovich theorem and Corollary 6.3.1 are not comparable with respect to the accessibility of solution, as we can see in [9], where two simple examples are given which demonstrate this fact.

Sometimes, the convergence of Newton's method can be established using, indistinctly, the Newton-Kantorovich theorem or Corollary 6.3.1. Again in [9], we can also see when each of these two results gives more accurate information on the solutions of $F(x) = 0$. Three cases are distinguished and an example that illustrates this is given, where we observe that, under the hypotheses of Corollary 6.3.1, the solution of $F(x) = 0$ are located in terms of the solutions of cubic polynomial (6.4), while, under the hypotheses of the Newton-Kantorovich theorem, the solutions of $F(x) = 0$ are located in terms of the solutions of quadratic polynomial (6.3).

### 6.3.2   $F''$ is a Center Hölder Operator

In [9, 10], the authors extend the semilocal convergence result obtained in Corollary 6.3.1 to a more general situation assuming, instead of condition (6.6), that $F''$ satisfies in $\Omega$ the following condition of Hölder type:

$$\|F''(x) - F''(x_0)\| \le K\|x - x_0\|^p, \ p \ge 0, \ \text{for all } x \in \Omega. \tag{6.7}$$

Observe that condition (6.7) is reduced to condition (6.6) if $p = 1$.

In particular, the authors give a new semilocal convergence result (Theorem 6.3.2, see below) for the convergence of Newton's method under the following conditions:

    (C1)   There exist $\Gamma_0 = [F'(x_0)]^{-1} \in \mathcal{L}(Y, X)$, for some $x_0 \in \Omega$, with $\|\Gamma_0\| \le \beta$ and $\|\Gamma_0 F(x_0)\| \le \eta$.

    (C2)   $\|F''(x_0)\| \le M$.

    (C3)   $\|F''(x) - F''(x_0)\| \le K\|x - x_0\|^p, \ p \ge 0, \ \text{for all } x \in \Omega$.

    (C4)   $\psi(\alpha) \le 0$, where $\psi$ is the function

$$\psi(t) = \frac{K}{(p+1)(p+2)}t^{2+p} + \frac{M}{2}t^2 - \frac{t}{\beta} + \frac{\eta}{\beta} \tag{6.8}$$

    and $\alpha$ is the unique positive solution of $\psi'(t) = 0$.

As for functions (6.3) and (6.4), we obtain function (6.4) by solving the problem of interpolation fitting given by conditions (C1)–(C2)–(C3).

Notice that Remark 6.3.1 is also satisfied when the function $\phi$ is substituted by the function $\psi$ (see [9]).

**Theorem 6.3.2** *Let $F : \Omega \subseteq X \to Y$ be a twice continuously differentiable operator defined on a nonempty open convex domain $\Omega$ of a Banach space $X$ with values in a Banach space $Y$. Suppose that conditions (C1)–(C2)–(C3)–(C4) are satisfied and $B(x_0, t^*) \subset \Omega$. Then, Newton's sequence, given by (6.1) and starting at $x_0$, converges to a solution $x^*$ of $F(x) = 0$. Moreover, $x_n, x^* \in \overline{B(x_0, t^*)}$, for all $n \in \mathbb{N}$, and $x^*$ is unique in $B(x_0, t^{**}) \cap \Omega$ if $t^* < t^{**}$ or in $\overline{B(x_0, t^*)}$ if $t^* = t^{**}$. Furthermore,*

$$\|x^* - x_n\| \le t^* - t_n, \quad n = 0, 1, 2, \dots,$$

*where $t_n = t_{n-1} - \frac{\psi(t_{n-1})}{\psi'(t_{n-1})}$, with $n \in \mathbb{N}$, $t_0 = 0$ and $\psi(t)$ is defined in (6.8).*

Moreover, some error estimates are given in [9] for Newton's method when Newton's real sequence is defined from functions (6.4) and (6.8).

### 6.3.3  F'' is a Center ω-Lipschitz Operator

In [7], the authors generalize all the semilocal convergence results obtained previously to a still more general situation assuming, instead of condition (6.7), that $F''$ satisfies in $\Omega$ the following condition:

$$\|F''(x) - F''(x_0)\| \leq \omega(\|x - x_0\|), \tag{6.9}$$

where $\omega : [0, +\infty) \rightarrow \mathbb{R}$ is a nondecreasing continuous function such that $\omega(0) = 0$.

Observe that condition (6.9) is reduced to conditions (6.6) or (6.7) if $\omega(z) = Kz$ or $\omega(z) = Kz^p$, respectively.

On the other hand, if we consider (C1)–(C2)–(6.9), we cannot obtain, by interpolation fitting, a real function $f$ for Theorem 6.2.1, as Kantorovich does, since (6.9) does not allow determining the class of functions where (C1)–(C2)–(6.9) can be applied. To solve this problem, we proceed without interpolation fitting, by solving an initial value problem, as we do in Sect. 6.2 to obtain the Kantorovich polynomial given in (6.3).

First of all, as our idea is to generalize the hypotheses of Huang and Gutiérrez by modifying conditions (B1)–(B2)–(6.6) and (C1)–(C2)–(C3), respectively, we first consider that there exists a real function $f \in \mathcal{C}^{(2)}([t_0, t'])$, with $t_0, t' \in \mathbb{R}$, which satisfies:

(H1)  There exists $\Gamma_0 = [F'(x_0)]^{-1} \in \mathcal{L}(Y, X)$, for some $x_0 \in \Omega$, with $\|\Gamma_0\| \leq -\dfrac{1}{f'(t_0)}$

and $\|\Gamma_0 F(x_0)\| \leq -\dfrac{f(t_0)}{f'(t_0)}$, and $\|F''(x_0)\| \leq f''(t_0)$.

(H2)  $\|F''(x) - F''(x_0)\| \leq f''(t) - f''(t_0)$, for $\|x - x_0\| \leq t - t_0, x \in \Omega$ and $t \in [t_0, t']$.

Next, we use the majorant principle to prove the semilocal convergence of Newton's method under general conditions (H1)–(H2). For this, we construct a majorizing sequence $\{t_n\}$ of Newton's sequence $\{x_n\}$ in the Banach space $X$. To obtain the sequence $\{t_n\}$, we use the previous real function $f(t)$ defined in $[t_0, t'] \subset \mathbb{R}$ as follows:

$$t_0 \text{ given}, \quad t_{n+1} = N_f(t_n) = t_n - \frac{f(t_n)}{f'(t_n)}, \quad n = 0, 1, 2, \dots \tag{6.10}$$

We have seen that Kantorovich constructs a majorizing sequence $\{s_n\}$ from the application of Newton's method to polynomial (6.3) and $s_0 = 0$, so that $\{s_n\}$ converges to the smallest positive root $s^*$ of the equation $p(s) = 0$. The convergence of the sequence is obvious, since the polynomial $p(s)$ is a decreasing convex function in $[s_0, s']$. Therefore, to construct a majorizing sequence $\{t_n\}$ from $f(t)$, it is necessary that the function $f(t)$ has at least one zero $t^*$, such that $t^* \in [t_0, t']$, and the sequence $\{t_n\}$ is increasing and convergent to $t^*$. When this happens, the semilocal convergence of Newton's sequence $\{x_n\}$ is guaranteed, in the Banach space $X$, from the convergence of the real sequence $\{t_n\}$ defined in (6.10).

### 6.3.3.1 Semilocal Convergence Result

To see the above-mentioned, we first study the convergence of the real sequence $\{t_n\}$ defined in (6.10). If conditions (H1)–(H2) are satisfied and there exists a root $\alpha \in (t_0, t')$ of $f'(t) = 0$ such that $f(\alpha) \leq 0$, then the equation $f(t) = 0$ has only one root $t^*$ in $(t_0, \alpha)$. Indeed, if $f(\alpha) < 0$, as $f(t_0) > 0$, then $f(t)$ has at least one zero $t^*$ in $(t_0, \alpha)$ by continuity. Besides, since $f''(t_0) \geq 0$, from (H2) it follows that $f''(t) \geq 0$ for $t \in (t_0, \alpha)$, so that $f'(t)$ is increasing and $f'(t) < 0$ for $t \in (t_0, \alpha)$. Also, as $f'(t_0) < 0$, $f(t)$ is decreasing for $t \in [t_0, \alpha)$. In consequence, $t^*$ is the unique root of $f(t) = 0$ in $(t_0, \alpha)$. On the other hand, if $f(\alpha) = 0$, then $\alpha$ is a double root of $f(t) = 0$ and we choose $t^* = \alpha$.

As a consequence, the convergence of the real sequence $\{t_n\}$ is guaranteed from the next theorem.

**Theorem 6.3.3** *Suppose that there exist $f$, such that conditions (H1)–(H2) are satisfied, and a root $\alpha \in (t_0, t')$ of $f'(t) = 0$ such that $f(\alpha) \leq 0$. Then, the sequence $\{t_n\}$, given in (6.10), is a nondecreasing sequence that converges to $t^*$.*

*Proof* As $f(t_0) > 0$, then $t_0 - t^* \leq 0$. By the mean value theorem, we obtain

$$t_1 - t^* = N_f(t_0) - N_f(t^*) = N_f'(\theta_0)(t_0 - t^*) \quad \text{with} \quad \theta_0 \in (t_0, t^*),$$

so that $t_1 < t^*$, since $N_f'(t) = \dfrac{f(t)f''(t)}{f'(t)^2} > 0$ in $[t_0, t^*)$.

On the other hand, we have

$$t_1 - t_0 = -\frac{f(t_0)}{f'(t_0)} \geq 0.$$

By mathematical induction on $n$, we obtain $t_n < t^*$ and $t_n - t_{n-1} \geq 0$, since $(t_{n-1}, t^*) \subset (t_0, t^*)$.

Therefore, we infer that sequence (6.10) converges to $r \in [t_0, t^*]$. Moreover, since $t^*$ is the unique root of $f(t) = 0$ in $[t_0, t^*]$, it follows that $r = t^*$. The proof is complete.

From the previous notation, we consider the known degree of logarithmic convexity of a real function $f$ (see [11]) as the function

$$L_f(t) = \frac{f(t)f''(t)}{f'(t)^2},$$

whose extension to Banach spaces can be used to construct majorizing sequences for Newton's method in the Banach space $X$. For this, we define the degree of logarithmic convexity in Banach spaces. So, we suppose that $F : \Omega \subseteq X \to Y$ is a twice continuously differentiable operator in $\Omega$ and there exists the operator $[F'(x)]^{-1} : Y \to X$. Moreover, since $F''(x) : X \times X \to Y$, it follows $F''(x)[F'(x)]^{-1}F(x) \in \mathcal{L}(X, Y)$ and

$$L_F(x) : \Omega \xrightarrow{\ F''(x)[F'(x)]^{-1}F(x)\ } Y \xrightarrow{\ [F'(x)]^{-1}\ } \Omega.$$

In addition, we have $L_F(x) = N'_F(x)$.

Now, it is easy to see that

$$x_{n+1} - x_n = \int_{x_{n-1}}^{x_n} N'_F(x)\, dx = \int_{x_{n-1}}^{x_n} L_F(x)\, dx,$$

where $\{x_n\}$ is Newton's sequence in the Banach space $X$, so that

$$\|x_{n+1} - x_n\| \le t_{n+1} - t_n, \quad n = 0, 1, 2, \ldots,$$

provided that $\|L_F(x)\| \le L_f(t)$ for $\|x - x_0\| \le t - t_0$.

Before we see that (6.10) is a majorizing sequence of (6.1) in the Banach space $X$, we give the following technical lemma that is used later.

**Lemma 6.3.1** *Suppose that there exist $f \in C^{(2)}([t_0, t'])$, with $t_0, t' \in \mathbb{R}$, such that (H1)–(H2) are satisfied and $\alpha \in (t_0, t')$, such that $f'(\alpha) = 0$, and $B(x_0, \alpha - t_0) \subset \Omega$. Then, for $x \in B(x_0, \alpha - t_0)$, the operator $L_F(x) = [F'(x)]^{-1}F''(x)[F'(x)]^{-1}F(x)$ exists and is such that*

$$\|L_F(x)\| \le \frac{f''(t)}{f'(t)^2}\|F(x)\| \quad for \quad \|x - x_0\| \le t - t_0. \tag{6.11}$$

*Proof* We start proving that the operator $[F'(x)]^{-1}$ exists for $x \in B(x_0, \alpha - t_0)$ and, for $t \in (t_0, \alpha)$ with $\|x - x_0\| \le t - t_0$, we also have

$$\|[F'(x)]^{-1}F'(x_0)\| \le \frac{f'(t_0)}{f'(t)} \quad \text{and} \quad \|[F'(x)]^{-1}\| \le -\frac{1}{f'(t)}. \tag{6.12}$$

From

$$\|I - \Gamma_0 F'(x)\| = \left\| -\Gamma_0 \int_{x_0}^{x} F''(z)\, dz \right\|$$

$$\le \|\Gamma_0\| \|F''(x_0)\| \|x - x_0\| + \|\Gamma_0\| \left\| \int_{x_0}^{x} (F''(z) - F''(x_0))\, dz \right\|$$

$$\leq -\frac{f''(t_0)}{f'(t_0)}(t - t_0)$$

$$+ \|\Gamma_0\| \left\| \int_0^1 \left( F'' (x_0 + \tau(x - x_0)) - F''(x_0) \right) (x - x_0)\, d\tau \right\|.$$

and

$$\|z - x_0\| = \tau \|x - x_0\| \leq \tau(t - t_0) = t_0 + \tau(t - t_0) - t_0 = u - t_0,$$

where $z = x_0 + \tau(x - x_0)$ and $u = t_0 + \tau(t - t_0)$, with $\tau \in [0, 1]$, it follows from (H2) that

$$\|I - \Gamma_0 F'(x)\| \leq -\frac{f''(t_0)}{f'(t_0)}(t - t_0) - \frac{1}{f'(t_0)} \int_{t_0}^t \left( f''(u) - f''(t_0) \right) du = 1 - \frac{f'(t)}{f'(t_0)} < 1,$$

since $f'(t)$ is increasing and $f'(t_0) \leq f'(t) \leq 0$. In consequence, by the Banach lemma, the operator $[F'(x)]^{-1}$ exists and is such that

$$\|[F'(x)]^{-1}\| \leq \frac{\|\Gamma_0\|}{1 - \|I - \Gamma_0 F'(x)\|} \leq -\frac{1}{f'(t)}.$$

In addition,

$$\|[F'(x)]^{-1} F'(x_0)\| \leq \frac{1}{1 - \|I - \Gamma_0 F'(x)\|} \leq \frac{f'(t_0)}{f'(t)}.$$

Therefore (6.12) holds.

On the other hand, if $x \in B(x_0, \alpha - t_0)$ and $t \in (t_0, \alpha)$ are such that $\|x - x_0\| \leq t - t_0$, we have

$$\|F''(x)\| \leq \|F''(x_0)\| + \|F''(x) - F''(x_0)\| \leq f''(t_0) + f''(t) - f''(t_0) = f''(t)$$

and (6.11) also holds. The proof is complete.

Next, from the following lemma, we see that (6.10) is a majorizing sequence of sequence (6.1) in the Banach space $X$.

**Lemma 6.3.2** *Under the hypotheses of Lemma 6.3.1, the following items are true for all $n = 0, 1, 2, \ldots$*

($i_n$)   $x_n \in B(x_0, t^* - t_0)$.

($ii_n$)   $\|\Gamma_0 F(x_n)\| \leq -\dfrac{f(t_n)}{f'(t_0)}.$

($iii_n$)   $\|x_{n+1} - x_n\| \leq t_{n+1} - t_n.$

*Proof* We prove $(i_n)$–$(iii_n)$ by mathematical induction on $n$. Firstly, $x_0$ given, it is clear that $x_1$ is well-defined and

$$\|x_1 - x_0\| = \|\Gamma_0 F(x_0)\| \leq -\frac{f(t_0)}{f'(t_0)} = t_1 - t_0 < t^* - t_0.$$

Then $(i_0)$–$(iii_0)$ hold.

We now suppose that $(i_j)$–$(iii_j)$ are true for $j = 0, 1, \ldots, n - 1$ and prove that the three items are also true for $j = n$.

As $x_n = x_{n-1} - [F'(x_{n-1})]^{-1} F(x_{n-1})$, it is clear that $x_n$ is well-defined, since the operator $[F'(x_{n-1})]^{-1}$ exists by Lemma 6.3.1. Moreover,

$$
\begin{aligned}
\|x_n - x_0\| &\leq \|x_n - x_{n-1}\| + \|x_{n-1} - x_{n-2}\| + \cdots + \|x_1 - x_0\| \\
&\leq t_n - t_{n-1} + t_{n-1} - t_{n-2} + \cdots + t_1 - t_0 \\
&< t^* - t_0,
\end{aligned}
$$

so that $x_n \in B(x_0, t^* - t_0)$ and $(i_n)$ holds.

After that, we consider $x = x_{n-1} + s(x_n - x_{n-1})$, with $s \in [0, 1]$, and $\|x - x_{n-1}\| = s\|x_n - x_{n-1}\| \leq s(t_n - t_{n-1})$. Therefore,

$$
\begin{aligned}
\|x - x_0\| &\leq \|x_{n-1} - x_0\| + s\|x_n - x_{n-1}\| \\
&\leq t_{n-1} - t_0 + s(t_n - t_{n-1}) \\
&= t - t_0,
\end{aligned}
$$

with $t = t_{n-1} + s(t_n - t_{n-1}) \in [t_{n-1}, t_n]$, and since $\|x - x_0\| \leq t - t_0 < t^* - t_0$, it is clear that $x \in B(x_0, t^* - t_0)$ for $x \in [x_{n-1}, x_n]$. From Lemma 6.3.1, we have that the operators $[F'(x)]^{-1}$ and $L_F(x)$ exist and

$$\|L_F(x)\| \leq \frac{f''(t)}{f'(t)^2} \|F(x)\| \quad \text{with} \quad t = t_{n-1} + s(t_n - t_{n-1}) \quad \text{and} \quad s \in [0, 1].$$

Besides,

$$\|L_F(x)\| \leq -\frac{f''(t)}{f'(t)} \frac{f'(t_0)}{f'(t)} \|\Gamma_0 F(x)\|,$$

simply by writing $L_F(x) = [F'(x)]^{-1} F''(x) [F'(x)]^{-1} F'(x_0) \Gamma_0 F(x)$ and applying (6.12).

Taking now into account the last inequality and Taylor's series, we write

$$\|\Gamma_0 F(x)\| = \left\| \Gamma_0 F(x_{n-1}) + \Gamma_0 F'(x_{n-1})(x - x_{n-1}) + \int_{x_{n-1}}^{x} \Gamma_0 F''(z)(x - z)\, dz \right\|$$

$$\leq (1 - s)\|\Gamma_0 F(x_{n-1})\| + \frac{1}{2}\|\Gamma_0\|\|F''(x_0)\|\|x - x_{n-1}\|^2$$

$$+ \|\Gamma_0\| \int_0^1 \|F''(x_{n-1} + \tau(x - x_{n-1})) - F''(x_0)\| \|x - x_{n-1}\|^2 (1 - \tau)\, d\tau.$$

As $\|x - x_{n-1}\| = s\|x_n - x_{n-1}\| \leq s(t_n - t_{n-1}) \leq t - t_{n-1}$, for $z = x_{n-1} + \tau(x - x_{n-1})$ with $\tau \in [0, 1]$, we have $\|z - x_0\| \leq u - t_0$, where $u = t_{n-1} + \tau(t - t_{n-1})$. Consequently,

$$\|\Gamma_0 F(x)\| \leq -(1 - s)\frac{f(t_{n-1})}{f'(t_0)} - \frac{1}{2}\frac{f''(t_0)}{f'(t_0)}(t - t_{n-1})^2 -$$

$$- \frac{1}{f'(t_0)} \int_0^1 \left( f''(t_{n-1} + \tau(t - t_{n-1})) - f''(t_0) \right)(t_n - t_{n-1})^2 (1 - \tau)\, d\tau$$

$$= -\frac{1}{f'(t_0)} \left( f(t_{n-1}) + f'(t_{n-1})(t - t_{n-1}) + \int_{t_{n-1}}^{t} f''(u)(t - u)\, du \right)$$

$$= -\frac{f(t)}{f'(t_0)}.$$

If we take $s = 1$ above, we obtain $x = x_n$, $t = t_n$ and $\|\Gamma_0 F(x_n)\| \leq -\frac{f(t_n)}{f'(t_0)}$. In addition, $(ii_n)$ holds and $\|L_F(x)\| \leq \frac{f(t)f''(t)}{f'(t)^2} = L_f(t)$.

Finally, to prove $(iii_n)$, just see that

$$\|x_{n+1} - x_n\| = \left\| \int_{x_{n-1}}^{x_n} L_F(x)\, dx \right\|$$

$$\leq \int_0^1 \|L_F(x_{n-1} + \tau(x_n - x_{n-1}))\| \|x_n - x_{n-1}\|\, d\tau$$

$$\leq \int_0^1 L_f(t_{n-1} + \tau(t_n - t_{n-1}))(t_n - t_{n-1})\, d\tau$$

$$= \int_{t_{n-1}}^{t_n} L_f(u)\, du$$

$$= t_{n+1} - t_n,$$

since $x = x_{n-1} + \tau(x_n - x_{n-1})$ with $\tau \in [0, 1]$ and

$$\|x - x_0\| \leq \|x_{n-1} - x_0\| + \tau\|x_n - x_{n-1}\| \leq t_{n-1} + \tau(t_n - t_{n-1}) - t_0 = t - t_0,$$

where $t = t_{n-1} + \tau(t_n - t_{n-1})$. The proof is complete.

Once we have seen that (6.10) is a majorizing sequence of (6.1), we are ready to prove the semilocal convergence of (6.1) in the Banach space $X$.

**Theorem 6.3.4 (See [7])** *Let* $F : \Omega \subseteq X \rightarrow Y$ *be a twice continuously differentiable operator defined on a nonempty open convex domain* $\Omega$ *of a Banach space* $X$ *with values in a Banach space* $Y$. *Suppose that there exist* $f \in C^{(2)}([t_0, t'])$, *with* $t_0, t' \in \mathbb{R}$, *such that (H1)–(H2) are satisfied, and a root* $\alpha \in (t_0, t')$ *of* $f'(t) = 0$ *such that* $f(\alpha) \leq 0$, *and* $B(x_0, t^* - t_0) \subset \Omega$. *Then, Newton's sequence, given by (6.1) and starting at* $x_0$, *converges to a solution* $x^*$ *of* $F(x) = 0$. *Moreover,* $x_n, x^* \in B(x_0, t^* - t_0)$, *for all* $n \in \mathbb{N}$, *and*

$$\|x^* - x_n\| \leq t^* - t_n, \quad n = 0, 1, 2, \ldots,$$

*where* $\{t_n\}$ *is defined in (6.10).*

*Proof* Observe that $\{x_n\}$ is convergent, since $\{t_n\}$ is a majorizing sequence of $\{x_n\}$ and convergent. Moreover, as $\lim_{n \to +\infty} t_n = t^*$, if $x^* = \lim_{n \to +\infty} x_n$, then $\|x^* - x_n\| \leq t^* - t_n$, for all $n = 0, 1, 2, \ldots$ Furthermore, from item $(ii_n)$ of the last lemma, we have $\|\Gamma_0 F(x_n)\| \leq -\dfrac{f(t_n)}{f'(t_0)}$, for all $n = 0, 1, 2, \ldots$ Then, by letting $n \to +\infty$ in the last inequality, it follows $F(x^*) = 0$ by the continuity of $F$. The proof is complete.

### 6.3.3.2 Uniqueness of Solution

After proving the semilocal convergence of Newton's method and locating the solution $x^*$, we prove that $x^*$ is unique. Note that if $f(t)$ has two real zeros $t^*$ and $t^{**}$ such that $t_0 < t^* \leq t^{**}$, then the uniqueness of solution follows from the next theorem.

**Theorem 6.3.5** *Under the hypotheses of Theorem 6.3.4, the solution* $x^*$ *is unique in* $B(x_0, t^{**} - t_0) \cap \Omega$ *if* $t^* < t^{**}$ *or in* $\overline{B(x_0, t^* - t_0)}$ *if* $t^* = t^{**}$.

*Proof* Suppose that $t^* < t^{**}$ and $y^*$ is another solution of $F(x) = 0$ in $B(x_0, t^{**} - t_0) \cap \Omega$. Then,

$$\|y^* - x_0\| \leq \rho(t^{**} - t_0) \quad \text{with} \quad \rho \in (0, 1).$$

We now suppose that $\|y^* - x_k\| \leq \rho^{2^k}(t^{**} - t_k)$ for $k = 0, 1, \ldots, n$. In addition,

$$
\begin{aligned}
\|y^* - x_{n+1}\| &= \left\| -\Gamma_n \left( F(y^*) - F(x_n) - F'(x_n)(y^* - x_n) \right) \right\| \\
&= \left\| -\Gamma_n \left( \int_0^1 \left( F''\left( x_n + \tau(y^* - x_n) \right) - F''(x_0) \right) (1 - \tau)(y^* - x_n)^2 \, d\tau \right.\right. \\
&\qquad \left.\left. + \frac{1}{2} F''(x_0)(y^* - x_n)^2 \right) \right\|.
\end{aligned}
$$

As $\|x_n + \tau(y^* - x_n) - x_0\| \leq t_n + \tau(t^{**} - t_n) - t_0$, it follows that

$$
\|y^* - x_{n+1}\| \leq -\frac{\mu}{f'(t_n)} \|y^* - x_n\|^2,
$$

where $\mu = \dfrac{1}{2} f''(t_0) + \displaystyle\int_0^1 \left( f''\left( t_n + \tau(t^{**} - t_n) \right) - f''(t_0) \right) (1 - \tau) \, d\tau$.

On the other hand, we also have

$$
\begin{aligned}
t^{**} - t_{n+1} &= -\frac{1}{f'(t_n)} \left( \int_{t_n}^{t^{**}} \left( f''(t) - f''(t_0) \right) (t^{**} - t) \, dt + \frac{1}{2} f''(t_0)(t^{**} - t_n)^2 \right) \\
&= -\frac{\mu}{f'(t_n)} (t^{**} - t_n)^2.
\end{aligned}
$$

Therefore,

$$
\|y^* - x_{n+1}\| \leq \frac{t^{**} - t_{n+1}}{(t^{**} - t_n)^2} \|y^* - x_n\|^2 \leq \rho^{2^{n+1}}(t^{**} - t_{n+1}),
$$

so that $y^* = x^*$.

If $t^* = t^{**}$ and $y^*$ is another solution of $F(x) = 0$ in $\overline{B(x_0, t^* - t_0)}$, then $\|y^* - x_0\| \leq t^* - t_0$. Proceeding similarly to the previous case, we can prove by mathematical induction on $n$ that $\|y^* - x_n\| \leq t^* - t_n$. Since $t^* = t^{**}$ and $\lim_n t_n = t^*$, the uniqueness of solution is now easy to follow. The proof is complete.

Note that the uniqueness of solution established in Theorem 6.3.5 includes the uniqueness of solution given by Huang (Theorem 6.3.1) and Gutiérrez (Theorem 6.3.2) when $f(t)$ is reduced to $\phi(t)$ or to $\psi(t)$, respectively, and $t_0 = 0$.

### 6.3.3.3   A Priori Error Estimates

We finish this section by seeing the quadratic convergence of Newton's method under conditions (H1)–(H2). First, if $f(t)$ has two real zeros $t^*$ and $t^{**}$ such that $t_0 < t^* \leq t^{**}$, we can then write

$$
f(t) = (t^* - t)(t^{**} - t)\ell(t)
$$

with $\ell(t^*) \neq 0$ and $\ell(t^{**}) \neq 0$. We then obtain the following theorem from Ostrowski's technique [14].

**Theorem 6.3.6** *Let* $f \in C^{(2)}([t_0, t'])$ *with* $t_0, t' \in \mathbb{R}$. *Suppose that* $f(t)$ *has two real zeros* $t^*$ *and* $t^{**}$ *such that* $t_0 < t^* \leq t^{**}$.

(a) *If* $t^* < t^{**}$, *then*

$$\frac{(t^{**} - t^*)\theta^{2^n}}{m_1 - \theta^{2^n}} \leq t^* - t_n \leq \frac{(t^{**} - t^*)\Delta^{2^n}}{M_1 - \Delta^{2^n}}, \quad n \geq 0,$$

*where* $\theta = \frac{t^*}{t^{**}}m_1$, $\Delta = \frac{t^*}{t^{**}}M_1$, $m_1 = \min\{Q_1(t); t \in [t_0, t^*]\}$, $M_1 = \max\{Q_1(t); t \in [t_0, t^*]\}$, $Q_1(t) = \frac{(t^{**}-t)\ell'(t)-\ell(t)}{(t^*-t)\ell'(t)-\ell(t)}$ *and provided that* $\theta < 1$ *and* $\Delta < 1$.

(b) *If* $t^* = t^{**}$, *then*

$$m_2^n t^* \leq t^* - t_n \leq M_2^n t^*,$$

*where* $m_2 = \min\{Q_2(t); t \in [t_0, t^*]\}$, $M_2 = \max\{Q_2(t); t \in [t_0, t^*]\}$, $Q_2(t) = \frac{(t^*-t)\ell'(t)-\ell(t)}{(t^*-t)\ell'(t)-2\ell(t)}$ *and provided that* $m_2 < 1$ *and* $M_2 < 1$.

*Proof* Let $t^* < t^{**}$ and denote $a_n = t^* - t_n$ and $b_n = t^{**} - t_n$ for all $n = 0, 1, 2, \ldots$ Then

$$f(t_n) = a_n b_n \ell(t_n), \quad f'(t_n) = a_n b_n \ell'(t_n) - (a_n + b_n)\ell(t_n)$$

and

$$a_{n+1} = t^* - t_{n+1} = t^* - t_n + \frac{f(t_n)}{f'(t_n)} = \frac{a_n^2 (b_n \ell'(t_n) - \ell(t_n))}{a_n b_n \ell'(t_n) - (a_n + b_n)\ell(t_n)}.$$

From $\dfrac{a_{n+1}}{b_{n+1}} = \dfrac{a_n^2 (b_n \ell'(t_n) - \ell(t_n))}{b_n^2 (a_n \ell'(t_n) - \ell(t_n))}$, it follows

$$m_1 \left(\frac{a_n}{b_n}\right)^2 \leq \frac{a_{n+1}}{b_{n+1}} \leq M_1 \left(\frac{a_n}{b_n}\right)^2.$$

In addition,

$$\frac{a_{n+1}}{b_{n+1}} \leq M_1^{2^{n+1}-1} \left(\frac{a_0}{b_0}\right)^{2^{n+1}} = \frac{\Delta^{2^{n+1}}}{M_1}, \qquad \frac{a_{n+1}}{b_{n+1}} \geq m_1^{2^{n+1}-1} \left(\frac{a_0}{b_0}\right)^{2^{n+1}} = \frac{\theta^{2^{n+1}}}{m_1}.$$

Taking then into account that $b_{n+1} = (t^{**} - t^*) + a_{n+1}$, it follows:

$$\frac{(t^{**} - t^*)\theta^{2^{n+1}}}{m_1 - \theta^{2^{n+1}}} \leq t^* - t_{n+1} \leq \frac{(t^{**} - t^*)\Delta^{2^{n+1}}}{M_1 - \Delta^{2^{n+1}}}.$$

If $t^* = t^{**}$, then $a_n = b_n$ and

$$a_{n+1} = \frac{a_n \left( a_n \ell'(t) - \ell(t) \right)}{a_n \ell'(t) - 2\ell(t_n)}.$$

Consequently, $m_2 a_n \leq a_{n+1} \leq M_2 a_n$ and

$$m_2^{n+1} t^* \leq t^* - t_{n+1} \leq M_2^{n+1} t^*.$$

The proof is complete.

From the last theorem, it follows that the convergence of Newton's method, under conditions (H1)–(H2), is quadratic if $t^* < t^{**}$ and linear if $t^* = t^{**}$.

Notice that the a priori error estimates established in Theorem 6.3.6 are exactly the same as those given by Gutiérrez in [9] when $f(t) = \phi(t)$ or $f(t) = \psi(t)$ with $t_0 = 0$.

### 6.3.3.4   Particular Cases

The way of getting polynomial (6.3) by solving an initial value problem has the advantage of being able to be generalized to conditions (H1)–(H2), in contrast to get polynomial (6.3) using interpolation, so that we can then obtain the real function $f$ for Theorem 6.2.1 under very general conditions. In the following, we see three different cases that can be deduced as particular cases of the general semilocal convergence result given in Theorem 6.3.4.

Firstly, if conditions (H1)–(H2) are reduced to (B1)–(B2)–(6.6), we are in the conditions of [12] and can find the real function $f$ by solving the following initial value problem:

$$\begin{cases} \phi''(t) = M + Kt, \\ \phi(0) = \dfrac{\eta}{\beta}, \quad \phi'(0) = -\dfrac{1}{\beta}, \end{cases}$$

whose solution is polynomial (6.4), since

$$f''(t) - f''(t_0) = K(t - t_0) = Kt \quad \text{and} \quad f''(t) = M + Kt.$$

Notice that polynomial (6.4) satisfies the hypotheses of Theorem 6.3.4 with $t_0 = 0$ and, consequently, the semilocal convergence of Newton's method can be guaranteed in the Banach space $X$ from Theorem 6.3.4.

Secondly, if conditions (H1)–(H2) are reduced to (C1)–(C2)–(C3), we are in the conditions of [9, 10] and can find the real function $f$ by solving the following initial value problem:

$$\begin{cases} \psi''(t) = M + Kt^p, \\ \psi(0) = \dfrac{\eta}{\beta}, \quad \psi'(0) = -\dfrac{1}{\beta}, \end{cases}$$

whose solution is function (6.8), since

$$f''(t) - f''(t_0) = K(t - t_0)^p = Kt^p \quad \text{and} \quad f''(t) = M + Kt^p.$$

Notice that function (6.8) satisfies the hypotheses of Theorem 6.3.4 with $t_0 = 0$ and, consequently, the semilocal convergence of Newton's method can also be guaranteed in the Banach space $X$ from Theorem 6.3.4.

Thirdly, if conditions (H1)–(H2) are reduced to (C1)–(C2)–(6.9), we are in the conditions of [7] that generalize the conditions given in [9, 10, 12] and find the real function $f$ by solving the following initial value problem:

$$\begin{cases} \varphi''(t) = M + \omega(t), \\ \varphi(0) = \dfrac{\eta}{\beta}, \quad \varphi'(0) = -\dfrac{1}{\beta}, \end{cases}$$

whose solution is given in the following theorem.

**Theorem 6.3.7** *Suppose that the function $\omega(t)$ is continuous for all $t \in [0, t']$, with $t' > 0$. Then, for any real numbers $\beta \neq 0$, $\eta$ and $M$, there exists only one solution $\varphi(t)$ of the last initial value problem in $[0, t']$; that is:*

$$\varphi(t) = \int_0^t \int_0^s \omega(z) \, dz \, ds + \frac{M}{2} t^2 - \frac{t}{\beta} + \frac{\eta}{\beta}. \tag{6.13}$$

To apply Theorem 6.3.4, the equation $\varphi(t) = 0$ must have at least one root greater than zero, so that we have to guarantee the convergence of the real sequence $\{t_n\}$ defined in (6.10) with $f(t) \equiv \varphi(t)$ and $t_0 = 0$ to this root. We then give the following properties of the function $\varphi(t)$ defined in (6.13), whose proofs are easy to follow.

**Theorem 6.3.8** *Let $\varphi$ and $\omega$ be the functions defined respectively in (6.13) and (6.9).*

(a) *There exists only one positive solution $\alpha > t_0$ of the equation*

$$\varphi'(t) = \int_0^t \omega(z) \, dz + Mt - \frac{1}{\beta} = 0, \tag{6.14}$$

*which is the unique minimum of $\varphi(t)$ in $[0, +\infty)$ and $\varphi(t)$ is nonincreasing in $[t_0, \alpha)$.*

(b) *If $\varphi(\alpha) \leq 0$, then the equation $\varphi(t) = 0$ has at least one root in $[0, +\infty)$. Moreover, if $t^*$ is the smallest root of $\varphi(t) = 0$ in $[0, +\infty)$, we have $0 < t^* \leq \alpha$.*

Taking into account the hypotheses of Theorem 6.3.8, function (6.13) satisfies the conditions of Theorem 6.3.4 and the semilocal convergence of Newton's method is then guaranteed in the Banach space $X$. In particular, we have the following theorem, whose proof follows immediately from Theorem 6.3.4.

**Theorem 6.3.9** *Let $F : \Omega \subseteq X \rightarrow Y$ be a twice continuously differentiable operator defined on a nonempty open convex domain $\Omega$ of a Banach space $X$ with values in a Banach space $Y$ and $\varphi(t)$ be the function defined in (6.13). Suppose that (C1)–(C2)–(6.9) are satisfied, there exists a root $\alpha > 0$ of (6.14), such that $\varphi(\alpha) \leq 0$, and $B(x_0, t^*) \subset \Omega$, where $t^*$ is the smallest root of $\varphi(t) = 0$ in $[0, +\infty)$. Then, Newton's sequence $\{x_n\}$, given by (6.1), converges to a solution $x^*$ of $F(x) = 0$ starting at $x_0$. Moreover, $x_n, x^* \in \overline{B(x_0, t^*)}$, for all $n \in \mathbb{N}$, and*

$$\|x^* - x_n\| \leq t^* - t_n, \quad n = 0, 1, 2, \ldots,$$

*where $t_n = t_{n-1} - \frac{\varphi(t_{n-1})}{\varphi'(t_{n-1})}$, with $n \in \mathbb{N}$, $t_0 = 0$ and $\varphi(t)$ defined in (6.13).*

*Remark 6.3.2* Note that the three real functions $\phi$, $\psi$ and $\varphi$, from which the majorizing sequences are defined, also have the property of the translation of the Kantorovich polynomial given in (6.3), see Remark 6.2.1, so they are independent of the value of $t_0$. For this reason, we always choose $t_0 = 0$, which simplifies considerably the expressions used.

Note that the choice of $x_0$ is fundamental, since it is not enough to satisfy condition (6.9) to guarantee the convergence of Newton's method. As it is well-known, the starting point $x_0$ should be close to the solution $x^*$. In addition, having the possibility that there are several points that satisfy condition (6.9) is an advantage to be able to select the starting point. Therefore, the condition

$$\|F''(x) - F''(y)\| \leq \omega(\|x - y\|), \quad \text{for all } x, y \in \Omega, \tag{6.15}$$

where $\omega : [0, +\infty) \rightarrow \mathbb{R}$ is a nondecreasing continuous function such that $\omega(0) = 0$, is more useful than condition (6.9).

*Example 6.3.2* As we have indicated in the introduction, an aim of this work is to modify condition (A2) on the operator $F''$, so that the domain of starting points for Newton's method is modified with respect to that of Kantorovich under conditions (A1)–(A3). In the following, we see a situation where Kantorovich's study is improved from conditions (C1)–(C2)–(6.9).

The situation presented is a particular case of a nonlinear integral equation of the following mixed Hammerstein type [8]:

$$x(s) = h(s) + \int_a^b G(s, t)H(x(t)) \, dt, \quad s \in [a, b],$$

where $-\infty < a < b < +\infty$, $h$, $G$ and $H$ are known functions and $x$ is the function to determine. Integral equations of this type appear very often in several applications to real world problems. For example, in problems of dynamic models of chemical reactors [1], vehicular traffic theory, biology and queuing theory [2]. The Hammerstein integral equations also appear in the electro-magnetic fluid dynamics and can be reformulated as two-point boundary value problems with certain nonlinear boundary conditions and in multi-dimensional analogues which appear as reformulations of elliptic partial differentiable equations with nonlinear boundary conditions (see [15] and the references given there).

In particular, we consider a nonlinear equation of the form:

$$x(s) = h(s) + \int_a^b G(s,t) \left(\lambda x(t)^3 + \delta x(t)^{2+p}\right) dt, \quad s \in [a,b], \tag{6.16}$$

$p \in [0,1]$ and $\lambda, \delta \in \mathbb{R}$, where $h$ is a continuous function in $[a,b] \times [a,b]$ and the kernel $G$ is the Green function.

Solving Eq. (6.16) is equivalent to solve $F(x) = 0$, where $F : \Omega \subset C([a,b]) \to C([a,b])$, $\Omega = \{x \in C([a,b]); x(s) \geq 0, s \in [a,b]\}$ and

$$[F(x)](s) = x(s) - h(s) - \int_a^b G(s,t) \left(\lambda x(t)^3 + \delta x(t)^{2+p}\right) dt,$$

where $s \in [a,b]$, $p \in [0,1]$ and $\lambda, \delta \in \mathbb{R}$.

In addition,

$$[F'(x)y](s) = y(s) - \int_a^b G(s,t) \left(3\lambda x(t)^2 + (2+p)\delta x(t)^{1+p}\right) y(t)\, dt,$$

$$[F''(x)(yz)](s) = - \int_a^b G(s,t) \left(6\lambda x(t) + (2+p)(1+p)\delta x(t)^p\right) z(t)y(t)\, dt.$$

Notice that condition (A2) of Kantorovich is not satisfied, since $\|F''(x)\|$ is not bounded in $\Omega$. Moreover, it is not easy to locate a domain where $\|F''(x)\|$ is bounded and contains a solution of $F(x) = 0$. Notice also that conditions (B3) and (C3) are not satisfied either, since $F''(x)$ is not Lipschitz-continuous or Hölder-continuous in $\Omega$, so that we cannot apply Theorems 6.3.1 and 6.3.2 to guarantee the semilocal convergence of Newton's method to a solution of (6.16).

Initially, we transform Eq. (6.16) into a finite dimensional problem and, later, we apply Newton's method to approximate a solution of this problem. Then, we approximate the integral of (6.16) by a Gauss-Legendre quadrature formula with $m$ nodes:

$$\int_a^b q(t)\, dt \simeq \sum_{i=1}^m w_i\, q(t_i),$$

where the nodes $t_i$ and the weights $w_i$ are determined. Now, if we denote the approximations of $x(t_i)$ and $h(t_i)$ by $x_i$ and $h_i$, respectively, with $i = 1, 2, \ldots, m$, then Eq. (6.16) is equivalent to the following nonlinear system of equations:

$$x_i = h_i + \sum_{j=1}^{m} b_{ij} \left( \lambda x_j^3 + \delta x_j^{2+p} \right), \quad i = 1, 2, \ldots, m,$$

where

$$b_{ij} = \begin{cases} w_j \dfrac{(b - t_i)(t_j - a)}{b - a} & \text{if } j \le i, \\[2mm] w_j \dfrac{(b - t_j)(t_i - a)}{b - a} & \text{if } j > i. \end{cases}$$

The last nonlinear system can be then written as follows:

$$F(\bar{x}) = \bar{x} - \bar{h} - B(\lambda \tilde{x} + \delta \hat{x}) = 0, \tag{6.17}$$

where $\bar{x} = (x_1, x_2, \ldots, x_m)^T, \bar{h} = (h_1, h_2, \ldots, h_m)^T, B = (b_{ij}), \tilde{x} = (x_1^3, x_2^3, \ldots, x_m^3)^T$ and $\hat{x} = (x_1^{2+p}, x_2^{2+p}, \ldots, x_m^{2+p})^T$. In view of what the domain $\Omega$ is for Eq. (6.16), we consider $F : \widetilde{\Omega} \subset \mathbb{R}^m \to \mathbb{R}^m$, where $\widetilde{\Omega} = \{(x_1, x_2, \ldots, x_m) \in \mathbb{R}^m; x_i \ge 0 \text{ for } i = 1, 2, \ldots, m\}$. In addition,

$$F(\bar{x}) = \begin{pmatrix} x_1 - h_1 - \sum_{j=1}^{m} b_{1j} \left( \lambda x_i^3 + \delta x_i^{2+p} \right) \\ x_2 - h_2 - \sum_{j=1}^{m} b_{2j} \left( \lambda x_i^3 + \delta x_i^{2+p} \right) \\ \vdots \\ x_m - h_m - \sum_{j=1}^{m} b_{mj} \left( \lambda x_i^3 + \delta x_i^{2+p} \right) \end{pmatrix},$$

$$F'(\bar{x}) = I - B(3\lambda D_1(\bar{x}) + (2 + p)\delta D_2(\bar{x})),$$

where $D_1(\bar{x}) = \text{diag}\{x_1^2, x_2^2, \ldots, x_m^2\}$ and $D_2(\bar{x}) = \text{diag}\{x_1^{1+p}, x_2^{1+p}, \ldots, x_m^{1+p}\}$, and

$$F''(\bar{x})\bar{y}\,\bar{z} = -B \left( \left( 6\lambda x_1 + (2 + p)(1 + p)\delta x_1^p \right) y_1 z_1, \ldots, \right.$$
$$\left. \left( 6\lambda x_m + (2 + p)(1 + p)\delta x_m^p \right) y_m z_m \right)^T,$$

where $\bar{y} = (y_1, y_2, \ldots, y_m)^T$ and $\bar{z} = (z_1, z_2, \ldots, z_m)^T$.

Moreover, provided that $\|B\| \left( 3|\lambda| \|D_1(\bar{x}_0)\| + (2 + p)|\delta| \|D_2(\bar{x}_0)\| \right) < 1$, we have

$$\|\Gamma_0\| \le \frac{1}{1 - \|B\| \left( 3|\lambda| \|D_1(\bar{x}_0)\| + (2 + p)|\delta| \|D_2(\bar{x}_0)\| \right)} = \beta,$$

$$\|\Gamma_0 F(\bar{x}_0)\| \le \frac{\|\bar{x}_0 - \bar{g} - B(|\lambda|\tilde{x}_0 + \delta\tilde{x}_0)\|}{1 - \|B\| \left( 3\lambda \|D_1(\bar{x}_0)\| + (2 + p)|\delta| \|D_2(\bar{x}_0)\| \right)} = \eta,$$

where $\bar{x}_0 = (\check{x}_1, \check{x}_2, \ldots, \check{x}_m)^T$, $\tilde{x}_0 = (\check{x}_1^3, \check{x}_2^3, \ldots, \check{x}_m^3)^T$ and $\hat{x}_0 = (\check{x}_1^{2+p}, \check{x}_2^{2+p}, \ldots, \check{x}_m^{2+p})^T$. Furthermore,

$$\|F''(\bar{x})\| = \sup_{\|\bar{y}\|=1, \|\bar{z}\|=1} \|F''(\bar{x})\overline{yz}\| \quad \text{with} \quad \|F''(\bar{x})\overline{yz}\| \leq \|B\|\|v(\bar{x}, \bar{y}, \bar{z})\|,$$

where $v(\bar{x}, \bar{y}, \bar{z}) = \left((6\lambda x_1 + (2+p)(1+p)\delta x_1^p)y_1 z_1, \ldots, (6\lambda x_m + (2+p)(1+p)\delta x_m^p)y_m z_m\right)^T$.

For the infinity norm, we have

$$\|v(\bar{x}, \bar{y}, \bar{z})\|_\infty = \left(6|\lambda|\|\bar{x}\|_\infty + (2+p)(1+p)|\delta|\|x_i\|_\infty^p\right)\|\bar{y}\|_\infty\|\bar{z}\|_\infty,$$

so that

$$\|F''(\bar{x})\|_\infty \leq \|B\|_\infty \left(6|\lambda|\|\bar{x}\|_\infty + (2+p)(1+p)|\delta|\|\bar{x}\|_\infty^p\right).$$

Consequently,

$$\|F''(\bar{x}_0)\|_\infty \leq \|B\|_\infty \left(6|\lambda|\|\bar{x}_0\|_\infty + (2+p)(1+p)|\delta|\|\bar{x}_0\|_\infty^p\right) = M,$$

$$\|F''(\bar{x}) - F''(\bar{x}_0)\|_\infty \leq \|B\|_\infty \left(6|\lambda|\|\bar{x} - \bar{x}_0\|_\infty + (2+p)(1+p)|\delta|\|\bar{x} - \bar{x}_0\|_\infty^p\right)$$

and then $\omega(z) = \|B\|_\infty \left(6|\lambda|z + (2+p)(1+p)|\delta|z^p\right)$.

Observe that in this case $\|F''(\bar{x})\|_\infty$ is not bounded in general, since the function $\chi(t) = 6|\lambda|t + (2+p)(1+p)|\delta|t^p$ is increasing. Therefore, condition (A2) of Kantorovich is not satisfied.

To solve the difficulty of applying Kantorovich's conditions, a common alternative is to locate the solutions in a domain $\widetilde{\Omega}_0 \subset \widetilde{\Omega}$ and look for a bound for $\|F''(\bar{x})\|_\infty$ in $\widetilde{\Omega}_0$ (see [5]). In the next example we see that we cannot use this alternative either, because a priori we cannot find a domain $\widetilde{\Omega}_0$ which contains solutions of the equation.

We consider the equation of type (6.16) given by

$$x(s) = 1 + \int_0^1 G(s, t)\left(x(t)^3 + \frac{1}{4}x(t)^{\frac{5}{2}}\right) dt, \quad s \in [0, 1]. \tag{6.18}$$

Once Eq. (6.16) is discretized, the solutions $\bar{x}^*$ of the corresponding nonlinear system given by (6.17) must satisfy

$$\|\bar{x}^*\|_\infty - 1 - \|B\|\left(\|\bar{x}^*\|_\infty^3 + \frac{1}{4}\|\bar{x}^*\|_\infty^{\frac{5}{2}}\right) \leq 0,$$

which does not imply restrictions on $\|\bar{x}^*\|_\infty$, so that we cannot locate a domain $\widetilde{\Omega}_0 \subset \widetilde{\Omega}$ where $\|F''(\bar{x})\|_\infty$ is bounded and contains a solution $\bar{x}^*$. In consequence,

we cannot guarantee the semilocal convergence of Newton's method to a discretized
solution of (6.18) from the Newton-Kantorovich theorem.

Now, we make clear the importance of the fact that the starting point $\bar{x}_0$ is
close to the solution $\bar{x}^*$. If we consider, for example, $m = 8$ and the starting
point $\bar{x}_0 = (2, 2, \ldots, 2)^T$, we cannot apply Theorem 6.3.9, since $\beta = 3.3821\ldots$,
$\eta = 0.9280\ldots, M = 1.6465\ldots, \alpha = 0.1698\ldots,$

$$\varphi(t) = (0.2744\ldots) - (0.2956\ldots)t + (0.8232\ldots)t^2 + (0.0308\ldots)t^{\frac{5}{2}} + (0.1235\ldots)t^3$$

and $\varphi(\alpha) = 0.2489\ldots > 0$.

However, if we take into account condition (6.15), that is satisfied for all $y \in \widetilde{\Omega}_0$,
instead of (6.9), that is just satisfied at the starting point $\bar{x}_0$, we can select other
starting point $\bar{x}_0$. For example, if we choose $\bar{x}_0 = (1, 1, \ldots, 1)^T$, we obtain in this
case $\beta = 1.7248\ldots, \eta = 0.2499\ldots, M = 0.8571\ldots, \alpha = 0.5236\ldots,$

$$\varphi(t) = (0.1449\ldots) - (0.5797\ldots)t + (0.4285\ldots)t^2 + (0.0308\ldots)t^{\frac{5}{2}} + (0.1235\ldots)t^3$$

and $\varphi(\alpha) = -0.0172\ldots \leq 0$. Therefore, we can guarantee the convergence
of Newton's method, starting at $\bar{x}_0 = (1, 1, \ldots, 1)^T$, from Theorem 6.3.9. In
addition, after five iterations, Newton's method converges to the solution $\bar{x}^* =
(x_1^*, x_2^*, \ldots, x_8^*)^T$ which is shown in Table 6.1. Moreover, since $t^* = 0.3596\ldots$
and $t^{**} = 0.6822\ldots$, the domains of existence and uniqueness of solution are
respectively

$$\{\bar{v} \in \widetilde{\Omega}; \|\bar{v} - \bar{1}\|_\infty \leq 0.3596\ldots\} \quad \text{and} \quad \{\bar{v} \in \widetilde{\Omega}; \|\bar{v} - \bar{1}\|_\infty < 0.6822\ldots\}.$$

*Remark 6.3.3* Finally, we first observe that applying condition (6.9), we have

$$\|F''(x) - F''(x_0)\| \leq \omega(\|x - x_0\|) \leq \omega(\|x\| + \|x_0\|), \quad x_0, x \in \Omega,$$

so that

$$\|F''(x)\| \leq \|F''(x_0)\| + \omega(\|x\| + \|x_0\|) = \tilde{\omega}(\|x\|), \quad x_0, x \in \Omega,$$

where $\tilde{\omega} : [0, +\infty) \to \mathbb{R}$ is a continuous monotonous function. Observe that
the previous new condition is milder than condition (6.9). In addition, this new

**Table 6.1** Numerical
solution of system (6.17)
associated to integral
equation (6.18)

| $i$ | $x_i^*$ | $i$ | $x_i^*$ |
|---|---|---|---|
| 1 | 1.021626... | 5 | 1.302053... |
| 2 | 1.105232... | 6 | 1.218581... |
| 3 | 1.218581... | 7 | 1.105232... |
| 4 | 1.302053... | 8 | 1.021626... |

condition also generalizes condition (A2) of the Newton-Kantorovich theorem. In next section, we study the semilocal convergence of Newton's method under this new condition.

## 6.4   $F''$ is an $\omega$-Bounded Operator

To generalize everything said so far, we replace condition (A2) with the following milder condition

$$\|F''(x)\| \leq \tilde{\omega}(\|x\|), \quad x \in \Omega, \tag{6.19}$$

where $\tilde{\omega} : [0, +\infty) \rightarrow \mathbb{R}$ is a continuous monotonous (nondecreasing or nonincreasing) function. Obviously, condition (6.19) generalizes (A2).

In this situation, if we consider (A1)–(6.19), we cannot obtain again, by interpolation fitting, a real function $f$ for Theorem 6.2.1, since (6.19) does not allow determining the class of functions where (6.19) can be applied. To solve this problem, we proceed differently, by solving an initial value problem, as we have done in the previous sections. For this, as $\tilde{\omega}$ is monotonous, we have

$$\|F''(x)\| \leq \tilde{\omega}(\|x\|) \leq \varpi(t; \|x_0\|, t_0),$$

where

$$\varpi(t; \|x_0\|, t_0) = \begin{cases} \tilde{\omega}(\|x_0\| - t_0 + t) \text{ if } \tilde{\omega} \text{ is nondecreasing, provided that} \\ \qquad \|x\| - \|x_0\| \leq \|x - x_0\| \leq t - t_0, \\ \tilde{\omega}(\|x_0\| + t_0 - t) \text{ if } \omega \text{ is nonincreasing, provided that} \\ \qquad \|x_0\| - \|x\| \leq \|x - x_0\| \leq t - t_0. \end{cases} \tag{6.20}$$

Observe that $\varpi(t; \|x_0\|, t_0)$ is a nondecreasing function.

As a result of the above, instead of (6.19), we consider

$$\|F''(x)\| \leq \varpi(t; \|x_0\|, t_0), \quad \text{when} \quad \|x - x_0\| \leq t - t_0, \tag{6.21}$$

where $\varpi : [t_0, +\infty) \rightarrow \mathbb{R}$ is a continuous monotonous function such that $\varpi(t_0; \|x_0\|, t_0) \geq 0$. The corresponding initial value problem to solve is then

$$\begin{cases} y''(t) = \varpi(t; \|x_0\|, t_0), \\ y(t_0) = \dfrac{\eta}{\beta}, \quad y'(t_0) = -\dfrac{1}{\beta}. \end{cases} \tag{6.22}$$

whose solution is given in the following theorem.

**Theorem 6.4.1** *We suppose that $\varpi(t; \|x_0\|, t_0)$ is continuous for all $t \in [t_0, t']$. Then, for any real numbers $\beta \neq 0$ and $\eta$, there exists only one solution $g(t)$ of initial value problem (6.22) in $[t_0, t']$, that is,*

$$g(t) = \int_{t_0}^{t} \int_{t_0}^{\theta} \varpi(\xi; \|x_0\|, t_0) \, d\xi \, d\theta - \frac{t - t_0}{\beta} + \frac{\eta}{\beta}, \tag{6.23}$$

*where $\varpi$ is the function defined in (6.20).*

Observe that (6.23) with $t_0 = 0$ is reduced to the Kantorovich polynomial given in (6.3) if $\varpi$ is constant.

### 6.4.1 Semilocal Convergence Result

By analogy with Kantorovich, if we want to apply the technique of majorizing sequence to our particular problem, the equation $g(t) = 0$, where $g$ is defined in (6.23), must have at least one root greater than $t_0$, so that we have to guarantee the convergence of the real sequence

$$t_0 \text{ given,} \quad t_{n+1} = N_g(t_n) = t_n - \frac{g(t_n)}{g'(t_n)}, \quad n = 0, 1, 2, \ldots, \tag{6.24}$$

to this root, for obtaining a majorizing sequence under conditions (A1)–(6.19). Clearly, the first we need is to analyse the function $g$ defined in (6.23). Then, we give some properties of the function $g$, whose proofs are easy to follow.

**Theorem 6.4.2** *Let $g$ and $\varpi$ be the functions defined in (6.23) and (6.20), respectively.*

(a) *There exists only one positive solution $\alpha > t_0$ of the equation*

$$g'(t) = \int_{t_0}^{t} \varpi(\xi; \|x_0\|, t_0) \, d\xi - \frac{1}{\beta} = 0, \tag{6.25}$$

*which is the unique minimum of $g$ in $[t_0, +\infty)$ and $g$ is nonincreasing in $[t_0, \alpha)$.*

(b) *If $g(\alpha) \leq 0$, then the equation $g(t) = 0$ has at least one root in $[t_0, +\infty)$. Moreover, if $t^*$ is the smallest root of $g(t) = 0$ in $[t_0, \infty)$, we have $t_0 < t^* \leq \alpha$.*

As we are interested in the fact that (6.24), where the function $g$ defined in (6.23), is a majorizing sequence of the sequence $\{x_n\}$ defined by Newton's method in the Banach space $X$, we establish the convergence of $\{t_n\}$ in the next result, whose proof follows similarly to that of Theorem 6.3.3.

**Theorem 6.4.3** *Let $\{t_n\}$ be the real sequence defined in (6.24), where the function $g$ is given in (6.23). Suppose that there exist a solution $\alpha > t_0$ of Eq. (6.25) such that*

$g(\alpha) \leq 0$. Then, the sequence $\{t_n\}$ is nondecreasing and converges to the root $t^*$ of the equation $g(t) = 0$.

The following is to prove that (6.24) is a majorizing sequence of the sequence $\{x_n\}$ and this sequence is well-defined, provided that $B(x_0, t^* - t_0) \subset \Omega$. Previously, from (A1), we observe that

$$\|x_1 - x_0\| = \|\Gamma_0 F(x_0)\| \leq \eta = t_1 - t_0 < t^* - t_0.$$

**Theorem 6.4.4** Let $g$ be the function defined in (6.23). Suppose that conditions (A1)–(6.19) are satisfied. Suppose also that $g(\alpha) \leq 0$, where $\alpha$ is a solution of Eq. (6.25) such that $\alpha > t_0$, and $B(x_0, t^* - t_0) \subset \Omega$. Then, $x_n \in B(x_0, t^* - t_0)$, for all $n \in \mathbb{N}$. Moreover, (6.24) is a majorizing sequence of the sequence $\{x_n\}$, namely

$$\|x_n - x_{n-1}\| \leq t_n - t_{n-1}, \quad \text{for all} \quad n \in \mathbb{N}.$$

*Proof* We prove the theorem from the next four recurrence relations (for $n \in \mathbb{N}$).

$(i_n)$   There exists $\Gamma_n = [F'(x_n)]^{-1}$ and $\|\Gamma_n\| \leq -\frac{1}{g'(t_n)}$.
$(ii_n)$   $\|F(x_n)\| \leq g(t_n)$.
$(iii_n)$   $\|x_{n+1} - x_n\| \leq t_{n+1} - t_n$.
$(iv_n)$   $\|x_{n+1} - x_0\| \leq t^* - t_0$.

We begin proving $(i_1)$–$(iv_1)$.

$(i_1)$: From $x = x_0 + \tau(x_1 - x_0)$ and $t = t_0 + \tau(t_1 - t_0)$, where $0 \leq \tau \leq 1$, it follows $\|x - x_0\| = \tau \|x_1 - x_0\| \leq \tau(t_1 - t_0) = t - t_0$, so that

$$\|I - \Gamma_0 F'(x_1)\| = \left\| \int_0^1 \Gamma_0 F''(x_0 + t(x_1 - x_0))(x_1 - x_0)\, dt \right\|$$

$$\leq \|\Gamma_0\| \int_0^1 \varpi(t_0 + \tau(t_1 - t_0); \|x_0\|, t_0)\, d\tau(t_1 - t_0)$$

$$= \beta \int_{t_0}^{t_1} \varpi(t; \|x_0\|, t_0)\, dt$$

$$= 1 - \frac{g'(t_1)}{g'(t_0)}$$

$$< 1,$$

since $\beta = -\frac{1}{g'(t_0)}$ and $\|x_1 - x_0\| \leq t_1 - t_0$. Then, from Banach's lemma, we obtain that there exists $\Gamma_1$ and $\|\Gamma_1\| \leq -\frac{1}{g'(t_1)}$.

$(ii_1)$: From Taylor's series, $\|x_1 - x_0\| \leq t_1 - t_0$ and the algorithm of Newton's method, we have

$$F(x_1) = \int_{x_0}^{x_1} F''(x)(x - x_0)\, dx = \int_0^1 F''(x_0 + \tau(x_1 - x_0))(1 + \tau)\, d\tau(x_1 - x_0)^2$$

and

$$\|F(x_1)\| \leq \int_0^1 \varpi\,(t_0 + \tau(t_1 - t_0); \|x_0\|, t_0)\,\tau\,d\tau(t_1 - t_0)^2$$

$$= \int_0^1 g''\,(t_0 + \tau(t_1 - t_0))\,(1 + \tau)\,d\tau(t_1 - t_0)^2$$

$$= g(t_1).$$

$(iii_1)$: $\|x_2 - x_1\| \leq \|\Gamma_1\|\|F(x_1)\| \leq -\frac{g(t_1)}{g'(t_1)} = t_2 - t_1$.

$(iv_1)$: $\|x_2 - x_0\| \leq \|x_2 - x_1\| + \|x_1 - x_0\| \leq t_2 - t_0 \leq t^* - t_0$.

If we now suppose that $(i_j)$–$(iv_j)$ are true for $j = 1, 2, \ldots, n$, we can prove that $(i_{n+1})$–$(iv_{n+1})$ are also true, so that $(i_n)$–$(iv_n)$ are true for all $n \in \mathbb{N}$ by mathematical induction. The proof is complete.

We are then ready to prove the following semilocal convergence result for Newton's method under conditions (A1)–(6.19).

**Theorem 6.4.5** *Let $F : \Omega \subseteq X \to Y$ be a twice continuously differentiable operator defined on a nonempty open convex domain $\Omega$ of a Banach space $X$ with values in a Banach space $Y$ and $g(t)$ be the function defined in (6.23). Suppose that (A1)–(6.19) are satisfied, there exists a root $\alpha > t_0$ of (6.25), such that $g(\alpha) \leq 0$, and $B(x_0, t^* - t_0) \subset \Omega$, where $t^*$ is the smallest root of $g(t) = 0$ in $[t_0, +\infty)$. Then, Newton's sequence $\{x_n\}$ converges to a solution $x^*$ of $F(x) = 0$ starting at $x_0$. Moreover, $x_n, x^* \in \overline{B(x_0, t^* - t_0)}$, for all $n \in \mathbb{N}$, and*

$$\|x^* - x_n\| \leq t^* - t_n, \quad n = 0, 1, 2, \ldots,$$

*where $\{t_n\}$ is defined in (6.24).*

*Proof* Observe that $\{x_n\}$ is convergent, since $\{t_n\}$ is a majorizing sequence of $\{x_n\}$ and convergent. Moreover, as $\lim\limits_{n \to +\infty} t_n = t^*$, if $x^* = \lim\limits_{n \to +\infty} x_n$, then $\|x^* - x_n\| \leq t^* - t_n$, for all $n = 0, 1, 2, \ldots$ Furthermore,

$$\|F'(x_n) - F'(x_0)\| = \left\| \int_0^1 F''\,(x_0 + \tau(x_n - x_0))\,d\tau(x_n - x_0) \right\|$$

$$\leq \int_0^1 \varpi\,(t_0 + \tau(t_n - t_0); \|x_0\|, t_0)\,d\tau(t^* - t_0)$$

$$\leq \varpi(t^*; \|x_0\|, t_0)(t^* - t_0),$$

since $\|x - x_0\| \leq t - t_0$ and $\|x_n - x_0\| \leq t^* - t_0$, so that

$$\|F'(x_n)\| \leq \|F'(x_0)\| + \varpi(t^*; \|x_0\|, t_0)(t^* - t_0),$$

and consequently, the sequence $\{\|F'(x_n)\|\}$ is bounded. Therefore, from

$$\|F(x_n)\| \leq \|F'(x_n)\|\|x_{n+1} - x_n\|,$$

it follows that $\lim_n \|F(x_n)\| = 0$, and, by the continuity of $F$, we obtain $F(x^*) = 0$. The proof is complete.

### 6.4.2 Uniqueness of Solution

Once we have proved the semilocal convergence of Newton's method and located the solution $x^*$, we prove the uniqueness of $x^*$. First, we note that if $\varpi(t_0; \|x_0\|, t_0) > 0$, then $g'$ is increasing and $g'(t) > 0$ in $(\alpha, +\infty)$, since $\varpi$ is nondecreasing and $\varpi(t; \|x_0\|, t_0) = g''(t) > 0$. Therefore, $g$ is strictly increasing and convex in $(\alpha, +\infty)$. The last guarantees that $g$ has two real zeros $t^*$ and $t^{**}$ such that $t_0 < t^* \leq t^{**}$. Second, if $\varpi(t_0; \|x_0\|, t_0) = 0$ and $\varpi(t'; \|x_0\|, t_0) > 0$, for some $t' > t_0$, then it takes place the same as in the previous case. Third, finally, if $\varpi(t_0; \|x_0\|, t_0) = 0$ and $\varpi(t; \|x_0\|, t_0) = 0$, for all $t > t_0$, then $g$ is lineal. Note that the latter is not restrictive because only the lineal case is eliminated. Observe that this case is trivial: if $g(t) = at + b$, then $N_g(t) = -\frac{b}{a}$, which is the solution of $g(t) = 0$.

**Theorem 6.4.6** *Under the conditions of Theorem 6.4.5, the solution $x^*$ is unique in $B(x_0, t^{**} - t_0) \cap \Omega$ if $t^* < t^{**}$ or in $B(x_0, t^* - t_0))$ if $t^* = t^{**}$.*

The proof of Theorem 6.4.6 is analogous to that of Theorem 6.3.5.

### 6.4.3 Improvement of the Domain of Starting Points, the Domains of Existence and Uniqueness of Solution and the Error Bounds

In this section, by means of a simple example, we show that we can improve the domain of starting points, the domains of existence and uniqueness of solution and the a priori error bounds for Newton's method if we use condition (6.19) instead of condition (A2).

Consider the equation $F(x) = 0$, where $F : \Omega = (0, a) \to \mathbb{R}$ and $F(x) = x^3 - a$ with $a > 1$. Then,

$$\|\Gamma_0\| = \frac{1}{3x_0^2} = \beta, \quad \|\Gamma_0 F(x_0)\| = \frac{|x_0^3 - a|}{3x_0^2} = \eta, \quad \|F''(x)\| = 6|x|.$$

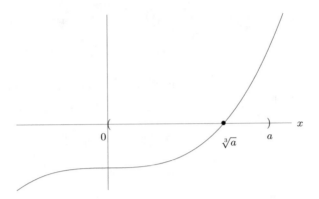

**Fig. 6.1** $F(x) = x^3 - a$

In consequence, we have $L = 6a$ for the Newton-Kantorovich theorem and $\varpi(t; \|x_0\|, t_0) = \varpi(t; \|x_0\|, 0) = \tilde{\omega}(t + \|x_0\|) = 6(t + |x_0|)$ for Theorem 6.4.5.

When analyzing the domain of starting points for Newton's method from the Newton-Kantorovich theorem and Theorem 6.4.5, we will only pay attention to the interval $(0, \sqrt[3]{a})$, since Newton's method always converges if we choose $x_0$ in the interval $(\sqrt[3]{a}, a)$, since $F$ is increasing and convex in $(\sqrt[3]{a}, a)$, see Fig. 6.1.

For the Newton-Kantorovich theorem, we need that $L\beta\eta \leq \frac{1}{2}$, which is equivalent to $3x_0^4 + 4ax_0^3 - 4a^2 \equiv d(x_0) \geq 0$, since $x_0 \in (0, \sqrt[3]{a})$. In addition, $x_0 \in (r^*, \sqrt[3]{a})$, where $r^*$ is such that $d(r^*) = 0$. For Theorem 6.4.5, we need that

$$0 \geq g(\alpha) = (5 - 4\sqrt{2})|x_0|^3 + |x_0^3 - a|,$$

where $g(t) = (t + |x_0|)^3 - 6|x_0|^2 t - |x_0|^3 + |x_0^3 - a|$ and $\alpha = (\sqrt{2} - 1)|x_0|$, which is equivalent to $4(1 - \sqrt{2})x_0^3 + a \leq 0$, since $x_0 \in (0, \sqrt[3]{a})$. Consequently, $x_0 \geq \sqrt[3]{\frac{a}{4(\sqrt{2}-1)}}$.

If we consider the particular case $a = 2011$, we obtain $x_0 \in (12.6026\ldots, \sqrt[3]{2011})$ for the Newton-Kantorovich theorem and $x_0 \in (10.6670\ldots, \sqrt[3]{2011})$ for Theorem 6.4.5. Therefore, we improve the domain of starting points by Theorem 6.4.5 with respect to the Newton-Kantorovich theorem.

Taking then, for example, $x_0 = 12.61$, we obtain $s^* = 0.01520\ldots$ and $s^{**} = 0.06387\ldots$ for the Newton-Kantorovich theorem, so that the domains of existence and uniqueness of solution are respectively

$$\{z \in (0, 2011); |z - x_0| \leq 0.01520\ldots\} \quad \text{and} \quad \{z \in (0, 2011); |z - x_0| < 0.06387\ldots\}.$$

For Theorems 6.4.5 and 6.4.6, we have that $t^* = 0.01229\ldots$ and $t^{**} = 9.96796\ldots$ are the roots of the equation $g(t) = 0$, so that the domains of existence and

**Table 6.2**  Absolute error and a priori error bounds for $x^* = \sqrt[3]{2011}$

| $n$ | $\lvert x^* - x_n \rvert$ | $\lvert t^* - t_n \rvert$ | $\lvert s^* - s_n \rvert$ |
|---|---|---|---|
| 0 | $1.2266\ldots \times 10^{-2}$ | $1.2290\ldots \times 10^{-2}$ | $1.52011\ldots \times 10^{-2}$ |
| 1 | $1.1936\ldots \times 10^{-5}$ | $1.1983\ldots \times 10^{-5}$ | $2.92236\ldots \times 10^{-3}$ |
| 2 | $1.1290\ldots \times 10^{-11}$ | $1.1421\ldots \times 10^{-11}$ | $1.56661\ldots \times 10^{-4}$ |

uniqueness of solution are respectively

$$\{z \in (0, 2011); \lvert z - x_0 \rvert \leq 0.01229\ldots\} \quad \text{and} \quad \{z \in (0, 2011); \lvert z - x_0 \rvert < 9.96796\ldots\}.$$

Therefore the domains of existence and uniqueness of solution that we have obtained from Theorems 6.4.5 and 6.4.6 are better than those obtained from the Newton-Kantorovich theorem.

Finally, we also obtain better error bounds from majorizing sequences; see Table 6.2, where $\{s_n\}$ denotes the majorizing sequence obtained from the Kantorovich polynomial given in (6.3) with $s_0 = 0$ and $\{t_n\}$ denotes the majorizing sequence obtained from (6.24) with the function $g$ defined in (6.23) and $t_0 = 0$.

### 6.4.4   A Priori Error Estimates

Next, we give some a priori error estimates without having to calculate previously the majorizing sequence $\{t_n\}$. For this, we use Ostrowski's technique above-mentioned.

If $g$ has two real positive zeros $t^*$ and $t^{**}$ such that $t^* \leq t^{**}$, we can then write

$$g(t) = (t^* - t)(t^{**} - t)\ell(t)$$

with $\ell(t^*) \neq 0$ and $\ell(t^{**}) \neq 0$. Next, we give a result which provides some a priori error estimates for Newton's method, whose proof is exactly the same as that of Theorem 6.3.6. Remember that we have written above how the function $\varpi$ should be for $g$ to have two real positive roots.

**Theorem 6.4.7**  *Suppose that the function $g$ defined in (6.23) has two real positive roots $t^*$ and $t^{**}$ such that $t_0 < t^* \leq t^{**}$.*

(a)  *If $t^* < t^{**}$, then*

$$\frac{(t^{**} - t^*)\theta^{2^n}}{m_1 - \theta^{2^n}} \leq t^* - t_n \leq \frac{(t^{**} - t^*)\Delta^{2^n}}{M_1 - \Delta^{2^n}}, \quad n \geq 0,$$

*where $\theta = \frac{t^*}{t^{**}}m_1$, $\Delta = \frac{t^*}{t^{**}}M_1$, $m_1 = \min\{Q_1(t); t \in [t_0, t^*]\}$, $M_1 = \max\{Q_1(t); t \in [t_0, t^*]\}$, $Q_1(t) = \frac{(t^{**} - t)\ell'(t) - \ell(t)}{(t^* - t)\ell'(t) - \ell(t)}$ and provided that $\theta < 1$ and $\Delta < 1$.*

*(b) If $t^* = t^{**}$, then*

$$m_2^n t^* \leq t^* - t_n \leq M_2^n t^*,$$

*where $m_2 = \min\{Q_2(t); t \in [t_0, t^*]\}$, $M_2 = \max\{Q_2(t); t \in [t_0, t^*]\}$, $Q_2(t) = \frac{(t^*-t)\ell'(t)-\ell(t)}{(t^*-t)\ell'(t)-2\ell(t)}$ and provided that $m_2 < 1$ and $M_2 < 1$.*

From the last theorem, it follows that the convergence of Newton's method, under conditions (A1)–(6.19), is quadratic if $t^* < t^{**}$ and linear if $t^* = t^{**}$.

*Remark 6.4.1* Note that real function (6.23), from which the majorizing sequences is defined, also has the property of the translation of the Kantorovich polynomial given in (6.3), see Remark 6.2.1, so it is independent of the value of $t_0$. For this reason, we can choose $t_0 = 0$, which simplifies considerably the expressions used and, as a consequence, the last results are independent of the value $t_0$.

**Acknowledgements** This work was supported in part by the project MTM2014-52016-C2-1-P of Spanish Ministry of Economy and Competitiveness.

# References

1. Bruns, D.D., Bailey, J.E.: Nonlinear feedback control for operating a nonisothermal CSTR near an unstable steady state. Chem. Eng. Sci. **32**, 257–264 (1977)
2. Deimling, K.: Nonlinear Functional Analysis. Springer, Berlin (1985)
3. Ezquerro, J.A., Hernández, M.A.: Generalized differentiability conditions for Newton's method. IMA J. Numer. Anal. **22**, 187–205 (2002)
4. Ezquerro, J.A., Hernández, M.A.: On an application of Newton's method to nonlinear operators with $\omega$-conditioned second derivative. BIT **42**, 519–530 (2002)
5. Ezquerro, J.A., Hernández, M.A.: Halley's method for operators with unbounded second derivative. Appl. Numer. Math. **57**, 354–360 (2007)
6. Ezquerro, J.A., González, D., Hernández, M.A.: Majorizing sequences for Newton's method from initial value problems. J. Comput. Appl. Math. **236**, 2246–2258 (2012)
7. Ezquerro, J.A., González, D., Hernández, M.A.: A general semilocal convergence result for Newton's method under centered conditions for the second derivative. ESAIM – Math. Model. Numer. Anal. **47**, 149–167 (2013)
8. Ganesh, M., Joshi, M.C.: Numerical solvability of Hammerstein integral equations of mixed type. IMA J. Numer. Anal. **11**, 21–31 (1991)
9. Gutiérrez, J.M.: A new semilocal convergence theorem for Newton's method. J. Comput. Appl. Math. **79**, 131–145 (1997)
10. Gutiérrez, J.M., Hernández, M.A.: An application of Newton's method to differential and integral equations. ANZIAM J. **42**, 372–386 (2001)
11. Hernández, M.A., Salanova, M.A.: Indices of convexity and concavity. Application to Halley method. Appl. Math. Comput. **103**, 27–49 (1999)
12. Huang, Z.: A note on the Kantorovich theorem for Newton iteration. J. Comput. Appl. Math. **47**, 211–217 (1993)

13. Kantorovich, L.V., Akilov, G.P.: Functional Analysis. Pergamon, Oxford (1982)
14. Ostrowski, A.M.: Solution of Equations in Euclidean and Banach Spaces. Academic, London (1943)
15. Rashidinia, J., Zarebnia, M.: New approach for numerical solution of Hammerstein integral equations. Appl. Math. Comput. **185**, 147–154 (2007)
16. Yamamoto, T.: On the method of tangent hyperbolas in Banach spaces. J. Comput. Appl. Math. **21**, 75–86 (1988)

# Chapter 7
# Complexity of an Homotopy Method at the Neighbourhood of a Zero

**J.-C. Yakoubsohn, J.M. Gutiérrez, and Á.A. Magreñán**

**Abstract** This paper deals with the enlargement of the region of convergence of Newton's method for solving nonlinear equations defined in Banach spaces. We have used an homotopy method to obtain approximate zeros of the considered function. The novelty in our approach is the establishment of new convergence results based on a Lipschitz condition with a $L$-average for the involved operator. In particular, semilocal convergence results (Kantorovich-type results), as well as local convergence results ($\gamma$-theory) are obtained.

## 7.1   Introduction

This paper is concerned with enlarging the region of convergence of the Newton's method using an homotopy method. Let us consider a mapping $f$ defined from an open set $\Omega$ of a Banach space $X$ with values in a Banach space $Y$. We will suppose that $f \in C^1(\Omega, Y)$, i.e., the (Fréchet) derivative $f'(x) \in \mathcal{L}(X, Y)$ for all $x \in \Omega$. From $x \in \Omega$ be such that $f'(x)$ is one-to-one and onto we define the Newton operator

$$N_f(x) = x - f'(x)^{-1}f(x). \qquad (7.1)$$

We also note

$$x_{k+1} = N_f(x_k), \quad k \geq 0, \qquad (7.2)$$

J.-C. Yakoubsohn
Institut de Mathématiques de Toulouse, Université Paul Sabatier, Toulouse, France

J.M. Gutiérrez (✉)
Departamento de Matemáticas y Computación, Universidad de La Rioja, Logroño, Spain
e-mail: jmguti@unirioja.es

Á.A. Magreñán
Departamento de TFG/TFM, Universidad Internacional de La Rioja, Logroño, Spain
e-mail: angel.magrenan@unir.net

© Springer International Publishing Switzerland 2016
S. Amat, S. Busquier (eds.), *Advances in Iterative Methods for Nonlinear Equations*, SEMA SIMAI Springer Series 10,
DOI 10.1007/978-3-319-39228-8_7

the Newton sequence when it is well defined. We are interested with the problem of approximating a regular solution $w$ of the non-linear equation

$$f(x) = 0 \tag{7.3}$$

using an homotopy in the form

$$h(x, t) = f(x) - tf(x_0), \tag{7.4}$$

where $x_0 \in \Omega$ is a given point and $t \in [0, 1]$. This is a geometric point of view of finding a solution of the non-linear equation (7.3). In fact, let us consider the segment line $S = \{tf(x_0) : t \in [0, 1]\}$ and the set $f^{-1}(S)$. Under the assumption that $f'(x_0)$ is one-to-one and onto then the implicit function theorem applies in a neighbourhood of $x_0$ and there exists a curve $x(t)$ solution of $f(x(t)) = tf(x_0)$ defined on $[1 - \delta, 1]$ for some $\delta > 0$. This curve is solution of the initial value problem

$$\dot{x}(t) = -Df(x(t))^{-1}f(x_0), \quad x(1) = x_0. \tag{7.5}$$

In general, this initial value problem has not solution on $[0, 1]$. In the case where it is possible, the goal is to follow numerically the curve $x(t)$, that is implicitly defined by $h(x(t), t) = 0$ thanks to the Newton operator associated to $h(., t)$. More precisely the homotopy method associated to (7.4) consists in the definition of a sequence, called homotopy sequence, $t_0 = 1 > t_1 > \cdots > t_k > \cdots > 0$ such that

$$x_{k+1} = N_{h(..t_{k+1})}(x_k), \ k \geq 0, \tag{7.6}$$

is an approximate zero of $x(t_{k+1})$ where $h(x(t_{k+1}), t_{k+1}) = 0$. The notion of approximate zero that we use is the following

**Definition 1** A $f$-regular ball is an open ball in which all the points $x$ are regular, i.e., $f'(x)$ is one-to-one and onto.

A point $x_0$ is a regular approximate zero of $f$ if there exists a $f$-regular ball containing a zero $w$ of $f$ and a sequence $(x_k)_{k \geq 0}$ that converges to $w$.

In the sequel we will say approximate zero for regular approximate zero. If the sequence $(x_k)_{k \geq 0}$ is the Newton sequence (7.2) (respectively homotopy sequence (7.6)) we call $x_0$ Newton approximate zero (respectively homotopy approximate zero).

We now describe the framework and the background material on which we state the results. We follow the approach taken by Wang Xinghua in several papers [27] and [28] where a Lipschitz condition with a $L$-average is introduced. We give a slightly different formulation of this notion.

**Definition 2** Let us consider $L(s)$ a positive non decreasing real function define on $[0, \upsilon[$ where $\upsilon > 0$. Let $z \in \Omega$ be given such that $f'(z)$ is continuous, one-to-one and onto. We say that the function $f'(z)^{-1}f'$ is $L$-center Lipschitz in $z$ if there exists a quantity

$$\gamma(f,z) \tag{7.7}$$

such that for all $x, y \in \Omega$ that satisfy $\gamma(f,z)(||x-z||+t||x-y||) \le \upsilon$ for all $t \in [0,1]$, one has

$$||f'(z)^{-1}(f'((1-t)x+ty)-f'(x))|| \le \int_{\gamma(f,z)||x-z||}^{\gamma(f,z)(||x-z||+t||y-x||)} L(s)ds. \tag{7.8}$$

The quantity $\gamma(f,z)$ introduced in (7.7) plays a key role in the development of our theory. For instance, the general case is described with $L(s) = 1$ and the analytic case by $L(s) = \dfrac{2}{(1-s)^3}$. To explain the motivation of this paper we need to state two results. The first one gives a point estimate of the radius of the ball around a point where the derivative of $f$ is one-to-one and onto.

**Lemma 1** *Let $z \in \Omega$ be such that $f'(z)$ is one-to-one and onto. Let $\upsilon$ such that*

$$\int_0^\upsilon L(s)ds = 1. \tag{7.9}$$

1. *Then for all $x \in B(z, \upsilon/\gamma(f,z))$ the derivative $f'(z)$ is one-to-one and onto. Moreover*

$$||f'(x)^{-1}f'(z)|| \le \frac{1}{1 - \int_0^{\gamma(f,z)||x-z||} L(s)ds}. \tag{7.10}$$

*We call $\gamma$-ball of $z$ the ball $B(z, \upsilon/\gamma(f,z))$.*
2. *For all $x \in B(z, \upsilon/\gamma(f,z))$ the quantity $\gamma(f,x)$ satisfies the inequality*

$$\int_0^{\gamma(f,x)||x-z||} L(s)ds \le \frac{\int_0^{\gamma(f,z)||x-z||} L(s)ds}{1 - \int_0^{\gamma(f,z)||x-z||} L(s)ds}. \tag{7.11}$$

According to the previous lemma we will suppose that there exists a non decreasing positive function $\varphi$ defined on $[0, \upsilon[$ such that $\varphi(0) = 1$ for all $x$ lying in the $\gamma$-ball of $z$ one has:

$$\gamma(f,x) \le \varphi(\gamma(f,z)||x-z||) \, \gamma(f,z). \tag{7.12}$$

The reason for that will be explained at Sect. 7.3. We also call $\alpha(f,z)$ the quantity

$$\alpha(f,z) = \gamma(f,z)\beta(f,z), \quad \text{with } \beta(f,z) = ||f'(z)^{-1}f(z)|| \tag{7.13}$$

and $\gamma(f,z)$ defined in (7.7). The next result that motivates this paper is a theorem of existence of a zero of $f$ obtained by the homotopy approach via the initial value problem (7.5). The conditions of existence are identical to those stated by the Newton-Kantorovich theorem, see Sect. 7.2. The following result generalizes the one given by Meyer in [14].

**Theorem 1** *Let $x_0 \in \Omega$ such that $f'(x_0)f$ be L-center Lipschitz in $x_0$. Let us suppose $\alpha(f,x_0) \leq \int_0^\upsilon L(s)sds$ and $B(x_0,\upsilon) \subset \Omega$ where $\alpha(f,x_0)$ is defined in (7.13) and $\upsilon$ in (7.9). Then the solution of the initial value problem (7.5) exists in the ball $B(x_0,\upsilon_1/\gamma(f,x_0))$ for all $t \in [0,1]$ where $\upsilon_1$ is defined at Theorem 3. Consequently $x(0)$ is a zero of $f$.*

A proof of this result appears in Sect. 7.4. We remark that theorem 1 does not apply for the points of the $\gamma$-ball of $x_0$ for which $\alpha(f,x_0) > \int_0^\upsilon L(s)sds$. This fact motivates the study of the following problem. Let us suppose the solution of the initial value problem (7.5) is in the $\gamma$-ball of $x_0$: what about the number $k$ of steps to get an approximate zero $x_k$ of $f = h(.,0)$? This number of steps describes the complexity of this homotopy method of finding an approximate zero of (7.3). In this paper we address this problem of complexity for an homotopy approximate zero. We now can explain and state the main result of this paper.

**Theorem 2** *Let $z \in \Omega$ be such that $f'(z)$ is one-to-one and onto. We consider $x_0$ lying in the $\gamma$-ball of $z$. We note $\mathsf{v}_0 = \gamma(f,z)\|x_0 - z\|$ and $0 \leq \mathsf{v} < \upsilon$. Let $\upsilon$ be defined as in (7.9) and $\upsilon_L$ be such that*

$$q(\upsilon_L) = 1 \tag{7.14}$$

*where*

$$q(u) = \frac{\int_0^u L(s)sds}{u(1 - \int_0^u L(s)ds)}. \tag{7.15}$$

*Let $c \geq 1$ and $g_a(t) = a - t + \int_0^t L(s)(t-s)ds$ where $a$ satisfies*

$$a \leq \min\left(\upsilon_L/c - \int_0^{\upsilon_L/c} L(s)(\upsilon_L/c - s)ds, \int_0^\upsilon L(s)sds\right), \tag{7.16}$$

*in order that the first real positive root of $g_a(u)$ is less than or equal to $\upsilon_L/c$. Let us denote*

$$A = \frac{\varphi(\mathsf{v})(\alpha(f,z) + \int_0^{\mathsf{v}_0} L(s)(\mathsf{v}_0 - s)ds + \mathsf{v}_0)}{(1 - \int_0^{\upsilon_L/c} L(s)ds)(1 - \int_0^{\mathsf{v}} L(s)ds)}$$

$$B = \frac{\int_0^{\upsilon_L/c} L(s)(\upsilon_L/c - s)ds + \upsilon_L/c}{1 - \int_0^{\upsilon_L/c} L(s)ds}$$

*where $\varphi$ is the function defined in (7.12) and assume that c satisfies*

$$\int_0^\upsilon L(s)s\,ds - B > 0. \tag{7.17}$$

*Let $x_0$ be such that the curve $x(t)$ solution of the initial problem (7.5) is contained in the $\gamma$-ball of z. Let us consider an homotopy time sequence $(t_k)_{k \geq 0}$ such that*

$$t_0 = 1,\ t_k > 0,\ t_{k-1} - t_k > t_k - t_{k+1} > 0,\quad k \geq 0 \quad and \quad \lim_{k \to 0} t_k = 0 \tag{7.18}$$

*with*

$$1 - t_1 = \frac{1 - \int_0^V L(s)\,ds}{\varphi(V)(\alpha(f,z) + \int_0^{V_0} L(s)(V_0 - s)\,ds + V_0)}a.$$

*Let $w_k$, $k \geq 0$, be such that $f(w_k) = t_k f(x_0)$. Then we have:*

*1. Each $w_k$ is a Newton approximate zero of $w_{k+1}$. Moreover*

$$\gamma(f,z)\varphi(V)\|w_{k+1} - w_k\| \leq a, k \geq 0. \tag{7.19}$$

*2. Each $x_k$ of the Newton sequence (7.6) is a Newton approximate zero of $w_k$, $k \geq 0$. Moreover*

$$\gamma(f,z)\varphi(V)\|x - w_k\| \leq \upsilon_L/c,\ k \geq 0. \tag{7.20}$$

*3. Let $N = \dfrac{\int_0^\upsilon L(s)s\,ds - B}{A}$. The number k of homotopy steps in order that $x_k$ is a Newton approximate zero of w is greater or equal to*

$$\left\lceil \frac{1-N}{1-t_1} \right\rceil \ if\ t_k := \max(0, 1 - k(1 - t_1)),\ k \geq 0.$$

$$\left\lceil \frac{\log N}{\log t_1} \right\rceil \ if\ t_k := t_1^k,\ k \geq 0.$$

$$\left\lceil \log_2\left(\frac{\log N}{\log t_1} + 1\right) \right\rceil \ if\ t_k = t_1^{2^k-1},\ k \geq 0.$$

In Sect. 7.2 we give the historical context of our study with some remarks to highlight Theorem 2. As it is impossible to do a complete review, our goal is to describe the main advances. Section 7.3 contains the $\alpha$-theorems and $\gamma$-theorems in the case of $L$-Lipschitz functions. Sections 7.4 and 7.5 are devoted to proof respectively Theorems 1 and 2. Finally, in Sect. 7.6 we formulate Theorem 2 when

$L(s) = 1$ and $L(s) = \dfrac{2}{(1-s)^3}$ which respectively correspond to the classical case studied by Kantorovich and to the *alpha*-theory initialized by Smale for the analytical case.


## 7.2 Survey and Remarks

In the literature there are two kind of results that give criteria to get a Newton approximate zero. The first ones show the existence of a root with quadratic convergence of the Newton sequence from a starting point: we call them $\alpha$-theorems. The second ones exhibit a ball around a root with quadratic convergence for the Newton sequence initialized from any point of this ball: we call them $\gamma$-theorems. We adopt the denomination given by Shub and Smale for polynomial systems. The $\gamma$-theorems suppose the existence of a root and they were the first to be discovered. Ypma presented in [30] the early history of the Newton method and analyze the known fact that Newton is not the only precursor. Raphson's name is often coupled with Newton's name. In fact the name of Newton-Raphson method has became very popular for numerically solving nonlinear scalar equations. Nevertheless, as the same Ypma suggests, the name of the mathematician Thomas Simpson should be included in the method because of his contributions. No it vain he established the method as known nowadays, except for rhetorical aspects. The doubling of significant digits is the main reason as observed by Newton and Raphson but curiously they gave no theoretical justification for this property. Lagrange shows in his treatise [13] on solving equations that these two methods are identical but presented in different points of view. This is the time of Lagrange that a French mathematician and astronomer, Jean-Raymond Mourraille, describes the geometric aspect of the method (tangent method) in [16], see also [5]: he sees the importance of initial conditions and he observes the local character of convergence to a root. The issue of quadratic convergence was understood after the works of Fourier [10] and Cauchy [4]. For instance Cauchy states the $\gamma$-ball $B(w, r)$ is contained in the real (respectively complex) set

$$\left\{ x_0 \ : \ |f(x_0)| \leq \frac{\min\limits_{|x-x_0|\leq r} |f'(x)|}{p \max\limits_{|x-x_0|<r} \|f''(x)\|} \right\}$$

with $p = 2$ in the real case and $p = 4$ in the complex case. All these works are related to the real roots of functions of one variable. The problem of the approximation of complex roots was extended by Cauchy who established the basis of the studies on the dynamics of complex variable functions. A century later, Milnor retraces in [15] the steps of the development of dynamics systems in the twentieth century since the precursors Fatou and Julia. But this, is other story. To return to Newton's method itself, Simpson in [21] generalize the method to solve a system of two equations with two unknowns. In 1891, Weierstrass generalizes in [26] Newton's method for the simultaneous calculation of all the roots of a polynomial to a system obtained via the symmetric functions of the roots given by Cardano's formulas. He uses a linear homotopy method to prove the fundamental theorem of algebra. Runge in [19], independently of Weierstrass, locally approximates the roots of a polynomial using Newton's method on the same system of symmetric functions. The first proof of the existence of a root of system with a finite number of equations thanks Newton method is given par Fine in 1916, see [9]. The result stated by Fine asserts the existence of a regular root with quadratic convergence of the Newton sequence under the condition

$$||f(x_0)|| \leq \frac{1}{n^{7/2}\mu^2\nu} \tag{7.21}$$

where $\mu$ (respectively $\nu$) is the maximum value of $||Df(x)||$ (respectively $||D^2f(x)||$) on the ball $B(x_0, ||Df(x_0)^{-1}f(x_0)||)$. This paper was ignored a long time after the Kantorovich work, see the state of the art by Ostrowski in 1966 in [18] where this result is unknown and also the interesting discussion by Ortega and Rheinboldt in [17].

But the analysis of Kantorovich was decisive in the modern period since the framework was extended to Banach spaces. In fact, the importance of Kantorovich's theory is based on the statement of a classical problem in Numerical Analysis in terms of Functional Analysis. The condition for the existence of a root is governed by the study of a real or complex sequence that "dominates" the initial Newton sequence. This majorant principle introduced by Kantorovich has been very successful and the number of papers on this subject is huge. The surveys of Yamamoto [29] and Galántai [11] give a good historical development of the main results obtained since the time of Kantorovich. Roughly speaking all these results suppose the existence of the derivative's bound over the domain $\Omega$. In our context, the quantity $\gamma(f, x_0)$ introduced in (7.7) represents this bound. In the eighties and nineties years, the polynomial and analytic case has been considered by Smale in [22, 23] and Shub and Smale in [20]. The new fact is that for analytic maps the $\alpha$-theorems (respectively $\gamma$-theorems) only use information at the initial point (respectively at the root). Under analyticity assumption, the quantity $\gamma(f, x_0)$ only depends of the evaluation of the all derivatives at a point $x_0$. More precisely

$$\gamma(f, x_0) = \sup_{k \geq 2} \left\| \frac{1}{k!}f'(x_0)^{-1}f^{(k)}(x_0) \right\|^{\frac{1}{k-1}}. \tag{7.22}$$

The local behavior of Newton's method is a severe limitation to approximate a root if one don't know a good starting point. In this context the homotopy methods are been massively used to find zeros of map, see for instance Ficken [8], Meyer [14], Ortega and Rheinboldt [17], Hirsch and Smale [12], Allgower and Georg [1], Chow et al. [6], Wacker [25] and the references within. More recently in the last decade, a special attention was paid to polynomial systems concerning the Smale's 17th

Problem: *Can a zero of n complex polynomial equations in n unknowns be found approximatively, on the average, in polynomial time with a uniform algorithm.* We refer to Beltrán and Pardo [3], Dedieu-Malajovich-Shub [7], Beltrán and Leykin in [2] and the references within.

## 7.3   $\alpha$ and $\gamma$ Theorems

We first state a Kantorovich-type result that proves, under a condition called semi-local at a point $x_0$ the existence of a regular zero of $f$ and gives also the behaviour of the Newton sequence. The power of this theory is to show that the behaviour of a Newton sequence defined in a Banach space is dominated by a real Newton sequence associated at a certain function of one variable. This function is universal in the sense that it depends only on the function $f'(x_0)^{-1}f$ and of a neighbourhood around $x_0$. We reformulate the result given by Wang Xinghua in [27] with a more precise estimate for the convergence of the Newton sequence.

**Theorem 3** *Let $x_0 \in \Omega$ such that $f'(x_0)^{-1}f$ be L-center Lipschitz in $x_0$. We introduce the real function*

$$g_{\alpha(f,x_0)}(u) := g(u) = \alpha(f,x_0) - u + \int_0^u L(s)(u-s)ds,$$

*where $\alpha(f,x_0)$ is defined in (7.13). Let us suppose $\alpha(f,x_0) \leq \int_0^\upsilon L(s)s\,ds$ where $\upsilon$ satisfies (7.9). Then*

1. *The function $g(u)$ is strictly convex and has two distinct real roots $\upsilon_1$ and $\upsilon_2$ and*

$$g(u) = (u - \upsilon_1)(u - \upsilon_2)l(u)$$

   *where $l(u) = \int_0^1 \int_0^1 tL((1-t)\upsilon_1 + t(1s)\upsilon_2 + tsu)ds\,dt$. Moreover the Newton sequence $s_0 = 0$, $s_{k+1} = N_g(s_k)$, $k \geq 0$, converges to $\upsilon_1$.*
2. *There exists a unique zero $w$ of $f$ in the ball $B(x_0, \upsilon/\gamma(f,x_0))$.*
3. *The Newton sequence $x_{k+1} = N_f(x_k)$, $k \geq 0$, is well defined and converges to $w$ in the closed ball $\bar{B}(x_0, u_1/\gamma)$. It is dominated by the sequence $(s_k)_{k \geq 0}$, i.e,*

$$||x_k - w|| \leq ||s_k - \upsilon_1||, \ k \geq 0.$$

4. *Moreover if $g(u) \geq \dfrac{\alpha(f,x_0)}{\upsilon_1 \upsilon_2}$ then*

$$||x_k - w|| \leq \frac{1}{\gamma(f,x_0)2^k} \left(\frac{\upsilon_1}{\upsilon_m}\right)^{2^k-1} \upsilon_1, \quad k \geq 0. \tag{7.23}$$

The proof of this $\alpha$-theorem is given in Appendix 1.

Traub and Woźniakowski give in [24] the optimal radius of the open ball around a regular root of $f$ in which the Newton sequence initialized from each point of this ball converges quadratically. We remember this $\gamma$-theorem stated by Wang Xinghua in [28].

**Theorem 4** *Let us suppose that $w$ is a regular zero of $f$ and that $f'(w)^{-1}f(x)$ is Lipschitz for all $x \in B(w, \upsilon/\gamma(f, w))$ where $\upsilon$ satisfies (7.9). Let $\upsilon_L$ defined in (7.15). Then for all $x \in B(w, \upsilon_L/\gamma(f, w))$ the Newton sequence $x_0 = x$, $x_{k+1} = N_f(x_k)$, $k \geq 0$, converges to $w$. Moreover*

$$||x_k - w|| \leq q(u)^{2^k - 1}||x_0 - w||$$

*where $u = \gamma(f, w)||x_0 - w|| < \upsilon_L$ and $q(u)$ is defined in (7.14).*

The proof is given in Appendix 2.

## 7.4 Homotopy and Existence of a Zero

We prove Theorem 1. To do that we use the following result:

**Theorem 5 (Lemma 1.1 in [14])** *Assume that an homotopy map $F(x, t)$ is continuous on $\Omega \times [0, 1]$. Suppose further that there exists a positive continuous function $H(u, t)$ such that $||F(x, t)|| \leq H(||x - x_0||, t)$ for all $(x, t) \in \Omega \times [0, 1]$. Let the maximum solution $u(t)$ of the initial value problem*

$$\dot{u}(t) = H(u, t), \quad u(0) = 0 \tag{7.24}$$

*be bounded and such that $B(x_0, u(1)) \subset \Omega$. Then there exists a solution $x(t)$ of*

$$\dot{x}(t) = F(x, t), \quad x(0) = x_0 \tag{7.25}$$

*which remains in $B(x_0, u(1))$ for $t \in [0, 1]$.*

*Proof of Theorem 1* Let us consider $F(x, t) = -\gamma(f, x_0)f'(x)^{-1}f(x_0)$. From Lemma 1 we have

$$\gamma(f, x_0)||f'(x)^{-1}f(x_0)|| \leq \frac{\alpha(f, x_0)}{1 - \int_0^{\gamma(f, x_0)||x - x_0||} L(s)ds}.$$

We consider $H(u, t) = \dfrac{\alpha(f, x_0)}{1 - \int_0^{\gamma(f,x_0)\|x-x_0\|} L(s)ds}$. In this case we remark the solution

of (7.24) becomes

$$-\dot{u}g'(u) = \alpha(f, x_0), \quad u(0) = 0,$$

with $g(u) = \alpha(f, x_0) - u + \int_0^u L(s)(u - s)ds$. Integrating this previous initial value problem between 0 and $t \in [0, 1]$ we obtain $-g(u(t)) + g(u(0)) = \alpha(f, x_0)t$, i.e.

$$\alpha(f, x_0)t - u(t) + \int_0^{u(t)} L(s)(u(t) - s)ds = 0. \tag{7.26}$$

The condition $\alpha(f, x_0) \leq \int_0^{\upsilon} L(s)sds$ where $\upsilon$ is defined in (7.9) ensures the existence of a solution $u(t)$ of (7.26). On the other hand $u(1) = \upsilon_1$ where $\upsilon_1$ is defined in Theorem 3. From the assumption the ball $B(x_0, \upsilon_1)$ is contained in $\Omega$. The assumptions of Theorem 5 are satisfied. Hence the solution $x(t)$ of the initial value problem (7.5) remains in the ball $B(x_0, \upsilon_1/\gamma(f, x_0))$. □

## 7.5   Proof of Theorem 2

All the proofs are based on the mean value theorem with integral remainder, see [31] for instance.

**Lemma 2** Let $f \in C^1(\Omega, Y)$. Then for any closed segment $[y, x] \in \Omega$ one has

$$f(y) = f(x) + \int_0^1 f'((1 - t)x + ty)(y - x)dt. \tag{7.27}$$

We first state a lemma concerning the properties of the integral remainder.

**Lemma 3** Let $u < \upsilon$.

1. The function $t \to \dfrac{1}{t^i} \int_0^t s^k L(s)ds$ is non decreasing when $k \geq i - 1$. In particular we are interested by the cases $(i, k) = (1, 0)$, $(i, k) = (2, 1)$ and $(i, k) = (1, 1)$.

2. The function $q(t)$ defined on $[0, \upsilon[$ by $q(0) = 0$ and $q(t) = \dfrac{\int_0^t L(s)sds}{t(1 - \int_0^t L(s)ds)}$ if $t \neq 0$ is continuous and strictly increasing.

3. $\displaystyle\int_0^1 \int_{tu}^u L(s)uds\, dt = \int_0^u sL(s)ds.$

4. $\displaystyle\int_0^1 \int_0^{tu} L(s)uds\, dt = \int_0^u L(s)(u - s)ds.$

*Proof* The first part follows from straightforward calculation of the derivative. In fact since $L$ is a non decreasing function and $k \geq i - 1$, we successively have:

$$
\left(\frac{1}{t^i}\int_0^t s^k L(s)ds\right)' = -\frac{i}{t^{i+1}}\int_0^t s^k L(s)ds + t^{k-i}L(t)
$$

$$
\geq L(t)\left(-it^{-i-1}\int_0^t s^k ds + t^{k-i}\right)
$$

$$
\geq L(t)t^{k-i}\left(-\frac{i}{k+1} + 1\right) \geq 0.
$$

The second part follows from the estimates $tL(0)/2 \leq q(t) \leq L(t)/2$ and from the first part. The third and fourth parts are proved by integration by parts using $F(t) = \int_{ta}^a L(s)$ and $G'(t) = 1$. So the proof is finished.    □

Let $w_k$ as defined in Theorem 2. The following lemma bound the quantity $\alpha(f,x)$ of a point $x$ closed to a $w_k$ in the $\gamma$-ball of $z$ thanks quantities involved a given point $x_0$ of this $\gamma$-ball and $w_k$.

**Lemma 4** *Let $x_0$ in the $\gamma$-ball of $z$. Let $v < \upsilon$, $u_k := \gamma(f,z)\|x - w_k\| < \upsilon_L/\varphi(v)$, $\gamma(f,z)\|z - w_k\| \leq v$. For all $x \in B(w_k, \upsilon_L/\varphi(v))$ we have:*

$$
\alpha(f,x) \leq \frac{\varphi(v)(\alpha(f,z) + \int_0^{v_0} L(s)(v_0 - s)ds + v_0)}{(1 - \int_0^v L(s)ds)(1 - \int_0^{u_k\varphi(v)} L(s)ds)}t_k
$$

$$
+ \frac{\int_0^{u_k\varphi(v)} L(s)(u_k\varphi(v) - s)ds + u_k\varphi(v)}{1 - \int_0^{u_k\varphi(v)} L(s)ds},  \tag{7.28}
$$

*where $v_0 = \gamma(f,z)\|x_0 - z\|$.*

*Proof* We know that $\alpha(f,x) = \gamma(f,x)\|f'(x)^{-1}f(x)\| \leq \gamma(f,z)\varphi(v)\|f'(x)^{-1}f(x)\|$. Let us bound $\|f'(x)^{-1}f(x)\|$.

$$
\|f'(x)^{-1}f(x)\| \leq \|f'(x)^{-1}f'(w_k)\| \, \|f'(w_k)^{-1}f(x)\|.
$$

Since $\upsilon_L < \upsilon$, we first know that for all $x \in B(w_k, \upsilon_L/\gamma(f,w_k))$ one has from Lemma 1:

$$
\|f'(x)^{-1}f'(w_k)\| \leq \frac{1}{1 - \int_0^{\gamma(f,w_k)\|x-w_k\|} L(s)ds} \leq \frac{1}{1 - \int_0^{u_k\varphi(v)} L(s)ds}.  \tag{7.29}
$$

Moreover

$$
f'(w_k)^{-1}f(x) = f'(w_k)^{-1}\left(f(w_k) + \int_0^1 (f'((1-t)w_k + tx) - f'(w_k)))(x - w_k)dt\right) + (x - w_k).
$$

Hence using $\gamma(f, w_k)||x - w_k|| \le u_k\varphi(v)$ and the inequality (7.8) we get

$$||f'(w_k)^{-1}f(x)|| \le ||f'(w_k)^{-1}f(w_k)|| + \int_0^1 \int_0^{t\gamma(f,w_k)||x-w_k||} L(s)dsdt||x - w_k|| + ||x - w_k||$$

$$\le ||f'(w_k)^{-1}f(w_k)|| + \frac{1}{\gamma(f,z)\varphi(v)} \int_0^1 \int_0^{tu_k\varphi(v)} L(s)dsdtu_k\varphi(v) + ||x - w_k||.$$

Since $\int_0^1 \int_0^{tu} L(s)dsdtu = \int_0^u L(t)(u-t)dt$ and $\gamma(f,x) \le \gamma(f,z)\varphi(v)$ it follows

$$\gamma(f,x)||f'(w_k)^{-1}f(x)||  \quad \le \gamma(f,z)\varphi(v)||f'(w_k)^{-1}f(w_k)||$$
$$+ \int_0^{u_k\varphi(v)} L(s)(u_k\varphi(v) - s)ds + u_k\varphi(v). \qquad (7.30)$$

We now look for a bound of $||f'(w_k)^{-1}f(w_k)||$.

$$\gamma(f,z)||f'(w_k)^{-1}f(w_k)||  \quad = \gamma(f,z)||f'(w_k)^{-1}f(x_0)||t_k$$
$$\le \gamma(f,z)\,||f'(w_k)^{-1}f'(z)||\,||f'(z)^{-1}f(x_0)||t_k$$
$$\le \frac{\gamma(f,z)||f'(z)^{-1}f(z)|| + \int_0^{v_0} L(s)(v_0 - s)ds + v_0}{1 - \int_0^v L(s)ds}t_k \qquad (7.31)$$

Combining the point estimates (7.29)–(7.31) we finally get

$$\alpha(f,x) \le \frac{\varphi(v)(\alpha(f,z) + \int_0^{v_0} L(s)(v_0 - s)ds + v_0)}{(1 - \int_0^v L(s)ds)(1 - \int_0^{u_k\varphi(v)} L(s)ds)}t_k$$
$$+ \frac{\int_0^{u_k\varphi(v)} L(s)(u_k\varphi(v) - s)ds + u_k\varphi(v)}{1 - \int_0^{u_k\varphi(v)} L(s)ds}.$$

So the proof is finished.                                                                 □

**Lemma 5** *Let us consider $c > 1$. Let $v$ be defined by (7.9) and $v_L$ as in (7.15). We denote for $k$ be given $u_k = \gamma(f,z)||x - w_k||$ and $v < v$. Let us suppose that the curve defined by $h(x(t), t) = 0$ and $x(0) = w_0$ is contained in the $\gamma$-ball of $z$. Let us suppose that $u_k\varphi(v) \le \frac{v_L}{c}$ and*

$$\gamma(f,z)\varphi(v)||w_k - w_{k+1}|| \le \frac{v_L}{c} - \frac{\int_0^{v_L/c} L(s)sds}{1 - \int_0^{v_L/c} L(s)ds}. \qquad (7.32)$$

*Then $y = N_{h(.,t_k)}(x)$ is a Newton approximate zero of $w_{k+1}$. Moreover one has $\gamma(f,z)\varphi(v)||y - w_{k+1}|| \le \frac{v_L}{c}.$*

*Proof* From Lemma 3 the function $q(t)$ is strictly increasing. Since $c > 1$ and from the definition of $v_L$ in (7.15), we first remark that $\dfrac{v_L}{c} > \dfrac{\int_0^{v_L/c} L(s)s\,ds}{1 - \int_0^{v_L/c} L(s)ds}$. We first apply Lemma 10 at the function $h(.,t_k)$ and $y = N_{h(.,t_k)}(x)$. We obtain

$$||y - w_k|| \le \frac{\int_0^{\gamma(f,w_k)||x-w_k||} L(s)s\,ds}{\gamma(f,w_k)||x-w_k||(1 - \int_0^{\gamma(f,w_k)||x-w_k||} L(s)ds)}||x - w_k||^2.$$

Since the function $q(t)$ is strictly increasing one has

$$||y - w_k|| \le \frac{\int_0^{u_k\varphi(v)} L(s)s\,ds}{u_k\varphi(v)(1 - \int_0^{u_k\varphi(v)} L(s)ds)}||x - w_k||$$

$$\le \frac{1}{\gamma(f,z)\varphi(v)} \frac{\int_0^{v_L/c} L(s)s\,ds}{1 - \int_0^{v_L/c} L(s)ds}.$$

Since $||y - w_{k+1}|| \le ||y - w_k|| + ||w_k - w_{k+1}||$ the conclusion follows easily from the assumption (7.32). So the proof is finished. $\qquad\square$

*Proof of Theorem 2* Let us proof the first part. More precisely let us prove for this value of $t_1$ that $w_k$ is a Newton approximate zero of $w_{k+1}$. To do that we will prove that $\gamma(f,z)\varphi(v)||w_{k+1}-w_k|| \le a$. Let $w_k$ and $w_{k+1}$ be such that $f(w_k) = t_k f(x_0)$ and $f(w_{k+1}) = t_{k+1}f(x_0)$. Since $f'(w_k)^{-1}(f(w_k) - f(w_{k+1})) = (t_k - t_{k+1})f'(w_k)^{-1}f(x_0)$ it follows from (7.31)

$$\gamma(f,z) ||f'(w_k)^{-1}(f(w_k) - f(w_{k+1}))|| = \gamma(f,z)||f'(w_k)^{-1}f(x_0)||(t_k - t_{k+1})$$

$$\le \frac{\alpha(f,z) + \int_0^{v_0} L(t)(v_0 - t)dt + v_0}{1 - \int_0^v L(s)ds}(t_k - t_{k+1}). \qquad (7.33)$$

On the other hand we have

$$f'(w_k)^{-1}(f(w_k) - f(w_{k+1})) = w_k - w_{k+1}$$

$$+ \int_0^1 f'(w_k)^{-1}(f'((1 - t)w_{k+1} + tw_k) - f'(w_k))(w_k - w_{k+1})dt.$$

Hence with $e_k = \gamma(f,z)||w_k - w_{k+1}||$ we successively have

$$\gamma(f,z)||f'(w_k)^{-1}(f(w_k) - f(w_{k+1}))|| \ge e_k - \int_0^1 \int_0^{t\gamma(f,w_k)||w_{k+1}-w_k||} L(s)e_k\,ds\,dt$$

$$\ge e_k - \frac{1}{\varphi(v)}\int_0^1 \int_0^{te_k\varphi(v)} L(s)e_k\varphi(v)ds\,dt.$$

From Lemma 3

$$\gamma(f,z)\|f'(w_k)^{-1}(f(w_k)-f(w_{k+1}))\| \geq e_k - \frac{1}{\varphi(\mathsf{v})}\int_0^{e_k\varphi(\mathsf{v})} L(s)(e_k\varphi(\mathsf{v})-t)dt. \quad (7.34)$$

We now combine the point estimates (7.34) and (7.33). From the definition of $t_1$ we obtain:

$$e_k\varphi(\mathsf{v}) - \int_0^{e_k\varphi(\mathsf{v})} L(s)(e_k\varphi(\mathsf{v})-t)dt \leq \frac{\varphi(\mathsf{v})(\alpha(f,z) + \int_0^{\mathsf{v}_0} L(t)(\mathsf{v}_0-t)dt + \mathsf{v}_0)}{1-\int_0^{\mathsf{v}} L(s)ds}(t_k - t_{k+1})$$

$$\leq \frac{\varphi(\mathsf{v})(\alpha(f,z) + \int_0^{\mathsf{v}_0} L(t)(\mathsf{v}_0-t)dt + \mathsf{v}_0)}{1-\int_0^{\mathsf{v}} L(s)ds}(1 - t_1) \quad (7.35)$$

$$\leq a.$$

The previous condition implies $g_a(e_k\varphi(\mathsf{v})) \geq 0$. But at this step we cannot deduce that $e_k\varphi(\mathsf{v})$ is less than the first positive root of $g_a(t)$. To prove that, let us show that $\alpha(h(t_{k+1},.),w_k) \leq a \leq \int_0^{\mathsf{v}} L(s)sds$. In fact we simultaneously have using $f(w_k) = t_k f(x_0)$ and (7.35):

$$\alpha(h(t_{k+1},.),w_k) = \gamma(f,w_k)\|f'(w_k^{-1}(f(w_k)-t_{k+1}f(x_0))\|$$

$$= \gamma(f,w_k)\|f'(w_k^{-1}f(x_0)\|(t_k - t_{k+1})$$

$$\leq \gamma(f,z)\varphi\|f'(w_k^{-1}f(x_0)\|(1 - t_1)$$

$$\leq a.$$

From Theorem 3, it follows that the Newton sequence associated to $h(t_{k+1},.)$ with starting point $w_k$ converges to $w_{k+1}$. In fact this Newton sequence is dominated by the real Newton sequence associated to $g_a(t)$ with starting to 0 which converges to the first positive root of $g_a(t)$. Since this root is less than $\mathsf{v}_L/c$, it follows that $e_k\varphi(\mathsf{v}) \leq \mathsf{v}_L/c$.

For the second part we proceed by induction. The inequality $\gamma(f,z)\varphi(\mathsf{v})\|x-w_k\|$ holds for $k = 0$. Now if this inequality holds for $k \geq 0$ be given then from the part one and Lemma 5 it is true at $k + 1$.

It remains to determine the number of homotopy steps in order that $\alpha(f,x_k) \leq \int_0^{\mathsf{v}} L(s)sds$. Using the point estimate (7.13), we need to have $At_k + B \leq \int_0^{\mathsf{v}} L(s)sds$, i.e, the index $k$ satisfies

$$t_k \leq \frac{\int_0^{\mathsf{v}} L(s)sds - B}{A} := N.$$

We first remark the three sequences $(1 - k(1 - t_1))_{k\geq 0}$, $(t_1^k)_{k\geq 0}$ and $(t_1^{2^k-1})_{k\geq 0}$ evidently satisfy the assumption (7.37). From that, a straightforward computation gives the number of homotopy steps for the three sequences considered. So the proof is finished. □

## 7.6 Study of Special Cases

**Theorem 6** *Let* $L(s) = 1$ *and* $c > 1 + \sqrt{13}/3 \sim 2.202$. *Let* $x_0$ *be such that* $1/2 < \alpha(f, x_0) < 1$. *Let us consider the sequence* $(t_k)_{k \geq 0}$ *with* $t_0 = 1$ *and*

$$1 - t_1 = \frac{2(1-v)^2}{9c^2(\alpha(f, z) + v_0^2/2 + v_0)}.$$

*Let us suppose*

$$t_0 = 1, \; t_k > 0, \; t_{k-1} - t_k > t_k - t_{k+1} > 0, \quad k \geq 0 \, and \, \lim_{k \to 0} t_k = 0. \tag{7.36}$$

*Let* $w_k$, $k \geq 0$, *be such that* $f(w_k) = t_k f(x_0)$. *We then have:*

1. $\gamma(f, z)\varphi(v)\|w_{k+1} - w_k\| \leq \min\left(\dfrac{2}{9c^2}, \dfrac{2(3c-2)}{9c^2}\right) = \dfrac{2}{9c^2} < 1/2, \; k \geq 0.$

2. $\gamma(f, z)\varphi(v)\|x - w_k\| \leq \dfrac{2}{3c}, \; k \geq 0.$

3. *Let* $N = \dfrac{(1-v)^2(9c^2 - 18c - 4)}{18c^2(\alpha(f, z) + v_0^2/2 + v_0)}$. *The number* $k$ *of homotopy steps in order that* $\alpha(f, x_k) \leq 1/2$ *is greater or equal to*

$$\left\lceil \frac{1 - N}{1 - t_1} \right\rceil \; if \; t_k^1 = 1 - k(1 - t_1).$$

$$\left\lceil \frac{\log N}{\log t_1} \right\rceil \; if \; t_k^2 = t_1^k.$$

$$\left\lceil \log_2\left(\frac{\log N}{\log t_1} + 1\right)\right\rceil \; if \; t_k^3 = t_1^{2^k - 1}.$$

The graphics in Fig. 7.1 illustrate the third part of Theorem 6 with $v_0 = 0$, $c = 5/2$ and $\alpha(f, z) = 1$.

**Theorem 7** *Let* $L(s) = \dfrac{2}{(1-s)^3}$ *and*

$$c > \left(\frac{5 - \sqrt{17}}{4}\right)\left(\sqrt{2} + 7/2 + \frac{\sqrt{20\sqrt{2} + 37}}{2}\right) \sim 1.96.$$

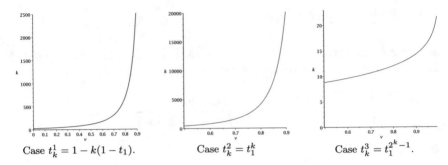

Case $t_k^1 = 1 - k(1 - t_1)$.    Case $t_k^2 = t_1^k$    Case $t_k^3 = t_1^{2^k - 1}$.

**Fig. 7.1** Number of steps $k$ with respect to $\upsilon$ in Theorem 6 with $v_0 = 0$, $\alpha(f, z) = 1$, $c = 5/2$. The parameter $\upsilon$ is defined in (7.9)

Let $x_0$ be such that $1/2 < \alpha(f, x_0) < 1$. Let us consider the sequence $(t_k)_{k \geq 0}$ with $t_0 = 1$ and

$$1 - t_1 = \frac{\left(5 - \sqrt{17}\right)\left(6\sqrt{17}c + 8c^2 - 5\sqrt{17} - 30c + 21\right)(1 - v_0)\left(2v^2 - 4v + 1\right)^2}{2c\left(\alpha(f, z)(1 - v_0) + v_0\right)(1 - v)\left(4c + \sqrt{17} - 5\right)^2}.$$

Let us suppose

$$t_0 = 1, \ t_k > 0, \ t_{k-1} - t_k > t_k - t_{k+1} > 0, \quad k \geq 0 \ and \ \lim_{k \to 0} t_k = 0. \tag{7.37}$$

Let $w_k$, $k \geq 0$, be such that $f(w_k) = t_k f(x_0)$. We then have:

1.

$$\gamma(f, z)\varphi(v)\|w_{k+1} - w_k\| \leq \frac{\left(5 - \sqrt{17}\right)\left(6\sqrt{17}c + 8c^2 - 5\sqrt{17} - 30c + 21\right)}{2c\left(4c + \sqrt{17} - 5\right)^2}$$

$$< 3 - 2\sqrt{2}, \quad k \geq 0.$$

2. $\gamma(f, z)\varphi(v)\|x - w_k\| \leq \dfrac{5 - \sqrt{17}}{4c}$, $k \geq 0$.

3. Let

$$N = \frac{\left[\begin{array}{l} 8\left(3 - 2\sqrt{2}\right)\left(8c^2 - 2\left(5 - \sqrt{17}\right)\left(2\sqrt{2} + 7\right)c \\ + \left(21 - 5\sqrt{17}\right)\left(2\sqrt{2} + 5\right)\right)(1 - v_0)\left((1 - v)^2 - 1/2\right)^2 \end{array}\right]}{\left(4c + \sqrt{17} - 5\right)^2 \left(\alpha(1 - v_0) + \alpha\right)(1 - v)}$$

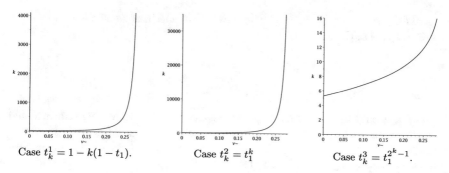

Case $t_k^1 = 1 - k(1 - t_1)$.     Case $t_k^2 = t_1^k$     Case $t_k^3 = t_1^{2^k - 1}$.

**Fig. 7.2** Number of steps $k$ with respect to $v$ in Theorem 7 with $v_0 = 0, \alpha(f, z) = 1$ and $c = 5/2$. The parameter $v$ is defined in (7.9)

*The number $k$ of homotopy steps in order that $\alpha(f, x_k) \leq 1/2$ is greater or equal to*

$$\left\lceil \frac{1 - N}{1 - t1} \right\rceil \text{ if } t_k^1 = 1 - k(1 - t_1).$$

$$\left\lceil \frac{\log N}{\log t_1} \right\rceil \text{ if } t_k^2 = t_1^k.$$

$$\left\lceil \log_2 \left( \frac{\log N}{\log t_1} + 1 \right) \right\rceil \text{ if } t_k^3 = t_1^{2^k - 1}.$$

The graphics in Fig. 7.2 illustrate the third part of Theorem 7 with $v_0 = 0, c = 5/2$ and $\alpha(f, z) = 1$.

**Acknowledgements** This work has been supported by the French project ANR-10-BLAN 0109 and the Spanish project MTM2011-28636-C02-01.

# Appendix 1: Proof of $\alpha$-Theorem

**Theorem 8** *Let us consider $x_0 \in \Omega$ and a function $G(x)$ defined on an open ball $B(x_0, r) \subset \Omega$ in $Y$. Let suppose that there exists a continuous real function $g :$ $[0, r[ \to [0, +\infty]$ such that $0 < ||G(x_0) - x_0|| \leq g(0)$. We assume that there exists $r_1 < r$ such that $g$ is non decreasing on the interval $[0, r_1]$ and that $r_1$ is the unique fix point of $g$ on $[0, r_1]$. We also suppose that*

1. *Let $x \in B(x_0, r_1)$ such that $y = G(x) \in B(x_0, r)$ and $G(y)$ well defined.*
2. *There exits two real positive numbers $t_1$ and $t_2$ such that $||G(x) - x|| \leq t_2 - t_1$ then one has $||G(y) - G(x)|| \leq g(t_2) - g(t_1)$.*

*Then the sequence* $x_{k+1} = G(x_k)$, $k \geq 0$, *is well defined and converges to a fix point* $w \in \bar{B}(x_0, r_1)$ *of G. Moreover*

$$||x_k - w|| \leq s_k - r_1, \quad k \geq 0,$$

*where* $s_0 = 0$ *and* $s_{k+1} = g(s_k)$, $k \geq 0$.

*Proof* It is easy to see that the sequence $(s_k)_{k \geq 0}$ is a non decreasing sequence and converges to $r_1$. Let us prove that the sequence $(x_k)_{k \geq 0}$ is a Cauchy's sequence. Since $||x_1 - x_0|| \leq s_1 - s_0 = g(0) \leq r_1$, we know that $x_1 \in \bar{B}(x_0, r_1)$. Let us suppose $||x_i - x_{i-1}|| \leq s_i - s_{i-1}$, $1 \leq i \leq k$. Then

$$||x_k - x_0|| \leq \sum_{i=1}^{k} ||x_i - x_{i-1}|| \leq \sum_{i=0}^{k-1} s_i - s_{i-1} = s_k \leq r_1.$$

Hence $x_k \in \bar{B}(x_0, r_1)$ and $||x_k - x_{k-1}|| \leq s_k - s_{k-1}$. Then $x_{k+1} = G(x_k)$ is well defined. Let us prove that $x_{k+1} \in B(x_0, r_1)$. Using the assumption we get:

$$||x_{k+1} - x_0|| \leq ||x_{k+1} - x_k|| + ||x_k - x_0|| \leq s_{k+1} - s_k + s_k = s_{k+1} \leq r_1.$$

In the same way this proves that the sequence $(x_k)_{k \geq 0}$ is a Cauchy's sequence. In fact

$$||x_p - x_q|| \leq \sum_{k=q}^{p-1} ||x_{k+1} - x_k|| \leq \sum_{k=q}^{p-1} s_{k+1} - s_k = s_p - s_q.$$

We also have $||x_p - x_0|| \leq r_1$. Hence there exists $w \in \bar{B}(x_0, r_1)$ which is the limit of the sequence $(x_k)$. $\square$

We show that the convergence of the Newton sequence defined by

$$x_{k+1} = N_f(x_k), \quad k \geq 0$$

towards a zero of $f$ can be reduced to the study of an universal Newton sequence

$$s_0 = 0, \quad s_{k+1} = N_g(s_k), \quad k \geq 0$$

where

$$g(u) = \alpha(f, x_0) - u + \int_0^u L(s)(u - s)ds.$$

To do that, we first study the behaviour of the sequence $(s_k)_{k \geq 0}$.

**Lemma 6** *Under the assumption of Theorem 3, we let* $\alpha := \alpha(f, x_0)$. *The function* $g(u)$ *is strictly convex on the interval* $[0, \Upsilon[$. *Let us suppose* $\alpha \leq \int_0^v sL(s)ds$. *Then* *we have:*

1. *The function $g(u)$ has two real roots $v_1$ and $v_2$ and*

$$g(u) = (u - v_1)(u - v_2) l(u)$$

*where $l(u) = \int_0^1 \int_0^1 t L((1-t)v_1 + t(1-s)v_2 + tsu)dsdt$. Moreover the sequence $s_0 = 0$, $s_{k+1} = N_g(s_k)$, $k \geq 0$, converges to $v_1$.*

2. *Let us suppose $l(u) \geq \dfrac{\alpha}{v_1 v_2}$. Then $g(u) \geq p(u) := \dfrac{\alpha}{v_1 v_2}(u - v_1)(u - v_2)$ on the interval $[0, v_1]$. We then have*

$$v_1 - s_k \leq v_1 - r_k \leq \frac{v_m}{2^k}\left(\frac{v_1}{v_m}\right)^{2^k}, \tag{7.38}$$

*where $(r_k)_{k \geq 0}$ is the Newton sequence associated to $p(u)$ and $v_m := \dfrac{v_1 + v_2}{2}$.*

*Proof* We have $g'(u) = -1 + \int_0^u L(s)ds$ and $g''(u) = L(u)$. Since $L(u) > 0$ the function $g(u)$ is strictly convex. Moreover, since the function $L$ is a non-decreasing positive function and $g'(0) = -1$, there exits a unique $v$ such that $g'(v) = 0$. Hence $v$ satisfies $\int_0^v L(s)ds = 1$. The condition $\alpha \leq \int_0^v L(s)sds$ implies $g(v) \leq 0$. Hence the function $g(u)$ has two non negative roots $v_1 := v_1(\alpha) \in [0, v]$ and $v_2 := v_2(\alpha) \in [v, \Upsilon[$.

The function $\alpha \to v_1(\alpha)$ is strictly increasing on the interval $[0, \int_0^v L(s)sds]$. In fact, the differentiation of the equation $g(v_1(\alpha)) = 0$ with respect $\alpha$ gives $v_1'(\alpha) = \dfrac{1}{1 - \int_0^{v_1(\alpha)} L(s)ds}$. This last quantity is non negative since $v_1(\alpha) < v$.

Let us prove $g(u) = (u - v_1)(u - v_2)l(u)$. In fact since $g(v_1) = 0$ we can write $g(u) = \int_0^1 g'((1-t)v_1 + tu)dt(u - v_1)$. The condition $g(v_2) = 0$ implies

$$\int_0^1 g'((1-t)v_1 + tu)dt = \int_0^1 t \int_0^1 g''((1-t)v_1 + t((1-s)v_2 + su)))dtds(u - v_2)$$

$$= \int_0^1 t \int_0^1 L((1-t)v_1 + t((1-s)v_2 + su)))dtds(u - v_2).$$

Under the condition $\alpha \leq \int_0^v sL(s)ds$, the strict convexity of $g(u)$ implies the convergence of the sequence $s_{k+1} = N_g(s_k)$ to $v_1$.

Let us prove the second part. We evidently have $g(u) \geq p(u)$, under the assumption $l(u) \geq \dfrac{\alpha}{v_1 v_2}$. The sequence $r_0 = 0$, $r_{k+1} = N_p(r_k)$ converges to $v_1$ and

$$v_1 - s_k \leq v_1 - r_k.$$

From Lemma 7 below the conclusion follows easily. □

**Lemma 7** *Let us consider the polynomial $p(u)$ as in Lemma 6. The Newton sequence $r_0 = 0$, $r_{k+1} = N_p(r_k)$, $k \geq 0$, converges to $v_1$. Moreover*

$$v_1 - r_k \leq \frac{v_m}{2^k} \left(\frac{v_1}{v_m}\right)^{2^k}, \quad k \geq 0$$

*with $v_m := \dfrac{v_1 + v_2}{2}$.*

*Proof* From $v_m^2 \geq v_1 v_2$, we first remark $p(u) \leq \frac{\alpha}{v_m^2}(v_m - u)^2$. Since $\dfrac{p(u)}{p'(u)} = \dfrac{(u - v_1)(u - v_2)}{2(u - v_m)}$ it follows:

$$v_m - r_{k+1} = v_m - r_k + \frac{p(r_k)}{p'(r_k)}$$

$$\geq v_m - r_k - \frac{(v_m - r_k)^2}{2(v_m - r_k)}$$

$$\geq \frac{v_m - r_k}{2}.$$

Next, the inequality $v_m - r_k \geq \dfrac{v_m}{2^k}$ holds for $k = 0$. The previous point estimate implies that $v_m - r_k \geq \dfrac{v_m}{2^k}$ holds for all $k \geq 0$.

The inequality of this lemma holds for $k = 0$. A straightforward computation gives

$$v_1 - r_{k+1} = v_1 - r_k + \frac{(r_k - v_1)(r_k - v_2)}{2(r_k - v_m)}$$

$$= \frac{(v_1 - r_k)^2}{2(v_m - r_k)}$$

$$\leq \frac{(v_1 - r_k)^2}{v_m} 2^{k-1}.$$

If $v_1 - r_k \leq \dfrac{v_m}{2^k} \left(\dfrac{v_1}{v_m}\right)^{2^k}$ then

$$v_1 - r_{k+1} \leq \frac{(v_1 - r_k)^2}{v_m} 2^{k-1} \leq \frac{v_m^2}{2^{2k}} \left(\frac{v_1}{v_m}\right)^{2^{k+1}} \frac{2^{k-1}}{v_m} = \frac{v_m}{2^{k+1}} \left(\frac{v_1}{v_m}\right)^{2^{k+1}}.$$

So the proof is finished.                                                                    $\square$

**Lemma 8** *Let us suppose that $f'(x_0)$ is invertible and $f'(x_0)^{-1}f(x)$ is Lipschitz in $x_0$. Let $\upsilon$ such that $\int_0^\upsilon L(s)ds = 1$ and $r = \dfrac{\upsilon}{\gamma(f,x_0)}$. Then $f'(x)$ is invertible in the ball $B(x,r)$ with the point estimate*

$$||f'(x)^{-1}f'(x_0)|| \leq \frac{1}{1 - \int_0^{\gamma(f,x_0)||x-x_0||} L(s)ds} = -\frac{1}{h'(\gamma(f,x_0)||x-x_0||)}.$$

*Proof* We have $||f'(x_0)^{-1}f'(x) - I|| \leq \displaystyle\int_0^{\gamma(f,x_0)||x-x_0||} L(s)ds$. From the Von Neumann lemma and the condition $\displaystyle\int_0^{\gamma(f,x_0)||x-x_0||} L(s)ds < \int_0^\upsilon L(s)ds = 1$, the conclusion follows easily. $\qquad\square$

**Lemma 9** *Let us suppose that $f'(x_0)$ is invertible and $f'(x_0)^{-1}f(x)$ is Lipschitz in $x_0$. Let $y = N_f(x)$ such that $\gamma(f,x_0)||x-x_0|| \leq u_1$ and $\gamma(f,x_0)||y-x|| \leq u_2 - u_1$ where $u_2 = N_h(u_1)$. Then*

$$\gamma(f,x_0)||f'(x_0)^{-1}f(y)|| \leq h(u_2).$$

*Proof* We have successively

$$\gamma(f,x_0)||f'(x_0)^{-1}f(y)|| = \gamma(f,x_0)||f'(x_0)^{-1}\left(f(x) + \int_0^1 f'((1-t)x + ty)(y-x)dt\right)||$$

$$\leq \gamma(f,x_0)||f'(x_0)^{-1}\left(\int_0^1 (f'((1-t)x+ty) - f'(x))(y-x)dt\right)||$$

$$\leq \int_0^1 \int_{\gamma(f,x_0)||x-x_0||}^{\gamma(f,x_0)(||x-x_0||+t||y-x||)} L(s)dsdt\gamma(f,x_0)||y-x||$$

$$\leq \int_0^1 \int_0^{\gamma(f,x_0)t||y-x||} L(s+\gamma(f,x_0)||x-x_0||)dsdt\gamma(f,x_0)||y-x||.$$

Since the function $L(t)$ is non decreasing and $\gamma(f,x_0)||x-x_0|| \leq u_1$ and $\gamma(f,x_0)||y-x|| \leq u_2 - u_1$ it follows

$$\gamma(f,x_0)||f'(x_0)^{-1}f(y)|| \leq \int_0^1 \int_0^{t(u_2-u_1)} L(s+u_1)dsdt(u_2-u_1)$$

$$= \int_0^1 \int_{u_1}^{u_1+t(u_2-u_1)} L(s)dsdt(u_2-u_1).$$

Using now that $h(u_1) = -h'(u_1)(u_2 - u_1)$ we obtain

$$\gamma(f,x_0)\|f'(x_0)^{-1}f(y)\| \leq \int_0^1 \int_{t_1}^{u_1+t(u_2-u_1)} L(s)ds(u_2-u_1)dt$$

$$\leq \int_0^1 \big(h'((1-t)u_1 + tu_2) - h'(u_1)\big)(u_2-u_1)dt$$

$$\leq h(u_1) + \int_0^1 h'((1-t)u_1 + tu_2)(u_2-u_1)dt$$

$$\leq h(u_2).$$

□

*Proof of Theorem 3* Lemmas 8 and 9 show that the Newton operator verifies the assumptions of Theorem 8. In fact we have first $\gamma(f,x_0)\|N_f(x_0) - x_0\| = \alpha(f,x_0)$. Moreover if $z = N_f(y)$, $y = N_f(x)$ and $\gamma(f,x_0)\|y-x\| \leq u_2 - u_1$ with $u_2 = N_f(u_1)$. Then combining Lemmas 8 and 9 it follows

$$\gamma(f,x_0)\|z-y\| = \gamma(f,x_0)\|f'(y)^{-1}f(y)\|$$

$$\leq \gamma(f,x_0)\|f'(y)^{-1}f'(x_0)\|\,\|f'(x_0)^{-1}f(y)\|$$

$$\leq -\frac{h(u_2)}{h'(u_2)}$$

$$\leq N_f(u_2) - N_f(u_1).$$

This proves Theorem 3.

## Appendix 2: Proof of $\gamma$-Theorem

To prove this theorem we first state:

**Lemma 10** *Let $x \in B(w, v_L/\gamma(f,w))$ where $v_L$ is defined as in Theorem 4. If $y = N_f(x)$ then*

$$\|y - w\| \leq \frac{\int_0^u L(s)sds}{u(1 - \int_0^u L(s)ds)}\|x - w\|$$

*where $u = \gamma(f,w)\|x - w\|$.*

*Proof* We have

$$y - w = x - w - f'(x)^{-1}f(x)$$

$$= f'(x)^{-1}f'(w)f'(w)^{-1}(f'(x)(x-w) - f(w) - \int_0^1 f'((1-t)w + tx)(x-w)dt)$$

$$= f'(x)^{-1}f'(w)f'(w)^{-1}(\int_0^1 (f'(x) - f'((1-t)w + tx))(x-w)dt).$$

Hence

$$||y - w|| \le ||f'(x)^{-1}f'(w)|| \, ||f'(w)^{-1}(\int_0^1 (f'(x) - f'((1-t)w + tx))(x-w)dt)||$$

$$\le \frac{1}{1 - \int_0^u L(s)ds} \int_0^1 \int_{tu}^u L(s)ds||x - w||dt.$$

From Lemma 3

$$\int_0^1 \int_{tu}^u L(s)ds\,dt\, u = \int_0^u L(s)s\,ds.$$

We finally get

$$\gamma(f, w)||y - w|| \le \frac{\int_0^u L(s)s\,ds}{1 - \int_0^u L(s)ds} = q(u)u.$$

So the proof is finished. □

*Proof of Theorem 4* We proceed by induction. It is true for $k = 0$. From Lemma 10 we have with $u_k = \gamma(f, w)||x_k - w||$:

$$||x_{k+1} - w|| \le \frac{q(u_k)}{u_k}\gamma(f, w)||x_k - w||^2.$$

From Lemma 3 the function $t \to q(t)/t$ is non decreasing.

Then under the condition $||x_k - w|| \le (q(u))^{2^k-1}||x_0 - w||$ we have $q(u_k)/u_k \le q(u)/u$. We then obtain:

$$||x_{k+1} - w|| \le \frac{q(u)}{u}\gamma(f, w)(q(u))^{2^{k+1}-2}||x_0 - w||^2$$

$$\le (q(u))^{2^{k+1}-1}||x_0 - w||.$$

The quantity $q(u)$ is less than 1 since $||x_0 - w|| \le r_w$ and the sequence $(x_k)$ converges to $w$. So the proof is finished. □

## Appendix 3: Separation Theorem

We now prove a separation theorem.

**Theorem 9** *Let $w$ be such that $f(w) = 0$ and $f'(w)^{-1}$ exists. Then for all zero $z \in U$ of $f$ and $z \neq w$ we have*

$$||z - w|| \geq \upsilon_L / \gamma(f, w).$$

*Proof* Since $f(w) = f(z) = 0$ one has $\int_0^1 f'((1-t)w + tz)(z-w)dt = 0$. Hence for

$$0 = ||f'(w)^{-1}f'(w)(z-w) - \int_0^1 f'(w)^{-1}(f'(w) - f'((1-t)w + tz))(z-w)dt||$$

$$\geq ||z - w|| - \int_0^1 \int_{t||z-w||}^{||z-w||} L(s)ds||z - w||dt$$

$$\geq ||z - w|| - \int_0^{||z-w||} L(s)s\,ds.$$

Hence $\frac{1}{||z-w||} \int_0^{||z-w||} L(s)s\,ds \geq 1$. From the definition of $r_w$ and that the function $t \to \frac{1}{t} \int_0^t L(s)s\,ds$ is non decreasing, we conclude to $||z - w|| \geq r_w$.

## References

1. Allgower, E.L., Georg, K.: Numerical Continuation Methods, vol. 33. Springer, Berlin (1990)
2. Beltrán, C., Leykin, A.: Certified numerical homotopy tracking. Exp. Math. **21**(1), 69–83 (2012)
3. Beltrán, C., Pardo, L.: Smale's 17th problem: average polynomial time to compute affine and projective solutions. J. Am. Math. Soc. **22**(2), 363–385 (2009)
4. Cauchy, A.: Sur la détermination approximative des racines d'une équation algébrique ou transcendante. In: Leçons sur le Calcul Différentiel, Buré frères, Paris, pp. 573–609 (1829)
5. Chabert, J.-L., Barbin, E., Guillemot, M., Michel-Pajus, A., Borowczyk, J., Djebbar, A., Martzloff, J.-C.: Histoire d'algorithmes: du caillou à la puce. Belin, Paris (1994)
6. Chow, S.N., Mallet-Paret, J., Yorke, J.A.: Finding zeroes of maps: homotopy methods that are constructive with probability one. Math. Comput. **32**(143), 887–899 (1978)
7. Dedieu, J.-P., Malajovich, G., Shub, M.: Adaptive step-size selection for homotopy methods to solve polynomial equations. IMA J. Numer. Anal. **33**(1), 1–29 (2013)
8. Ficken, F.: The continuation method for functional equations. Commun. Pure Appl. Math. **4**(4), 435–456 (1951)
9. Fine, H.B.: On Newton's method of approximation. Proc. Natl. Acad. Sci. U. S. A. **2**(9), 546 (1916)
10. Fourier, J.-B.J.: Analyse des équations déterminées, vol. 1. Firmin Didot, Paris (1831)
11. Galántai, A.: The theory of Newton's method. J. Comput. Appl. Math. **124**(1), 25–44 (2000)

12. Hirsch, M.W., Smale, S.: On algorithms for solving f(x)= 0. Commun. Pure Appl. Math. **32**(3), 281–312 (1979)
13. Lagrange, J.L., Poinsot, L.: Traité de la résolution des équations numériques de tous les degrés: avec des notes sur plusieurs points de la théorie des équations algébriques. chez Courcier, Paris (1806)
14. Meyer, G.H.: On solving nonlinear equations with a one-parameter operator imbedding. SIAM J. Numer. Anal. **5**(4), 739–752 (1968)
15. Milnor, J.: Dynamics in One Complex Variable (AM-160). Princeton University Press, Princeton (2011)
16. Mouraille, J.-R.: Traité de la résolution des équations en général. Jean Mossy éditeur, Marseille (1768)
17. Ortega, J.M., Rheinboldt, W.C.: Iterative Solution of Nonlinear Equations in Several Variables, No. 30. Siam, Philadelphia (2000)
18. Ostrowski, A.M.: Solution of Equations and Systems of Equations. Academic, New York (1966)
19. Runge, C.: Separation und approximation der wurzeln von gleichungen. In: Enzyklopadie der Mathematichen Wissenschaften, vol. 1, pp. 405–449 (1899)
20. Shub, M., Smale, S.: On the complexity of Bézout's theorem I - Geometric aspects. J. AMS **6**(2), 459–501 (1993)
21. Simpson, T.: Essays on Several Curious and Useful Subjects: In Speculative and Mix'd Mathematicks, Illustrated by a Variety of Examples. H. Woodfall, London (1740)
22. Smale, S.: The fundamental theorem of algebra and complexity theory. Bull. Am. Math. Soc. **4**(1), 1–36 (1981)
23. Smale, S.: Newton's method estimates from data at one point. In: The Merging of Disciplines: New Directions in Pure, Applied, and Computational Mathematics (Laramie, Wyo., 1985), pp. 185–196 (1986)
24. Traub, J.F., Woźniakowski, H.: Convergence and complexity of Newton iteration for operator equations. J. ACM **26**(2), 250–258 (1979)
25. Wacker, H.: A Summary of the Development on Imbedding Methods. Academic, New York, London (1978)
26. Weierstrass, K.: Neuer beweis des satzes, dass jede ganze rationale function einer verdnderlichen dargestellt werden kann als ein product aus linearen functionen derselben verdnderlichen. Mathematische Werke tome 3, 251–269 (1891)
27. Xinghua, W.: Convergence of Newton's method and inverse function theorem in banach space. Math. Comput. Am. Math. Soc. **68**(225), 169–186 (1999)
28. Xinghua, W.: Convergence of Newton's method and uniqueness of the solution of equations in banach space. IMA J. Numer. Anal. **20**(1), 123–134 (2000)
29. Yamamoto, T.: Historical developments in convergence analysis for Newton's and Newton-like methods. J. Comput. Appl. Math. **124**(1), 1–23 (2000)
30. Ypma, T.J.: Historical development of the Newton-Raphson method. SIAM Rev. **37**(4), 531–551 (1995)
31. Zeidler, E.: Nonlinear Functional Analysis and Its Applications I: Fixed-Point Theorems. Springer-Verlag, New York (1986)

# Chapter 8
# A Qualitative Analysis of a Family of Newton-Like Iterative Process with $R$-Order of Convergence At Least Three

**M.A. Hernández-Verón and N. Romero**

**Abstract** This work is focused on the study of iterative processes with R-order at least three in Banach spaces. We begin analyzing the semilocal convergence of a family of Newton-like iterative process. The most known iterative processes with R-order of convergence at least three are included in this family. In the study of iterative processes, there are two important points to bear in mind: the accessibility, which is analyzed by the convergence conditions required by the iterative process and the efficiency, which depends on the order of convergence and the operational cost in each step. These concepts are analyzed for the family of Newton-like iterative process. We obtain significant improvements from the study performed. Finally, considerations about the family of iterative processes are done and some numerical examples and applications to boundary-value problem are given.

## 8.1 Introduction

It is well-known that solving equations of the form

$$F(x) = 0, \tag{8.1}$$

where $F$ is a nonlinear operator, $F : \Omega \subseteq X \to Y$, defined on a non-empty open convex domain $\Omega$ of a Banach space $X$, with values in a Banach space $Y$, is a classical problem that appears in several areas of engineering and science. This problem can represent a differential equation, a boundary value problem, an integral equation, etc. Although some equations can be solved analytically, we usually look for numerical approximations of the solutions, since finding exact solutions is usually difficult. To approximate a solution of Eq. (8.1) we normally consider

M.A. Hernández-Verón • N. Romero (✉)

Departamento de Matemáticas y Computación, Universidad de La Rioja, Logroño (La Rioja) 26004, Spain

e-mail: mahernan@unirioja.es; natalia.romero@unirioja.es

© Springer International Publishing Switzerland 2016

S. Amat, S. Busquier (eds.), *Advances in Iterative Methods for Nonlinear Equations*, SEMA SIMAI Springer Series 10, DOI 10.1007/978-3-319-39228-8_8

iterative schemes. An iterative process is defined by an algorithm such that, from an initial approximation $x_0$, a sequence $\{x_n\}$ is constructed satisfying $\lim x_n = x^*$, where $F(x^*) = 0$. In particular, we consider one-point iterative processes in this form:

$$x_{n+1} = G(x_n), \quad \text{for all} \quad n \geq 0, \tag{8.2}$$

i.e., the new approach, $x_{n+1}$ which is determined by the iteration function $G$, depends only on the previous approximation $x_n$.

There are two fundamental aspects in the study of iterative processes: the convergence of the generated sequence $\{x_n\}$ and their speed of convergence. The convergence properties depend on the choice of the distance $\| \cdot \|$, but for a given distance, the speed of convergence of a sequence $\{x_n\}$ is characterized by the speed of convergence of the sequence of non-negative numbers $\|x^* - x_n\|$. An important measure of the speed of convergence is the R-order of convergence [28]. It is known that a sequence $\{x_n\}$ converges to $x^*$ with R-order at least $\tau$, with $\tau > 1$, if there are constants $C \in (0, \infty)$ and $\gamma \in (0, 1)$ such that $\|x^* - x_n\| \leq C\gamma^{\tau^n}, n = 0, 1, \ldots$. In this work, we study one-point iterative processes with R-order at least three. These processes are not used frequently in practice due to their operational cost, but there exist situations in such that the rise in the speed of convergence justify their use. For instance, these processes have been successfully used in problems where a fast convergence is required, such as stiff systems [14]. For instance, the case of quadratic equations, in which operator $F''$ is constant, is also a situation favorable to the implementation of these iterative processes [8].

We introduce a theory, the most general possible, relative to one-point iterative processes with R-order of convergence at least three. To that, we extend to Banach spaces a result due to W. Gander [12] which characterizes the form that the iterative processes have with R-order of convergence at least three in the scalar case:

Let $t^*$ be a simple zero of a real function $f$ and $H$ a function such that $H(0) = 1$, $H'(0) = \dfrac{1}{2}$ and $|H''(t)| < \infty$. The iteration $t_{n+1} = G(t_n) = t_n - H(L_f(t_n))\dfrac{f(t_n)}{f'(t_n)}$, with $L_f(t) = \dfrac{f(t)f''(t)}{f'(t)^2}$, has R-order of convergence at least three, where $L_f$ is the "degree of logarithmic convexity" (see [22]).

In this paper we consider $H$ in the following form:

$$H(z) = 1 + \frac{1}{2}z + \sum_{k \geq 2} A_k z^k, \quad \{A_k\}_{k \geq 2} \subset R^+, \tag{8.3}$$

where $\{A_k\}_{k \geq 2}$ is a non-increasing sequence such that $\sum_{k \geq 2} A_k z^k < +\infty$, for $\|z\| < r$.

The most known one-point iterative processes with $R$-order of convergence at least three satisfy this result. For instance, for the Chebyshev [16], the super-Halley [1] and the Halley [17] methods, the expressions of the function $H$ are the following:

$$H(L_f(t_n)) = 1 + \frac{1}{2}L_f(t_n), \quad H(L_f(t_n)) = 1 + \frac{1}{2}L_f(t_n) + \sum_{k\geq 2}\frac{1}{2}L_f(t_n)^k, \quad H(L_f(t_n)) =$$
$$1 + \frac{1}{2}L_f(t_n) + \sum_{k\geq 2}\frac{1}{2^k}L_f(t_n)^k, \text{ respectively.}$$

In view of the expression of the well known one-point iterative processes with $R$-order of convergence at least three, we can generalize these methods. Thus, we can obtain new iterative processes with $R$-order of convergence at least three by observing the sequential development of powers of the functions $H$ that have these methods.

In the scalar case, we can consider

$$\begin{cases} t_{n+1} = G(t_n) = t_n - H(L_f(t_n))\dfrac{f(t)}{f'(t)} \\ H(L_f(t_n)) = 1 + \dfrac{1}{2}L_f(t_n) + \displaystyle\sum_{k\geq 2} A_k L_f(t_n)^k, \quad \{A_k\}_{k\geq 2} \subset R^+, \end{cases} \qquad (8.4)$$

where $\{A_k\}_{k\geq 0}$ is a real decreasing sequence with $\displaystyle\sum_{k\geq 2} A_k t^k < +\infty$ for $|t| < r$, for what require that $|L_f(t)| < r$ for the well definition of $H$.

Our goal is to generalize these iterative processes in Banach spaces and study its main features.

The paper is organized as follows. In Sect. 8.2, we introduce a family of iterative processes with $R$-order of convergence at least three in Banach spaces which includes the most well-known iterative processes. We also include their algorithms in series development. In Sect. 8.3, we present a study of the general semilocal convergence theorem for the new family. We also include information about the existence and uniqueness of solution and a result on the a priori error estimates that leads to the third $R$-order of convergence of the iterative processes of the family. In Sect. 8.4, we analyze the accessibility domain for the constructed family, i.e., the set of starting points from which the iterative processes converge to a solution of an equation from any point of the set. In particular, we study the accessibility domains by means of: Attraction basins, obtained for a particular equation; Regions of accessibility obtained from the convergence conditions required to the iterative processes for a particular equation; Domains of parameters also given from the convergence conditions. We see that the domain of parameters of the Newton's method is less restrictive than these of processes of the family. So, in Sect. 8.5, we construct the hybrid iterative method that uses the Newton's method as predictor method and iterative processes of the family as corrector methods, so that it takes advantage of the domain of parameters of the predictor method and the speed of convergence of the corrector method. So, from the same starting points of the Newton's method, the predictor–corrector iterative methods converge with the same rate of convergence iterative processes of the family. In consequence, the accessibility of iterative processes of the family is improved by means of the Newton's method.

Finally, in Sect. 8.6, from the ideas of improving the speed of convergence and reducing the computational cost of one-point iterations, a new uniparametric family of multi-point iterations is constructed in Banach spaces to solve nonlinear equations. The semilocal convergence and the $R$-order of convergence of the new iterations are analyzed under Kantorovich type conditions. The section is finished with a semilocal convergence result under milder convergence conditions, so that the semilocal convergence is studied under the classical Newton-Kantorovich conditions [24]. Moreover, some numerical tests are given.

## 8.2 A Family of Iterative Processes with $R$-Order of Convergence At Least Three

To extend family (8.4) to Banach spaces, firstly it is necessary the extension of the degree of logarithmic convexity to operators defined in Banach spaces.

**Definition 1 ([18])** Let $F$ be a nonlinear twice Fréchet-differentiable operator in an open convex non-empty subset $\Omega$ of a Banach space $X$ in another Banach space $Y$ and let $\mathscr{L}(X, Y)$ be the set of bounded linear operators from $X$ into $Y$. If $x_0 \in \Omega$ and $\Gamma_0 = [F'(x_0)]^{-1} \in \mathscr{L}(Y, \Omega)$ exists, it is defined the "degree of logarithmic convexity" operator as $L_F : \Omega \rightarrow \mathscr{L}(\Omega, \Omega)$, where for a given $x_0 \in \Omega$, it corresponds the linear operator $L_F(x_0) : \Omega \rightarrow \Omega$ such that

$$L_F(x_0)(x) = [F'(x_0)]^{-1} F''(x_0) [F'(x_0)]^{-1} F(x_0)(x), \quad x \in \Omega.$$

From this definition, we consider the family of one-point iterative processes in the following form:

$$\begin{cases} x_{n+1} = G(x_n) = x_n - H(L_F(x_n))\Gamma_n F(x_n) \\ H(L_F(x_n)) = I + \dfrac{1}{2}L_F(x_n) + \displaystyle\sum_{k \geq 2} A_k L_F(x_n)^k, \quad \{A_k\}_{k \geq 2} \subset R^+, \end{cases} \quad (8.5)$$

where $\{A_k\}_{k \geq 2}$ is a non-increasing real sequence where $\displaystyle\sum_{k \geq 2} A_k t^k < +\infty$ for $|t| < r$.

Besides, we denote $L_F(x_n)^k$, $k \in N$, the composition $L_F(x)^k = \overbrace{L_F(x) \circ \cdots \circ L_F(x)}^{k}$ that is a linear operator in $\Omega$, and $I = L_F(x)^0$.

Taking into account the following result given in [5], it follows easily that, if $L_F(x_n)$ exists and $\|L_F(x_n)\| < r$, $n \geq 0$, then (8.6) is well defined:

*Let X be a Banach space and $T : X \to X$ a lineal and bounded operator, $T \in \mathcal{L}(X,X)$. If $\sum_{n=0}^{\infty} A_n t^n$, where $A_n \in R^+$, is convergent with radius of convergence r, then $\sum_{n=0}^{\infty} A_n T^n \in \mathcal{L}(X,X)$ is well-defined, provided that $\|T\| < r$,*

Notice that this family of one-point iterative processes is well defined in Banach spaces, when the operator $H$ is. It has that the operator $H$ is,

$$H(L_F(\_)) : \Omega \to \mathcal{L}(\Omega, \Omega) \to \mathcal{L}(\Omega, \Omega)$$

where it is associated to each $x_n$ a "polynomial" in $L_F(x_n)$, that is, $H(L_F(x_n)) = \sum_{k \geq 0} A_k L_F(x_n)^k$, with $A_0 = 1$ and $A_1 = 1/2$. We assume that $H$ is analytical in a neighborhood of zero.

On the other hand, it is clear that this family of one-point iterative processes, in the scalar case, has R-order of convergence at least three, since $H(0) = 1$, $H'(0) = \frac{1}{2}$ and $|H''(0)| < \infty$, (see [18]).

In Sect. 8.3, we prove this fact in Banach spaces. Observe that, the most well-known one-point iterative processes with R-order of convergence at least three admit a representation as the one given in algorithm (8.6):

Chebyshev [16]:

$$H(L_F(x_n)) = I + \frac{1}{2}L_F(x_n). \tag{8.6}$$

Chebyshev-like method [9]:

$$H(L_F(x_n)) = I + \frac{1}{2}L_F(x_n) + \frac{1}{2}L_F(x_n)^2. \tag{8.7}$$

Euler [8]:

$$H(L_F(x_n)) = 1 + \frac{1}{2}L_F(x_n) + \sum_{k \geq 2}(-1)^k 2^{k+1}\frac{1/2}{k+1}L_F(x_n)^k. \tag{8.8}$$

Halley [7]:

$$H(L_F(x_n)) = I + \frac{1}{2}L_F(x_n) + \sum_{k \geq 2}\frac{1}{2^k}L_F(x_n)^k. \tag{8.9}$$

Super-Halley [19]:

$$H(L_F(x_n)) = I + \frac{1}{2}L_F(x_n) + \sum_{k \geq 2}\frac{1}{2}L_F(x_n)^k. \tag{8.10}$$

Ostrowski [18]:

$$H\big(L_F(x_n)\big) = 1 + \frac{1}{2}L_F(x_n) + \sum_{k\geq 2}(-1)^k\frac{-1/2}{k}L_F(x_n)^k. \tag{8.11}$$

The Logarithmic method [1]:

$$H\big(L_F(x_n)\big) = 1 + \frac{1}{2}L_F(x_n) + \sum_{k\geq 2}\frac{1}{(k+1)!}L_F(x_n)^k. \tag{8.12}$$

The Exponential method [1]:

$$H\big(L_F(x_n)\big) = 1 + \frac{1}{2}L_F(x_n) + \sum_{k\geq 2}\frac{1}{k+1}L_F(x_n)^k. \tag{8.13}$$

## 8.3   Semilocal Convergence

Regarding the study of convergence, we can obtain results on global, semilocal or local convergence, in terms of the required conditions for the domain of $F$, the initial guess $x_0$ or the solution $x^*$ respectively. An interesting result of semilocal convergence for one-point iterative process of $R$-order of converge at least three in Banach spaces is given in [29], where the strongest required assumptions are:

Let us suppose that $F : \Omega \subseteq X \to Y$, $X, Y$ Banach spaces, $\Omega$ a nonempty open convex subset of $X$, $F$ is a twice continuously Fréchet-differentiable operator, which satisfies the following conditions:

(C1)   There exists a point $x_0 \in \Omega$ where the operator $\Gamma_0 = [F'(x_0)]^{-1} \in \mathscr{L}(Y, X)$ is defined and $\|\Gamma_0\| \leq \beta$.

(C2)   $\|\Gamma_0 F(x_0)\| \leq \eta$.

(C3)   There exists $M \geq 0$ such that $\|F''(x)\| \leq M$,   $\forall x \in \Omega$.

(C4)   There exists $N \geq 0$ such that $\|F'''(x)\| \leq N$,   $\forall x \in \Omega$.

Later [18], the condition (C4) is replaced by the milder condition

(C4')   For some fixed $K \geq 0$, $\|F''(x) - F''(y)\| \leq K\|x - y\|$,   $\forall x, y \in \Omega$.

Observe that condition (C4') means that $F''$ is Lipschitz continuous in $X$. The conditions (C1)–(C3) and (C4') for the operator $F$ are usually considered to prove the $R$-order of convergence at least three of iterative processes in Banach spaces.

### 8.3.1   Semilocal Convergence Under Kantorovich's Conditions

In general, to give a result of semilocal convergence for one-point iterative process with $R$-order of convergence at least three, it is necessary consider the following well-known Kantorovich conditions [24]: (C1)–(C3) and (C4'). In [18], we provide a result of semilocal convergence under these assumptions. To establish the convergence of the sequence $\{x_n\}$ it is enough to prove that it is a Cauchy sequence, since the sequence belongs to a Banach space. We also provide the uniqueness of the solution and the $R$-order of convergence. For this, first we define the following real functions:

$$
\begin{aligned}
&v(t) = \sum_{k \geq 2} 2A_k t^{k-2} \\
&h(t) = 1 + \frac{1}{2}t\big(1 + tv(t)\big), \quad \varphi(t) = 12 + 6t - 6h(t)(1 + 2t) + 3h(t)^2 t(2t - 1) \\
&f(t) = \frac{1}{1 - th(t)}, \\
&g(t, s) = \frac{t^2}{2}\left[1 + (1 + t)v(t) + \frac{t}{4}(1 + tv(t))^2\right] + \frac{s}{6}.
\end{aligned}
$$

$$(8.14)$$

**Theorem 1 ([18])**  *Let $F$ be a nonlinear twice Fréchet-differentiable operator under the previous conditions (C1)–(C3) and (C4'). We assume that*

$$a_0 = M\beta\eta < r, \quad a_0 h(a_0) < 1 \quad and \quad b_0 = K\beta\eta^2 < \varphi(a_0). \tag{8.15}$$

*Then, if $B(x_0, R) \subset \Omega$, where $R = \dfrac{h(a_0)\eta}{1 - f(a_0)g(a_0, b_0)}$, the family of iterative processes (8.6) starting in $x_0$, converges to a solution $x^*$ of Eq. (8.1) with R-order of convergence at least three. In this case, the solution $x^*$ and the iterations $x_n$ belong to $\overline{B(x_0, R)}$ and $x^*$ is a unique solution of (8.1) in $B(x_0, \frac{2}{M\beta} - R) \cap \Omega$. Moreover, the following a priori estimates of the error are also obtained*

$$
\| x^* - x_n \| < h(a_0\gamma^{\frac{3^n-1}{2}})\eta \,\frac{\gamma^{\frac{3^n-1}{2}}\Delta^n}{1 - \gamma^{3^n}\Delta} < \left(\gamma^{\frac{1}{2}}\right)^{3^n}\frac{R}{\gamma^{\frac{1}{2}}},
$$

*with $\gamma = f(a_0)^2 g(a_0, b_0) \in (0, 1)$ and $\Delta = \dfrac{1}{f(a_0)}.$*

Sometimes, (C4') is weakened by the condition:

$$\|F''(x) - F''(y)\| \leq K\|x - y\|^p, \quad \forall x, y \in \Omega, \ K \geq 0, \ p \in (0, 1],$$

this condition means that $F''$ is $(K, p)$-Hölder continuous in $\Omega$ [2].

### 8.3.2   Semilocal Convergence for Operators with Second Derivative $\omega$-Conditioned

According to the above, the number of equations that can be solved is limited. For instance, we cannot analyze the convergence of iteration (8.6) to a solution of equations where sums of operators, which satisfy Lipschitz or Hölder conditions, are involved, as for example in the following nonlinear integral operator of mixed Hammerstein type [13]

$$F(x)(s) = x(s) - u(s) + \sum_{i=1}^{m} \int_a^b k_i(s,t)\ell_i(x(t))\, dt, \quad s \in [a,b], \tag{8.16}$$

where $-\infty < a < b < \infty$, $u$, $\ell_i$, $k_i$, for $i = 1, 2, \ldots, m$, are known functions, $\ell_i''(x(t))$ is $(L_i, p_i)$-Hölder continuous in $\Omega$, for $i = 1, 2, \ldots, m$, and $x$ is a solution to be determined. This type of operator appear, for instance, in dynamic models of chemical reactors, see [4].

A generalization of the Lipschitz and Hölder conditions is this one:

(C4")   $\|F''(x) - F''(y)\| \leq \omega(\|x - y\|), \forall x, y \in \Omega$, where $\omega : R^+ \to R^+$ is a nondecreasing continuous function such that $\omega(0) = 0$,

which reduces to the Lipschitz and Hölder ones if $\omega(z) = Kz$ and $\omega(z) = Kz^p$ respectively. Besides, condition (C4") allows us to consider situations like the one for the operator in (8.16).

Moreover, we say that $\omega$ is $q$-quasi-homogeneous if,

$$\omega(\tau z) \leq \tau^q \omega(z), \quad \text{for all} \quad \tau \in [0, 1],$$

for some $q \geq 0$. Observe that if $\omega$ is $q$-quasi-homogeneous then it is also $s$-quasi-homogeneous with $s \in (0, q)$.

On the other hand, notice that the $R$-order of convergence depends on (C4), or depends on these milder assumptions. In any way, under conditions (C1)–(C3) and (C4'), or a milder assumption, (C4"), we can always guarantee the quadratic $R$-order of convergence [20].

In the following result, we establish the semilocal convergence of the family of iterative processes (8.6) under conditions (C1)–(C3), (C4") and

(C5)   There exists a continuous nondecreasing function $\varphi : [0, 1] \to R^+$ such that $\omega(\tau z) \leq \varphi(\tau)\omega(z)$, for all $\tau \in [0, 1]$ and $z \in R^+$.

We denote by

$$T = \int_0^1 \varphi(\tau)(1 - \tau)\, d\tau.$$

So, the property that $\omega$ is $q$-quasi-homogeneous is a particular instance of $(C5)$ for $\varphi(t) = t^q$. Also, we have

$$T = \int_0^1 \tau^q (1 - \tau) d\tau = \frac{1}{(q + 1)(q + 2)}.$$

We consider the following real function:

$$g(t, s) = \frac{t}{2} \left( t\nu(t) + h(t)^2 - 1 \right) + \frac{s}{(q + 2)(q + 1)}, \tag{8.17}$$

which is the same as that given in (8.14) for $q = 1$.

If $\omega$ is $q$-quasi-homogeneous, then the $R$-order of convergence is at least $q + 2$. Notice that $\omega(z)$ is 1-quasi-homogeneous under the Lipschitz condition and $\omega(z)$ is $p$-quasi-homogeneous under the Hölder condition with exponent $p$, so that we then obtain $R$-order of convergence at least three and $q + 2$ respectively.

**Theorem 2 ([21])** *Let F be a nonlinear twice Fréchet-differentiable operator under all conditions $(C1)$–$(C3)$, $(C4'')$ and $(C5)$. We assume that $a_0 = M\beta\eta < r$, $a_0 h(a_0) < 1$ and $b_0 = \beta\omega(\eta)\eta < \kappa_0$, where*

$$\kappa_0 = \frac{1}{2}(1 + q)(2 + q)\left(2 + a_0 - 4a_0\, h(a_0) - a_0\, h(a_0)^2 + 2\, a_0^2\, h(a_0)^2 - a_0^2\, l(a_0)\right).$$

*If $B(x_0, R\eta) \subset \Omega$, where $R = \frac{h(a_0)}{1 - f(a_0)g(a_0, b_0)}$, then the family of iterative processes (8.6) starting at $x_0$, converges to a solution $x^*$ of Eq. (8.1) with $R$-order of convergence at least $q + 2$. In this case, the solution $x^*$ and the iterations $x_n$ belong to $\overline{B(x_0, R\eta)}$ and $x^*$ is unique in $B(x_0, \frac{2}{M\beta} - R\eta) \cap \Omega$. Moreover, the following a priori estimates of the error are also obtained*

$$\|x^* - x_n\| < \gamma^{\frac{(q+2)^n - 1}{q+1}} \Delta^n \frac{h\left(\gamma^{\frac{(q+2)^n - 1}{q+1}} a_0\right)\eta}{1 - \gamma^{(q+2)^n}\Delta} < \left(\gamma^{\frac{1}{q+1}}\right)^{(q+2)^n} \frac{R\eta}{\gamma^{\frac{1}{q+1}}}.$$

*with $\gamma = f(a_0)^2 g(a_0, b_0) \in (0, 1)$ and $\Delta = \dfrac{1}{f(a_0)}$.*

This condition $(C4'')$ has another important feature. Note that conditions Lipschitz or Hölder also need to consider an appropriate domain to be checked, which may need a prelocation of the solution. However, the condition $(C4'')$ does not need this prelocation.

### 8.3.3    Semilocal Convergence Under Weak Conditions

The usual required assumptions to establish semilocal convergence of one-point iterative methods (8.2) with $R$-order of convergence at least three in Banach spaces, are the Kantorovich conditions (C1)–(C3) and (C4'). Notice that condition (C3) requires that $F''$ is bounded in $\Omega$. However, this condition is sometimes difficult to prove in practice. There are many situations in which $F''$ is not bounded; for instance, in some problems where nonlinear differential or integral equations are involved. In these problems, we have previously to locate the solution in a suitable domain where a solution $x^*$ is and where $F''$ is bounded.

For instance, if we consider the nonlinear integral equations of Fredholm-type and second class, [6]:

$$x(s) = f(s) + \lambda \int_a^b k(s,t)x(t)^n \, dt, \quad n \in N, \quad n \geq 3, \tag{8.18}$$

with $x \in \Omega \subseteq X = C[a,b]$, where $C[a,b]$ is the space of continuous functions defined in $[a,b]$, $\lambda$ a real number, $k(\cdot,\cdot)$ a continuous function in $[a,b] \times [a,b]$ and $f \in X$ a given function such that $f(s) > 0$, $s \in [a,b]$. We consider the equivalent problem to (8.18), i. e., to solve $F(x)(s) = 0$, with

$$F(x)(s) = x(s) - f(s) - \lambda \int_a^b k(s,t)x(t)^n \, dt, \quad s \in [0,1].$$

The first and the second Fréchet derivatives of the previous operator are

$$[F'(x)y](s) = y(s) - n\lambda \int_0^1 k(s,t)x(t)^{n-1}y(t) \, dt$$

and

$$[F''(x)yz](s) = -n(n-1)\lambda \int_0^1 G(s,t)x(t)^{n-2}z(t)y(t) \, dt.$$

Observe that it is not simple to obtain an upper bound of $F''$ that is independent of $\|x\|$. Then, if we use the aforesaid Kantorovich conditions to study the convergence of an iterative process of the family (8.6) to a solution $x^*$, we need to locate previously the solution $x^*$ in a suitable domain $\Omega$ where $F''$ is bounded. For instance, if we consider the space $X = C[0,1]$ and choose the initial function $x_0(s) = 1$, $f(s) = x_0(s)$, the kernel $k(s,t) = e^{s+t}$, $\lambda = 3/100$ and $n = 3$. Taking into account that $x^*$ in $C[0,1]$, and using the max-norm, we have

$$\|x^*\| - 1 - |\lambda| \max_{s \in [0,1]} \left| \int_0^1 e^{s+t} \, dt \right| \|x^*\|^3 \leq 0$$

i.e., $\|x^\star\| \le \theta_1 = 1.328674\ldots$ and $\|x^\star\| \ge \theta_2 = 1.746584\ldots$, where $\theta_1$ and $\theta_2$ are the positive roots of the real equation

$$z - 1 - 3z^3 e(e-1)/100 = 0. \tag{8.19}$$

Now, from (8.18), if we look for a solution $x^\star$ such that $\|x^\star\| \le \theta_1$, we can then consider $\Omega = B(0, \theta) \subseteq C[0,1]$, with $\theta \in (\theta_1, \theta_2)$ and obtain a domain $\Omega$ where a solution $x^\star$ is and $F''$ is bounded. However, if we look for a solution $x^{\star\star}$ such that $\|x^{\star\star}\| \ge \theta_2$, we cannot fix the domain $\Omega$. To avoid this problem, the condition (C3) must be relaxed. Then, we consider the milder one:

(C3') $\|F''(x_0)\| \le \alpha,$

where $F''$ is only bounded in the initial iteration $x_0$.

In [21], by using majorizing sequences [24] and assuming that the operator $F$ is twice Fréchet differentiable and $F''$ satisfies a Lipschitz type condition but it is unbounded, we provide the semilocal convergence of iterative process given by (8.6). Moreover, the $R$-order of convergence at least three is proved and some error estimates in terms of the majorizing sequence are given.

**Theorem 3 ([21])** *Let $F : \Omega \subseteq X \to Y$ be a twice Fréchet-differentiable nonlinear operator defined on a nonempty open convex subset $\Omega$ of a Banach space $X$ with values in another Banach space $Y$. Suppose (C1), (C2), (C3') and (C4') and $B(x_0, s^\star) \subseteq \Omega$, where $s^\star$ is the smallest positive root of $p$, with $p(t) = \frac{K}{6}t^3 + \frac{\alpha}{2}t^2 - \frac{1}{\beta}t + \frac{\eta}{\beta}$. If*

$$\eta \le \frac{1}{3K^2}\left[\beta(\alpha^2 + 2K/\beta)^{3/2} - \beta\alpha^3 - 3K\alpha\right]$$

*holds, then $p$ has two positive roots, $s^\star$, $s^{\star\star}$, $(s^\star \le s^{\star\star})$ and the sequence $\{x_n\}$ defined by (8.6) starting at $x_0$ converge to $x^\star$, solution of (8.1) in $\overline{B(x_0, s^\star)}$. If $s^\star < s^{\star\star}$ the solution is unique in $\overline{B(x_0, s^{\star\star})} \cap \Omega$ and methods (8.6) starting at $x_0 \in \Omega$, have $R$-order of convergence at least three. If $s^\star = s^{\star\star}$ the solution is unique in $B(x_0, s^\star)$.*

As we have mentioned earlier, the Lipschitz condition (C4') also depends on the considered domain. Notice that condition (C4") can be avoided, since for a starting fixed point $x_0$, we can consider: $\|F''(x)\| \le \|F''(x_0)\| + \tilde{\omega}(\|x - x_0\|) \le \|F''(x_0)\| + \tilde{\omega}(\|x\| + \|x_0\|) \le \omega(\|x\|)$. In order to consider more general situation as the aforesaid example (8.18) given in [10], we relax the previous convergence conditions to avoid the problem of pre-location. We then reconsider the convergence of (8.6) in Banach spaces by assuming only (C1), (C2) and

(C3") $\|F''(x)\| \le \omega(\|x\|), \forall x \in \Omega$, where $\omega : R^+ \cup \{0\} \to R^+ \cup \{0\}$ is a continuous real function such that $\omega(0) \ge 0$ and $\omega$ is a monotone function. The equation

$$4t - 2h(\beta\eta\varphi(t))((1 + \beta\eta\varphi(t))t + \eta) - \beta\eta\varphi(t)h(\beta\eta\varphi(t))^2(t - 2\eta) = 0$$

has at least one positive root, where

$$\varphi(t) = \begin{cases} \omega(\|x_0\| + t) & \text{if } \omega \text{ is non-decreasing,} \\ \omega(\|x_0\| - t) & \text{if } \omega \text{ is non-increasing.} \end{cases}$$

We denote the smallest root of the previous equation by $R$. Notice that $R$ must be less than $\|x_0\|$ if $\omega$ is non-increasing.

We define the scalar function:

$$\tilde{g}(t) = h(t) \left( 1 + \frac{t}{2} h(t) \right) - 1. \tag{8.20}$$

**Theorem 4 ([11])** *Let $F : \Omega \subseteq X \to Y$ be a twice continuously differentiable operator on a non-empty open convex domain $\Omega$ satisfying conditions $(C1), (C2), (C3'')$,*

$$a_0 h(a_0) < 1, \quad f(a_0)^2 \tilde{g}(a_0) < 1 \tag{8.21}$$

*and $B(x_0, R) \subseteq \Omega$, then methods (8.6), starting from $x_0$, generate a sequence $\{x_n\}$ which converges to a solution $x^* \in \overline{B(x_0, R)}$ of Eq. (8.1). Besides, the solution $x^*$ is unique in $B(x_0, \tilde{R})$, where $\tilde{R}$ is the biggest positive root of the equation*

$$\beta \int_0^1 \int_0^1 \varphi(s(R + t(\xi - R))) \, ds(R + t(\xi - R)) \, dt = 1. \tag{8.22}$$

As a particular case of the previous result, Theorem 4, if $\|F''(x)\| \leq M$, we can obtain a result of semilocal convergence [10]. However, for a $F''$ bounded operator this result still need a pre-location.

## 8.4 A Study of the Accessibility Domain for the Iterative Processes of the Family

An important aspect to consider when studying the applicability of an iterative method is the set of starting points that we can take into account, so that the iterative method converges to a solution of an equation from any point of the set, what we call accessibility of the iterative method. We can observe this experimentally by means of the attraction basin of the iterative method. The attraction basin of an iterative method is the set of all starting points from which the iterative method converges to a solution of a particular equation, once we fix some tolerance or a maximum number of iterations.

We can also study the accessibility of an iterative method from the convergence conditions required to the iterative method. Looking for some parallelism with the attraction basins, we can also consider the following experimental form of studying the accessibility of an iterative method. Therefore, if we consider a particular equation, we know that a given point $x \in \Omega$ has associated certain parameters of convergence. If the parameters of convergence satisfy the convergence conditions, we color the point $x$; otherwise, we do not. So, the region that is finally colored is what we call a region of accessibility of the iterative method.

Finally, we can study the accessibility from the domain of parameters. In this case, we only consider the convergence conditions of the iterative process considered and for one equation either. This is a theoretical and general study.

### 8.4.1   Attraction Basins

We can see the aforesaid relation between the $R$-order of convergence and the conditions on the starting point by means of the attraction basins associated to the iterative methods when they are applied to solve a particular equation [26, 32]. The set of all starting points from which an iterative process converges to a solution of the equation can be shown by means of the attraction basins.

Let $F(z) = \cos z - 1/5 = 0$, where $F : C \to C$. We identify the attraction basins of two solutions $z^* = \arctan(2\sqrt{6})$ and $z^{**} = -\arctan(2\sqrt{6})$ of the previous equation when we approximate these roots by the Chebyshev (8.6), the Super-Halley (8.10), the Halley (8.9) and logarithmic (8.12) methods, so that we can compare the attraction basins of the third order methods with the one of the Newton method. To do this, we take a rectangle $D \subset C$ to represent the regions and such that iterations start at every $z_0 \in D$. In every case, it is considered a grid of $512 \times 512$ points in $D$ and these points are chosen as $z_0$. We use the rectangle $[-1.75, 1.75] \times [-2, 2]$ which contains both solutions. The chosen iterative method, starting in $z_0 \in D$, can converge or diverge to any solution. In all the cases, the tolerance $10^{-3}$ and the maximum of 25 iterations are used. We do not continue if the required tolerance is not obtained with 25 iterations and we then decide that the iterative method does not converge to any solution starting from $z_0$.

In Figs. 8.1 and 8.2 the attraction basins of the Chebyshev, the Super-Halley, the Halley and logarithmic methods are respectively shown when the solutions $z^*$ and $z^{**}$ are approximated in the rectangle $D$. The pictures of the attraction basins are painted by the following strategy. It is assigned a color to each attraction basin according to the root at which an iterative method converges, starting from $z_0$. The color is made lighter or darker according to the number of iterations needed to reach the root with the fixed precision. In particular, it is assigned cyan for the positive root $z^*$ and magenta for the negative root $z^{**}$. Finally, the black color is assigned if the method does not converge to any solution with the fixed tolerance and maximum number of iterations. The graphics have been generated with Mathematica 5.1. For other strategies, reference [32] can be consulted and the references there given.

Observe in Figs. 8.1 and 8.2 that the Super-Halley is the most demanding of the four third-order methods that have considered, since the black color is more plentiful for the third-order methods. We can also observe that the lighter areas are greater

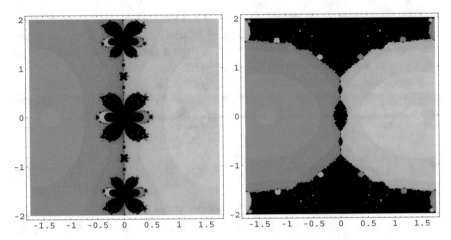

**Fig. 8.1** Attraction basins of the Chebyshev (*left*) and the Super-Halley (*right*) methods

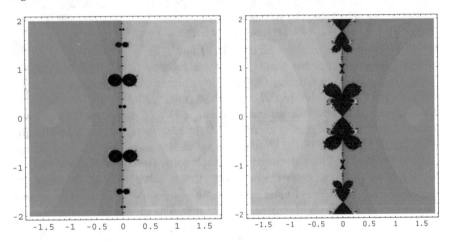

**Fig. 8.2** Attraction basins of the Halley (*left*) and Logarithmic (*right*) methods

for the third-order methods than for Newton's method, Fig. 8.7, as a consequence of the cubic $R$-order of convergence versus the quadratic $R$-order of convergence.

## 8.4.2 Regions of Accessibility

As we have indicate previously, the region of accessibility of an iterative method provides the domain of starting points from which we have guaranteed the convergence of the iterative method for a particular equation. In other words, the region of accessibility represents the domain of starting points that satisfy the convergence conditions required by the iterative method that we want to apply.

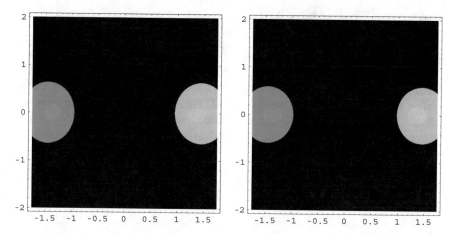

**Fig. 8.3** Region of accessibility of the Chebyshev (*left*) and Super-Halley (*right*) methods

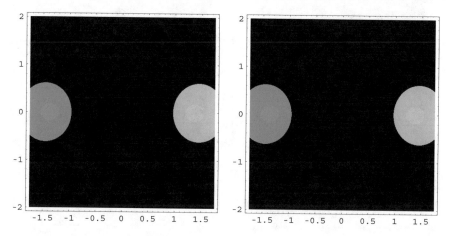

**Fig. 8.4** Region of accessibility of the Halley (*left*) and logarithmic (*right*) methods

We can graphically observe this with an example. If we consider the previous complex equation $F(z) = \cos z - 1/5 = 0$, we can respectively see in Figs. 8.3 and 8.4, according to conditions (8.15) required in Theorem 1, the accessibility regions associated with Chebyshev (8.6), Super-Halley (8.10), Halley (8.9) and logarithmic methods (8.12).

In view of Figs. 8.3 and 8.4, it is clear that the greatest region of accessibility is associated with the Chebyshev method, see Fig. 8.3. Similarly, as we analyze the domains of parameters, the Chebyshev method is the best performing in terms of accessibility.

### 8.4.3 Domains of Parameters

These regions represent the domains of accessibility of the different methods starting from the parameters $a_0$ and $b_0$ which are obtained from the starting point. Notice that the domain of parameters of the Chebyshev (8.6), Super-Halley (8.10), Halley (8.9) and logarithmic methods (8.12) represented in Figs. 8.5 and 8.6, are bounded and are delimited by the coordinated axes and the curve $b = \varphi(a_0)$. These domains of parameter are obtained to consider convergence conditions given in (8.15). For these methods, the function $\varphi$, given in (8.14), is reduced to $\varphi(t) = 3(t + 2)(2t - 1)(t^2 + 2t - 4)/4$ for the Chebyshev method, $\varphi(t) = 6(t^3 + 6t^2 - 12t + 4)/((t-2)^2)$ for the Super-Halley method, for the Halley method

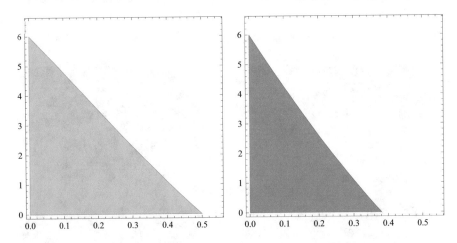

**Fig. 8.5** Domains of parameters of the Chebyshev (*left*) and the Super-Halley (*right*) methods

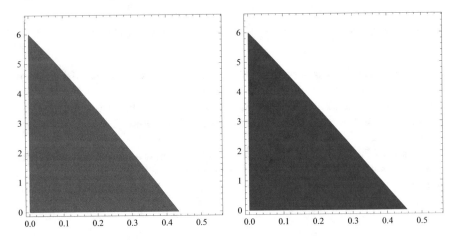

**Fig. 8.6** Domains of parameters of the Halley (*left*) and logarithmic (*right*) methods

and $\varphi(t) = 3(2t^4 - 9t^3 + 32t^2 - 32t + 8)/(4(t-1)^2)$ and for logarithmic method $\varphi(t) = ((3 + 30t - 24e^t t + 6t^2 + e^{2t}(-3 + 6t))/t$ (see [17]). In these cases, the condition $a_0 h(a_0) < 1$ follows from $b < \varphi(a_0)$, since $\varphi(a_0) > 0$.

## 8.5 Improvement of the Accessibility: Hybrid Method

It is well-known that the higher the $R$-order of convergence, the more restrictive the conditions on the starting point. For instance, this fact can be seen when one looks at the basin of attraction or accessibility region, Fig. 8.7 or domains of parameters Fig. 8.8 of the Newton method and compared with one third-order method.

To achieve the region of accessibility and the domain of parameter associated to the Newton's method we consider the semilocal convergence theorem:

**Theorem 5 ([15])** *Let F be a nonlinear twice Fréchet-differentiable operator on a nonempty open convex domain $\Omega$ under the previous conditions (C1)–(C3) If $B(x_0, R_N) \subseteq \Omega$, where $R_N = \frac{2(1-a)}{2-3a}\eta$ and $a = M\beta\eta$, and*

$$a = M\beta\eta < 1/2, \tag{8.23}$$

*then Eq. (8.1) has a solution $z^*$ and the Newton method converges to $z^*$ with R-order of convergence at least two.*

In Fig. 8.7 appear the attraction basin and the accessibility region when it is considered complex equation $F(z) = \cos z - 1/5 = 0$ according to Kantorovich convergence conditions given in Theorem 5, and the domain of parameters of the Newton method, respectively.

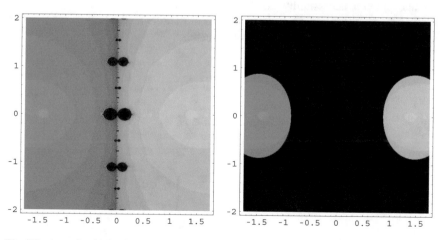

**Fig. 8.7** Attraction basin (*left*) and region of accessibility (*right*) of the Newton method

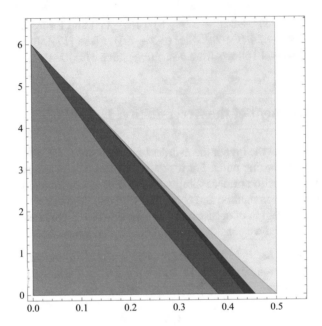

**Fig. 8.8** Domains of parameters of the Newton's method (*yellow*) and some third-order method in family (8.6)

Figure 8.8 shows the overlap of the graphics of parameter domains of the Newton's method (yellow) and some third-order method. Notice that, the domain of parameters of the Newton's method is an infinite vertical band colored in yellow.

Clearly, the accessibility domains of the Newton method is better comparing with any method of third order in all three cases that we are considering: Attraction basins, region of accessibility and domains of parameters.

To improve the accessibility domain of the third order methods (8.6), we consider the following predictor–corrector iterative method:

$$\begin{cases} \begin{cases} x_0 \in \Omega, \\ x_n = G_1(x_{n-1}), \quad n = 1, 2, \dots, N_0, \end{cases} \\ \begin{cases} z_0 = x_{N_0}, \\ z_n = G_2(z_{n-1}), \quad n \in N, \end{cases} \end{cases} \tag{8.24}$$

from any two one-point iterative processes:

$$\begin{cases} x_0 \in \Omega, \\ x_n = G_1(x_{n-1}), \quad n \in N, \end{cases} \quad \text{and} \quad \begin{cases} z_0 \in \Omega, \\ z_n = G_2(z_{n-1}), \quad n \in N. \end{cases}$$

where $x_0$ satisfies condition (8.23), while $z_0 = x_{N_0}$ satisfies the two conditions given in (8.15). For method (8.24) is convergent, we must do the following:

1. Find $x_0$, so that predictor method is convergent.
2. From the convergence of predictor method, calculate the value $N_0$ such that $x_{N_0}$ is a good starting point from which the convergence of corrector method (8.6) is guaranteed.

In short, we use the Newton's method for a finite number of steps $N_0$, provided that the starting point $x_0$ satisfies condition (8.23), until $z_0 = x_{N_0}$ satisfy conditions given in (8.15), and we then use methods (8.6) instead of the Newton's method. The key to the problem is therefore to guarantee the existence of $N_0$.

We also denote (8.24) by $(G_1, G_2)$ and, following the notation given in [27] for other combinations of methods, (8.24) is called hybrid method.

Then, for an initial couple $(a, b)$ in the domain of parameters of the Newton method from Kantorovich conditions (8.23), we try to guarantee the existence of a couple $(a_{N_0}, b_{N_0})$ in the domain of parameters of iteration (8.6), after a certain number of iterations $N_0$, so that the last couple satisfy conditions (8.15), thus can be considered as the initial one $(a_{N_0}, b_{N_0})$ for iteration (8.6) and such that the $R$-order of convergence at least three keeps (see Fig. 8.9).

With the goal of constructing a simple modification of (8.6) which converges from the same starting points as the Newton method, we consider the following

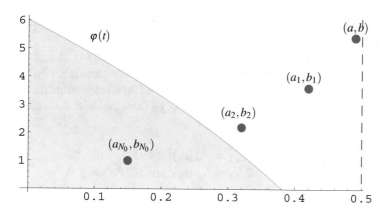

**Fig. 8.9** Evolution of the couples $(a_i, b_i)$ in the domains of parameters of the Newton method and an iterative process in family (8.6)

hybrid methods $(N, R_3)$, where $R_3$ denotes a third order method of family (8.6):

$$\begin{cases} \begin{cases} x_0 \in \Omega, \\ x_n = x_{n-1} - [F'(x_{n-1})]^{-1}F(x_{n-1}), \quad n = 1, 2, \ldots, N_0, \end{cases} \\ \begin{cases} z_0 = x_{N_0}, \\ y_{k-1} = z_{k-1} - [F'(z_{k-1})]^{-1}F(z_{k-1}), \\ z_k = y_{k-1} + \frac{1}{2}L_F(z_{k-1})H(L_F(z_{k-1}))(y_{k-1} - z_{k-1}), \quad k \in N, \end{cases} \end{cases} \tag{8.25}$$

where $L_F$ and $H$ are defined as in (8.6). Now, we consider in (8.25) that $x_0$ only satisfies (8.23), while $z_0 = x_{N_0}$ satisfies (8.15).

Now, we define the following real functions

$$f_N(t) = \frac{1}{1-t} \quad \text{and} \quad g_N(t) = \frac{t}{2(1-t)} \tag{8.26}$$

and the sequence $\{\alpha_n\}$ which guarantees the convergence of the Newton method:

$$\begin{cases} \alpha_0 = a, \quad \alpha_{n+1} = \alpha_n f_N(\alpha_n)g_N(\alpha_n), \quad n \geq 0, \\ \beta_0 = b, \quad \beta_{n+1} = \beta_n f_N(\alpha_n)g_N(\alpha_n)^2, \quad n \geq 0. \end{cases} \tag{8.27}$$

Observe that sequence $\{\alpha_n\}$ is strict decreasing when (8.23) is satisfied. On the other hand, the sequence $\{\beta_n\}$ is not necessary to prove the convergence of the Newton method, but it is essential to locate a valid starting point for iteration (8.6), see [18]. We can observe that the sequences $\{\alpha_n\}$ and $\{\beta_n\}$ are strictly decreasing to zero if $a < 1/2$.

In a similar way, see [18], from general conditions (C1)–(C3) and (C4') for iteration (8.6) and starting point $z_0$, we can consider the parameters $\tilde{a}_{N_0} = M\beta\eta$ and $\tilde{b}_{N_0} = K\beta\eta^2$ from $z_0$ and construct a system of recurrence relations. Thus, we can guarantee the semilocal convergence of (8.6) from the strict decreasing of the real sequences:

$$\begin{cases} a_0 = \tilde{a}_{N_0}, \quad a_{n+1} = a_n f(a_n)^2 g(a_n, b_n), \quad n \geq 0, \\ b_0 = \tilde{b}_{N_0}, \quad b_{n+1} = b_n f(a_n)^3 g(a_n, b_n)^2, \quad n \geq 0, \end{cases}$$

provided that (8.15) is satisfied for $\{a_n\}$ and $\{b_n\}$.

The idea is then to apply hybrid iteration (8.25) to approximate a solution $z^*$ of Eq. (8.1). To apply this iteration we take into account that the couple $(a, b)$ does not satisfy the convergence conditions given in (8.15), since, if it does, we apply corresponding iteration (8.6).

Now, we see that whenever the Newton method can be applied, iteration (8.6) can also be done. Firstly, as $\{\alpha_n\}$ is a strictly decreasing sequence to zero, there always exists $N_1 \in N$ such that $\alpha_{N_1} < r$. Secondly, as $t\psi(t) = \phi(t)$ is an increasing function with $\phi(0) = 0$, there also exists a number $N_2 \in N$ such that $\alpha_{N_2}\psi(\alpha_{N_2}) < 1$. Moreover, there exists $N_3 \in N$ such that $\alpha_{N_3} < \theta$, where $\theta$ is the smallest positive root of $\varphi(t) = 0$. Finally, if $b_0 \geq \varphi(\alpha_0)$ and $\{b_n\}$ is also a strictly decreasing sequence to zero, we can then guarantee the existence of $N_4 \in N$ such that $N_4 \geq N_3$ and

$$b_{N_4} < \varphi(\alpha_0) < \varphi(\alpha_{N_3}) < \varphi(\alpha_{N_4}),$$

since $\varphi$ is a decreasing function and $\{\varphi(\alpha_n)\}$ is a strictly increasing sequence. Observe then $\varphi(\alpha_{N_3}) > 0$. In consequence, we can take $N_0 = \max\{N_1, N_2, N_4\}$, choose $z_0 = x_{N_0}$ and apply iteration (8.6) from the starting point $z_0 = x_{N_0}$ to guarantee the convergence of (8.25).

Notice that condition $a < r$ could be omit in the case that we have the possibility of choosing a method of family (8.6), since $r$ could be $+\infty$, that is when the series that appears in (8.6) is finite. For instance, this happens with the Chebyshev method. In the case of the Super-Halley method, we have $r = 1$ and $a_0 < 0.380778...$, so that $a < r$ also holds.

We establish the semilocal convergence of (8.25) to a solution of (8.1) and obtain domains where the solution is located. From now on, we write sequence (8.25) in the following way:

$$w_n = \begin{cases} x_n, & \text{if } n \leq N_0, \\ z_{n-N_0}, & \text{if } n > N_0. \end{cases}$$

**Theorem 6 ([9])** *Let $X$ and $Y$ be two Banach spaces and $F : \Omega \subseteq X \to Y$ a twice Fréchet-differentiable operator on a nonempty open convex domain $\Omega$. We suppose that conditions (C1)–(C3) and (C4') are satisfied for some $x_0 \in \Omega$. If condition (8.23) holds and $B(x_0, R_N + R) \subseteq \Omega$, then there exists $N_0 \in N$ such that sequence $\{w_n\}$, given by (8.25), and starting from $w_0$, converges to a solution $z^*$ of Eq. (8.1). In this case, the solution $z^*$ and the iterations $w_n$ belong to $\overline{B(x_0, R_N + R)}$.*

Notice that we can apply (8.25) provided that the conditions of Theorem 6 hold, since there always exists $N_0$. However, if we are able to estimate a priori the value of $N_0$, then we can improve the algorithm, since it is not necessary to verify the conditions given in (8.15) in each step. In the following result we give an estimate of the value $N_0$.

**Theorem 7 ([9])** *We suppose that the hypothesis of the previous theorem are satisfied and condition (8.23) holds, but not (8.15), for some $x_0 \in \Omega$ satisfying (C1)–(C3). Let $z_0 = x_{N_0}$ with $N_0 = \max\{N_1, N_2, N_3, N_4\}$, $N_1 = 1 + \left[\frac{\log r - \log a}{\log(f(a)g(a))}\right]$, $N_2 = 1 + \left[\frac{-\log(a\psi(a))}{\log(f(a)g(a))}\right]$, $N_3 = 1 + \left[\frac{\log \theta - \log a}{\log(f(a)g(a))}\right]$ and $N_4 = 1 + \left[\frac{\log \lambda - \log b}{\log(f(a)g(a)^2)}\right]$, when they are positive, and consider them null in other cases, where $f$, $\varphi$ and $g$ are defined in (8.14), $\lambda = \varphi(\alpha_{N_3})$, $\theta$ is the smallest positive root of $\varphi(t) = 0$ and $[t]$ the integer part of the real number $t$. Then, $z_0$ satisfies (8.15).*

If we take again into account the complex equation $F(z) = \cos z - 1/5 = 0$ and conditions (8.23) and (8.15), we can see in Fig. 8.10 the regions of accessibility of

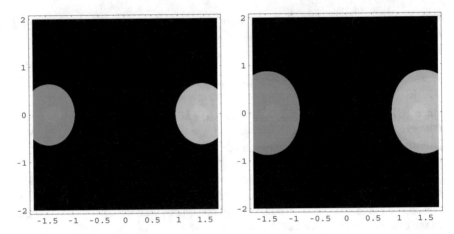

**Fig. 8.10** Region of accessibility of the Chebyshev (*left*) and the hybrid ($N, Ch$) (*right*) methods

the Chebyshev method (8.6) and the hybrid method ($N, Ch$). Notice that the domain of starting points of ($N, Ch$) coincides with the one of the Newton method, Fig. 8.7, although the intensity of the color changes, it is lighter for the hybrid method, since the number of iterations used to reach the solution of equation for the given tolerance is inferior.

### 8.5.1 Application

We apply the results obtained previously to obtain domains of existence and uniqueness of solutions when we use (8.25) to solve the following boundary-value problem [23, 25]:

$$\begin{cases} \dfrac{d^2x(t)}{dt^2} = e^{x(t)}, \\ x(0) = x(1) = 0, \end{cases} \tag{8.28}$$

where $x \in C^2([0, 1])$, $t \in [0, 1]$.

To obtain a numerical approximation of the solution of problem (8.28) we discretize the problem. We then introduce the points $t_i = ih$, $i = 0, 1, \ldots, n$, with $h = 1/n$ and $n$ a suitable integer. We have to determinate the values $x(t_i) = x_i$ which denote the approximations of the solution $x(t)$ in $t_i$. A standard approximation of the second derived in these points is

$$x_i'' \approx \frac{x_{i-1} - 2x_i + x_{i+1}}{h^2}, \quad x_i = x(t_i), \quad i = 1, 2, \ldots, n-1.$$

If we assume that $x_i$ satisfies Eq. (8.28), then

$$x_{i-1} - 2x_i + x_{i+1} - h^2 e^{x_i} = 0, \quad i = 1, 2, \dots, n - 1. \tag{8.29}$$

Moreover, since $x_0$ and $x_n$ are determinated by the boundary conditions, we have $x_0 = 0 = x_n$.

We then introduce the following notation. If

$$\mathbf{x} = \begin{pmatrix} x_1 \\ x_2 \\ \vdots \\ x_{n-2} \\ x_{n-1} \end{pmatrix}, \quad \Phi(\mathbf{x}) = h^2 \begin{pmatrix} e^{x_1} \\ e^{x_2} \\ \vdots \\ e^{x_{n-2}} \\ e^{x_{n-1}} \end{pmatrix}, \quad A = \begin{pmatrix} -2 & 1 & 0 & \cdots & 0 \\ 1 & -2 & 1 & \ddots & \vdots \\ 0 & \ddots & \ddots & \ddots & 0 \\ \vdots & \ddots & 1 & -2 & 1 \\ 0 & \cdots & 0 & 1 & -2 \end{pmatrix},$$

(8.29) can be written as

$$F(\mathbf{x}) \equiv A\mathbf{x} - \Phi(\mathbf{x}) = 0,$$

with $F : \Omega \subset R^{n-1} \to R^{n-1}$. We consider $\Omega = \{\mathbf{x} \in R^{n-1} \mid \|\mathbf{x}\| < \ln 8\}$ as the domain from which we obtain domains of existence and uniqueness of solutions for the equation $F(\mathbf{x}) = 0$. Moreover, we choose the norms

$$\|\mathbf{x}\|_\infty = \max_{1 \le i \le n-1} |x_i| \quad \text{and} \quad \|A\| = \sup_{\mathbf{x} \in R^{n-1}, \mathbf{x} \ne 0} \frac{\|A\mathbf{x}\|_\infty}{\|\mathbf{x}\|_\infty} = \max_{1 \le i \le n-1} \left( \sum_{j=1}^{n-1} |a_{ij}| \right).$$

Notice that the operator $F'$ is given by

$$F'(\mathbf{x})(\mathbf{u}) = A\mathbf{u} - h^2 \begin{pmatrix} e^{x_1} & 0 & \cdots & 0 \\ 0 & e^{x_2} & \cdots & 0 \\ \vdots & \vdots & \ddots & \vdots \\ 0 & 0 & \cdots & e^{x_{n-1}} \end{pmatrix} \mathbf{u}, \quad \forall \, \mathbf{u} \in R^{n-1},$$

where $\mathbf{u} = (u_1, u_2, \dots, u_{n-1})^T$, and $F''(\mathbf{x})$ is the following bilinear operator

$$F''(\mathbf{x})(\mathbf{u}, \mathbf{v}) = -h^2 \begin{pmatrix} e^{x_1} u_1 v_1 \\ e^{x_2} u_2 v_2 \\ \vdots \\ e^{x_{n-1}} u_{n-1} v_{n-1} \end{pmatrix}, \quad \forall \, \mathbf{u}, \mathbf{v} \in R^{n-1}.$$

To establish the convergence of (8.25) to a solution of (8.28), we take $5t(t-1)/2$ as initial function. If $n = 10$, then the initial approximation is $w_i^{(0)} = 5t_i(t_i - 1)/2$, with $i = 1, 2, \ldots, 9$. In this case, $a = 0.489219\ldots$ and $b = 0.253904\ldots$

We now try to use the Halley method given in (8.9) to approximate a solution of problem (8.28). Observe then that the conditions $a < r = 2$ and $a\psi(a) < 1$ hold, but we cannot take that starting point for the Halley method since the other two conditions $\varphi(a) > 0$ and $b < \varphi(a)$, which guarantee the convergence of the method, are not satisfied. However, for the hybrid method $(N, H)$, we have $N_1 = N_2 = 0$, $N_3 = 2$ and $N_4 = 0$, so that $N_0 = 2$. Therefore, after two iterations by the Newton method, we can consider the starting point $z_0 = x_2$ for the Halley method and apply $(N, H)$. Similarly, we can proceed with the Super-Halley method given in (8.10) and the method of family (8.6) such that $A_0 = 1$ and $A_1 = A_2 = 1/2$ to obtain respectively $N_0 = 3$ and $N_0 = 4$ for the corresponding hybrid methods.

In Table 8.1, we show the evolution of iterations $\mathbf{w_n} = (w_1^{(n)}, w_2^{(n)}, \ldots, w_9^{(n)})^T$ generated by the three methods when it is used the stop criterion $\|\mathbf{w_n} - z^*\|_\infty < C \times 10^{-150}$ ($C$ = constant) and using 150 significant digits. We have denoted by $(N, H)$, $(N, Chl)$ and $(N, SH)$ the hybrid methods given by (8.25) when the Halley, the Chebyshev-like and the Super-Halley methods are respectively used as methods of $R$-order at least three.

From the above we can observe that we have extended the region of accessibility of the three methods of $R$-order at least three until the region of accessibility of the Newton method.

Finally, from the initial approximation indicated above, we consider the vector shown in Table 8.2 as the numerical solution of (8.28), which has been obtained when the stop criterion indicated previously is used and after two iterations of the Newton method and four iterations of the Halley method.

**Table 8.1** Errors $\|\mathbf{w_n} - z^*\|_\infty$

| $n$ | Newton | $(N, H)$, $N_0 = 2$ | $(N, Ch - l)$, $N_0 = 3$ | $(N, S - H)$, $N_0 = 4$ |
|---|---|---|---|---|
| 0 | $5.11377\ldots \times 10^{-1}$ | $5.11377\ldots \times 10^{-1}$ | $5.11377\ldots \times 10^{-1}$ | $5.11377\ldots \times 10^{-1}$ |
| 1 | $7.62191\ldots \times 10^{-3}$ | $7.62191\ldots \times 10^{-3}$ | $7.62191\ldots \times 10^{-3}$ | $7.62191\ldots \times 10^{-3}$ |
| 2 | $2.10652\ldots \times 10^{-6}$ | $2.10652\ldots \times 10^{-6}$ | $2.10652\ldots \times 10^{-6}$ | $2.10652\ldots \times 10^{-6}$ |
| 3 | $1.57140\ldots \times 10^{-13}$ | $8.65986\ldots \times 10^{-20}$ | $1.57140\ldots \times 10^{-13}$ | $1.57140\ldots \times 10^{-13}$ |
| 4 | $8.70901\ldots \times 10^{-28}$ | $5.77528\ldots \times 10^{-60}$ | $4.05973\ldots \times 10^{-41}$ | $8.70901\ldots \times 10^{-28}$ |
| 5 | $2.67280\ldots \times 10^{-56}$ | | $6.76285\ldots \times 10^{-124}$ | $6.90369\ldots \times 10^{-84}$ |
| 6 | $2.51705\ldots \times 10^{-113}$ | | | |

**Table 8.2** Numerical solution of (8.28)

| $i$ | $z_i^*$ | $i$ | $z_i^*$ | $i$ | $z_i^*$ |
|---|---|---|---|---|---|
| 1 | $-4.14043\ldots \times 10^{-2}$ | 4 | $-1.09159\ldots \times 10^{-1}$ | 7 | $-9.57302\ldots \times 10^{-2}$ |
| 2 | $-7.32143\ldots \times 10^{-2}$ | 5 | $-1.13622\ldots \times 10^{-1}$ | 8 | $-7.32143\ldots \times 10^{-2}$ |
| 3 | $-9.57302\ldots \times 10^{-2}$ | 6 | $-1.09159\ldots \times 10^{-1}$ | 9 | $-4.14043\ldots \times 10^{-2}$ |

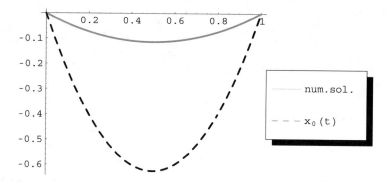

**Fig. 8.11** Interpolation of the numerical solution of (8.28)

By applying the hybrid method $(N, H)$, we obtain the existence and uniqueness domains $B(\mathbf{w_0}, 0.995959\ldots)$ and $B(\mathbf{w_0}, 1.19443\ldots)$, respectively. Moreover, by interpolating the numerical solution given in Table 8.2, which passes through the points $(t_i, z_i^*)$, $i = 1, 2, \ldots, 9$, we obtain the solution which appears in Fig. 8.11.

## 8.6   An Improvement of the Efficiency

In the study of iterative processes, there are two important points to pay attention: the speed of convergence, which is analyzed by the $R$-order of convergence, and the computational cost needed to compute $x_{n+1}$ from $x_n$, which is generally analyzed by taking into account the number of evaluations of $F$, $F'$, $F''\ldots$ that are necessary to obtain $x_{n+1}$ from $x_n$. From the previous ideas, to classify iterative processes, the efficiency index [31] is defined by the value $EI = p^{1/d}$, where $p$ is the order of convergence and $d$ the number of evaluations of $F$, $F'$, $F''\ldots$ in each step. In particular, this index is usually considered in the analysis of scalar equations, where the computational cost of the successive derivatives is not very different.

For one-point iterative methods, it is known that the $R$-order of convergence is a natural number. Moreover, one-point iterations of the form $x_{n+1} = \phi_1(x_n)$, $n \geq 0$, with order of convergence $d$, depend explicitly on the first $d - 1$ derivatives of $F$. This implies that their efficiency index is $EI = d^{1/d}$, $d \in N$. The best situation for this index is obtained when $d = 3$, (see Fig. 8.12). However, in general situations, this case is not considered as the most favorable one, but $d = 2$, namely Newton's method, even though its efficiency index is worse. It is due to the efficiency index does not consider several determinants. For example, in the case of nonlinear systems of dimension greater than one or non-finite situations (Banach spaces), the computational cost is high for computation of the corresponding operator $F''$, whereas Newton's method only uses $F$ and $F'$.

**Fig. 8.12** Efficiency index
$EI = d^{1/d}$

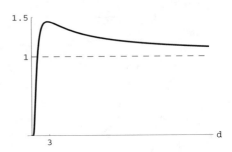

To improve the optimum efficiency index, $EI = 3^{1/3}$, we reduce computations of $F$ and $F'$, avoid the computation of $F''$ and increase the order of convergence. From the restrictions above-mentioned for the one-point methods, we construct a uniparametric family of multi-point iterations, namely, $x_{n+1} = \Psi(x_n^1, x_n^2, x_n^3, \ldots)$, with $x_n^i = \Psi_i(x_{n-1}^i)$, $i \geq 1$, $n \geq 0$ [31], where the second derivative $F''$ is not used, that depends on a parameter and is such that $EI = 4^{1/4}$. This efficiency index is the same as the one of Newton's method, but for some values of the parameter involved in the family, the efficiency index is $EI = 4^{1/3}$, which is better than Newton's method one and even the most favorable situation of one-point methods ($EI = 3^{1/3}$).

To do this, we use the following multi-point iteration:

$$\begin{cases} x_0 \text{ given,} \\ x_{n+1} = \Psi_1(x_n, u_n, v_n), \quad n \geq 0, \\ u_n = \mu_1(x_n), \quad v_n = \mu_2(x_n). \end{cases}$$

The first aim consists of avoiding the computation of $F''$ in these cubically $R$-convergent iterations. In [16], Hernández has been done this for Chebyshev's method in Banach spaces, but the order of convergence is not increased by the technique used there. In addition, as the second aim is to increase the order of convergence, we use a modification of the technique used in [16].

Observe first that if $A_0 = 1$ and $A_1 = A_2 = 1/2$, then (8.6) has order of convergence at least four when it is applied to solve quadratic equations (see [8]). Then, to extend this result to any equation and avoiding the computation of $F''$, from Taylor's formula, we have:

$$F'(u_n) = F'(x_n) + F''(x_n)(u_n - x_n) + \int_{x_n}^{u_n} F'''(x)(u_n - x)\, dx,$$

$$F'(v_n) = F'(x_n) + F''(x_n)(v_n - x_n) + \int_{x_n}^{v_n} F'''(x)(v_n - x)\, dx,$$

where $u_n = x_n + \theta_1(y_n - x_n)$, $v_n = x_n + \theta_2(y_n - x_n)$, $y_n = x_n - F(x_n)/F'(x_n)$, $\theta_1, \theta_2 \in [0, 1]$ and $\theta_1 \neq \theta_2$, and consequently

$$F'(v_n) - F'(u_n) \approx -(\theta_2 - \theta_1)F''(x_n)F(x_n)/F'(x_n),$$

$$L_F(x_n) \approx \left(\frac{-1}{\theta_2 - \theta_1}\right)\frac{F'(v_n) - F'(u_n)}{F'(x_n)} = \psi(u_n, v_n).$$

With this modification of the technique followed in [16], the parameters $\theta_1$ and $\theta_2$ are considered so that we can eliminate the computation of $F''$ and obtain order of convergence at least four.

Now, (8.6) is written in the way:

$$\begin{cases} x_0 \text{ given,} \\ x_{n+1} = \phi(x_n) = x_n - H(\psi(x_n))(F(x_n)/F'(x_n)), \quad n \geq 0, \end{cases} \tag{8.30}$$

where

$$\psi(x_n) = \left(\frac{-1}{\theta_2 - \theta_1}\right)\frac{F'(x_n - \theta_2(F(x_n)/F'(x_n))) - F'(x_n - \theta_1(F(x_n)/F'(x_n)))}{F'(x_n)},$$

$H$ is defined in (8.6) with $A_0 = 1$ and $A_1 = A_2 = 1/2$. Observe that if (8.30) is well defined, namely $\|\psi(x)\| < r$, only the computations of $F$ and $F'$ are required.

On the other hand, in the scalar case the Schröeder's characterization [30] can be used to obtain the order of convergence of (8.30). So, if $x^*$ is a solution of (8.1), then it is easy to prove that

$$\phi_3(x^*) = x^*, \quad \phi_3'(x^*) = \phi_3''(x^*) = 0, \quad \phi_3'''(x^*) = 0 \quad \text{if} \quad \theta_1 + \theta_2 = 2/3.$$

Consequently, the following family of iterations:

$$\begin{cases} x_0 \text{ given,} \\ y_n = x_n - F(x_n)/F'(x_n), \\ G(x_n) = \frac{-3}{2(1-3\theta)}\frac{F'(x_n + \frac{2-3\theta}{3}(y_n - x_n)) - F'(x_n + \theta(y_n - x_n))}{F'(x_n)}, \quad \theta \in [0, 1/3), \\ x_{n+1} = x_n - H(G(x_n))(F(x_n)/F'(x_n)), \quad n \geq 0, \end{cases} \tag{8.31}$$

where $H(z)$ is defined in (8.3) with $A_0 = 1$ and $A_1 = A_2 = 1/2$, has order of convergence at least four and the computations of $F$ and $F'$ are only used. The corresponding efficiency index is $EI = 4^{1/4}$, the same one as for Newton's method. But if $\theta = 0$, then $EI = 4^{1/3}$, which is better than Newton's one and any one-point iterative process.

*Example 1* We can now observe the behaviour of some iterations of (8.31) when they are applied to approximate the solution $x^* = 1.93345\ldots$ of the Kepler's equation given by $F(x) = x - \sin x - 1 = 0$. Two methods are applied: the

**Table 8.3** Errors $\|x_n - x^*\|_\infty$ for Kepler's equation

| $n$ | Newton's method | Jarratt's method | The modified Euler method |
|---|---|---|---|
| 1 | 5.47903... | 5.80236... | 5.80294... |
| 2 | **4.12344**... | 1.40564... | 1.69131... |
| 3 | 2.62048... | $1.81252... \times 10^{-2}$ | $6.66613... \times 10^{-3}$ |
| 4 | **2.03342**... | $5.54837... \times 10^{-9}$ | $2.35076... \times 10^{-11}$ |
| 5 | $1.74269... \times 10^{-1}$ | $5.00598... \times 10^{-35}$ | $3.62880... \times 10^{-45}$ |
| 6 | **1.23779**$... \times 10^{-2}$ | | |
| 7 | $5.21929... \times 10^{-5}$ | | |
| 8 | **9.38826**$... \times 10^{-10}$ | | |
| 9 | $3.03777... \times 10^{-10}$ | | |
| 10 | **3.18048**$... \times 10^{-38}$ | | |

**Table 8.4** The computational order of convergence $\rho$

| $n$ | Jarratt's method | The modified Euler method |
|---|---|---|
| 2 | 3.06885... | 4.49111... |
| 3 | 3.44737... | 3.51498... |
| 4 | 3.99818... | 4.00023... |
| 5 | 3.99999... | 4.00000... |

ones obtained from (8.31) when the third-order methods considered in (8.6) are the Super-Halley and the Euler methods, along with the value $\theta = 0$, which are respectively the Jarratt method [3] and the modified Euler method. From the starting point $x_0 = 15$, we have obtained the errors given in Table 8.3, where 50 significant digits and the stopping criterion $\|x_n - x^*\|_\infty < C \times 10^{-50}$ are used. Observe in Table 8.3 that iterations (8.31) are competitive if they are compared with the usual fourth-order iteration, the two steps of the Newton's method, whose values are written in bold. In addition, the computational cost is reduced.

To finish, we can see in Table 8.4 that the computational order of convergence (see [33])

$$\rho \approx \frac{\ln |(x_n - x^*)/(x_{n-1} - x^*)|}{\ln |(x_{n-1} - x^*)/(x_n - x^*)|} \tag{8.32}$$

of both methods is closed to four, so that it agrees with the order of convergence at least four of the methods.

Now, taking now into account that the series $\sum_{k\geq 0} A_k z^k$, has radius of convergence $r$, then

$$\tilde{H}(\tilde{G}(x)) = \sum_{k\geq 0} A_k \tilde{G}(x, y)^k \in \mathcal{L}(\Omega, \Omega)$$

if $\|\tilde{G}(x)\| < r$. So, iteration (8.31) can be extended to Banach spaces as follows:

$$
\begin{cases}
x_0 \text{ given,} \\
y_n = x_n - \Gamma_n F(x_n), \\
G(x_n) = \Gamma_n \left( F'\left(x_n + \frac{2-3\theta}{3}(y_n - x_n)\right) - F'(x_n + \theta(y_n - x_n)) \right), \theta \in [0, 1/3), \\
x_{n+1} = y_n + H(G(x_n))(y_n - x_n), \quad n \geq 0,
\end{cases}
$$

(8.33)

where $\Gamma_n = [F'(x_n)]^{-1}$, $H(z) = \sum_{k \geq 1} A_k z^k$, $A_1 = A_2 = 1/2$, $A_k \in R^+$, for $k \geq 3$.

In all the cases, the operator $H : \mathcal{L}(\Omega) \rightarrow \mathcal{L}(\Omega)$ is represented by the corresponding analytical operator; as for example, $H(G(x)) = \left[I - \frac{1}{2}G(x)\right]^{-1}$ for the Halley method, $H(G(x)) = I + \frac{1}{2}G(x)\left[I - G(x)\right]^{-1}$ for the Super-Halley method, etc.

For finite situations, $H(z) = \sum_{k=1}^{m_0} A_k z^k$, the most commonly used, we realize a study of the computational cost for iteration (8.33). Starting at $x_n$, for the finite dimensional case, the computation of the $(n + 1)$-step of (8.33) proceeds as follows:

1. Stage: Compute one $LR$-decomposition of $F'$ by the Gaussian elimination.
2. Stage: Solve the linear system: $F(x_n) + F'(x_n)\alpha_n = 0$.
3. Stage: Solve the linear systems: $F'(x_n)\beta_n - F'\left(x_n + \frac{2-3\theta}{3}\alpha_n\right) + F'(x_n + \theta\alpha_n) = 0$.
4. Stage: Calculate:

$$
\gamma_n = \frac{3}{2(1 - 3\theta)}\beta_n \left(\frac{1}{2}I + \frac{3}{2(1 - 3\theta)}\beta_n \left(\frac{1}{2}I + \frac{3}{2(1 - 3\theta)}\beta_n \right.\right.
$$
$$
\left.\left. \times \left(A_3 I + \left(\cdots\left(A_{m_0-1}I + \frac{3}{2(1 - 3\theta)}A_{m_0}\beta_n\right)\right)\right)\right)\right)\alpha_n.
$$

5. Stage: Set $x_{n+1} = y_n + \gamma_n$.

Observe that the linear systems considered above have the same associated matrix and then we only need one $LR$-decomposition of the matrix $F'(x_n)$ in each step.

The efficiency of higher order methods such as (8.33) is compared with its classical predecessor in Table 8.5. Observe that methods given in (8.33) are always superior.

**Table 8.5** Computational cost, order and efficiency index

| Method | Order | Ev. of $F$ | Ev. of $F'$ | Ev. of $F''$ | LR | EI |
|---|---|---|---|---|---|---|
| The Halley method | 3 | 1 | 1 | 1 | 2 | $3^{1/3}$ |
| The Super-Halley method | 3 | 1 | 1 | 1 | 2 | $3^{1/3}$ |
| (8.33) with $\theta = 1/4$ and $m_0 = 10$ | 4 | 1 | 3 | 0 | 1 | $4^{1/4}$ |
| (8.33) with $\theta = 2/3$ and $m_0 = 4$ | 4 | 1 | 2 | 0 | 1 | $4^{1/3}$ |

**Table 8.6** Computational cost, order and efficiency index

| Method | Order | Ev. of $F$ | Ev. of $F'$ | LR | EI |
|---|---|---|---|---|---|
| Two steps of Newton's method | 4 | 2 | 2 | 2 | $4^{1/4}$ |
| (8.33) with $\theta = 0$ and $m_0 = 2$ | 4 | 1 | 2 | 1 | $4^{1/3}$ |
| (8.33) with $\theta = 2/3$ and $m_0 = 5$ | 4 | 1 | 2 | 1 | $4^{1/3}$ |

On the other hand, iterations (8.33) must be compared with Newton's method (if two steps of Newton's method as one step of a fourth order method is considered). As we can see in Table 8.6, iterations (8.33) are competitive when they are even compared with a classical method of order four.

To obtain a semilocal convergence result for iterations (8.33) with at least $R$-order of convergence four, the following conditions of the Kantorovich type are usually required: Let $x_0 \in \Omega$ and suppose that there exists $\Gamma_0 = [F'(x_0)]^{-1} \in \mathscr{L}(Y, X)$ at $x_0$. We also assume that $F$ has continuous third-order Fréchet-derivative on $\Omega$ and

(i)   $\|\Gamma_0\| \le \beta$.
(ii)  $\|y_0 - x_0\| = \|\Gamma_0 F(x_0)\| \le \eta$.
(iii) $\|F''(x)\| \le M, \quad x \in \Omega$.
(iv)  $\|F'''(x)\| \le N, \quad x \in \Omega$.
(v)   $\|F'''(x) - F'''(y)\| \le K\|x - y\|. \quad x, y \in \Omega$.

We now denote $a_0 = M\beta\eta$, $b_0 = N\beta\eta^2$, $c_0 = K\beta\eta^3$, $d_0 = f(a_0)\ell(a_0, b_0, c_0)$ where $v$, $h$ and $f$ are given in (8.14) and

$$
\ell(t, s, r) = \frac{t^3}{2}\left(1 + \frac{1}{4}(1 + tv(t))^2\right) + \left(\frac{1}{4} + \frac{9\theta^2 - 6\theta + 2}{12}(1 - 3\theta)\right) ts
$$
$$
+ \left(1 + \frac{(2 - 3\theta)^3}{18} + \frac{3\theta^3}{2(1 - 3\theta)}\right)\frac{r^3}{12}
$$
$$
+ \left(t + \frac{3\theta + (2 - 3\theta)}{2(1 - 3\theta)}\right)\sum_{k \ge 3} A_k t^k.
$$

$$(8.34)$$

The semilocal convergence of sequence (8.33) now follows from the next theorem, which is also used to draw conclusions about the existence of a solution and the domain in which it is located, along with some a priori error bounds, which lead to iteration (8.33) converges with $R$-order of convergence at least four.

**Theorem 8** *Let $X$ and $Y$ be two Banach spaces and $F : \Omega \subseteq X \to Y$ a three times Fréchet differentiable operator on a non-empty open convex subset $\Omega$. Let $x_0 \in \Omega$ and suppose that all conditions (i)–(v) hold. If $a_0 < \frac{3r}{2(1-3\theta)}$, conditions*

$$
a_0 h(a_0) < 1 \quad and \quad f(a_0)^2 \ell(a_0, b_0, c_0) < 1 \tag{8.35}
$$

*are satisfied and $B(x_0, R\eta) \subseteq \Omega$, then the sequence $\{x_n\}$, given by (8.33) and starting at $x_0$, converges with R-order of convergence at least four to a solution $x^*$ of Eq. (8.1), the solution $x^*$ and the iterates $x_n$, $y_n$ belong to $\overline{B}(x_0, R\eta)$. Moreover, the following a priori error estimates are given*

$$\|x^* - x_n\| \le h(a_0)\,\eta\,\gamma^{\frac{4^n-1}{3}}\frac{\Delta^n}{1-\gamma^{4^n}\Delta}, \quad n \ge 0, \tag{8.36}$$

*where $\gamma = f(a_0)^2 \ell(a_0, b_0, c_0)$ and $\Delta = 1/f(a_0)$. Moreover, the solution $x^*$ of Eq. (8.1) is unique in $\Omega_0 = B(x_0, \frac{2}{M\beta} - R\eta) \cap \Omega$ provided that $R < 2/a_0$.*

## 8.6.1   Application

We now provide a numerical test where we apply the convergence result previously obtained. We consider the following boundary value problem:

$$\begin{cases} \dfrac{d^2x(t)}{dt^2} + \lambda x(t)^{1+p} + \mu x(t) = 0, & p \in (0,1], \quad \lambda, \mu \in R, \\ x(0) = x(1) = 0, \end{cases} \tag{8.37}$$

where $x \in C^2([0,1])$, $t \in [0,1]$. To obtain a numerical solution of (8.37), we first discretize the problem. Similarly to how we proceeded in Sect. 8.5.1, Eq. (8.37) can be written as a nonlinear system. So, we define the operator $F : R^{n-1} \to R^{n-1}$ by

$$F(\overline{x}) = A\overline{x} + \Phi(\overline{x}), \tag{8.38}$$

where

$$A = \begin{pmatrix} -2 & 1 & 0 & \cdots & 0 \\ 1 & -2 & 1 & \ddots & \vdots \\ 0 & \ddots & \ddots & \ddots & 0 \\ \vdots & \ddots & 1 & -2 & 1 \\ 0 & \cdots & 0 & 1 & -2 \end{pmatrix}, \quad \Phi(\overline{x}) = h^2 \begin{pmatrix} \lambda x_1^{1+p} + \mu x_1 \\ \lambda x_2^{1+p} + \mu x_2 \\ \vdots \\ \lambda x_{n-2}^{1+p} + \mu x_{n-2} \\ \lambda x_{n-1}^{1+p} + \mu x_{n-1} \end{pmatrix}$$

and $\overline{x} = (x_1 \ldots, x_{n-1})^t$. Notice that the operator $F'$ satisfies a Hölder continuity condition

$$F'(\overline{x})(\overline{u}) = \left(A + h^2\left(\mu D_0(\overline{x}) + \lambda(1+p)D_p(\overline{x})\right)\right)\overline{u}, \quad \forall \overline{u} \in R^{n-1},$$

where we have denoted by $D_k(\bar{x})$ the diagonal matrix with the components of the vector $(x_1^k, \ldots, x_{n-1}^k)^t$, and $F''(\bar{x})$ is a bilinear operator defined by

$$F''(\bar{x})(\bar{u}, \bar{v}) = \lambda(1 + p)ph^2 \begin{pmatrix} x_1^{p-1} u_1 v_1 \\ x_2^{p-1} u_2 v_2 \\ \vdots \\ x_{n-1}^{p-1} u_{n-1} v_{n-1} \end{pmatrix}, \quad \forall \bar{u}, \bar{v} \in R^{n-1}.$$

Initially, we consider Theorem 8. Observe that this theorem is important from the point of view of the $R$-order of convergence and the error estimates. If we choose $p = 1$, $\lambda = 6/5$ and $\mu = 0$ in (8.37), the boundary value problem is reduced to the following:

$$\begin{cases} \dfrac{d^2 x(t)}{dt^2} + \dfrac{6}{5} x(t)^2 = 0, \\ x(0) = x(1) = 0. \end{cases} \tag{8.39}$$

We now approximate the solution of the equation $F(\bar{x}) = 0$ by Jarratt's method [3], which is (8.33) with $A_0 = 1$, $A_k = 1/2$, for all $k \in N$, and $\theta = 0$. To prove the convergence of this method to a solution of the equation, we choose $n = 10$ and the initial approximation $\bar{u}_0 = 50 \sin(\pi \bar{x})$, so that

$$\bar{u}_0 = \begin{pmatrix} 15.4508\ldots \\ 29.3892\ldots \\ 40.4508\ldots \\ 47.5528\ldots \\ 50 \\ 47.5528\ldots \\ 40.4508\ldots \\ 29.3892\ldots \\ 15.4508\ldots \end{pmatrix},$$

and after two iterations by Jarratt's method, we have

$$\bar{u}_2^{(2)} = \begin{pmatrix} 3.00096\ldots \\ 5.90333\ldots \\ 8.42350\ldots \\ 10.1647\ldots \\ 10.7862\ldots \\ 10.1647\ldots \\ 8.42350\ldots \\ 5.90333\ldots \\ 3.00096\ldots \end{pmatrix}.$$

We then choose $\bar{u}_2^{(2)}$ as the initial iteration $x_0$ for Theorem 8, so that the hypotheses of Theorem 8 are now satisfied. Taking into account 50 significant digits, the following values of the involved parameters are obtained:

$$\beta = 11.4670\ldots, \qquad \eta = 0.934856\ldots, \qquad M = 0.024, \qquad N = 0, \qquad K = 0,$$

so that, the domain of existence and uniqueness of solutions of equation $F(\bar{x}) = 0$ are respectively $B(x_0, 1.14054\ldots)$ and $B(x_0, 6.20093\ldots)$. The numerical solution of (8.39), shown in Table 8.7, is obtained after five iterations by Jarratt's method.

Next, in Fig. 8.13, the approximations (continuous lines) and the numerical solution (discontinuous line) of (8.39) are interpolated to obtain the approximated solution of (8.39).

Moreover, Table 8.8 shows in the first two columns the error estimates $\|\bar{u}_n^{(i)} - x^*\|_\infty$ obtained with the stopping criterion $\|\bar{u}_n^{(i)} - x^*\|_\infty < C \times 10^{-50}$ $(i = 1, 2)$, where $\bar{u}_n^{(1)}$ and $\bar{u}_n^{(2)}$ denote respectively the approximations obtained by Newton's method and Jarratt's method. To see the behaviour of these error estimates, we have compared Jarratt's method with the two steps of Newton's method, whose $R$-order of convergence is at least four. To obtain the error estimates of the two steps of Newton's method, we have considered the ones given by Kantorovich in [24] for Newton's method, and they are written in bold, and for Jarratt's method, the ones given in (8.36). Notice that Newton's method starts at the fourth iteration, since the convergence conditions are satisfied from this iteration (see [24]). Observe that

**Table 8.7** Numerical solution of (8.39)

| $i$ | $x_i^*$ | $i$ | $x_i^*$ | $i$ | $x_i^*$ |
|---|---|---|---|---|---|
| 1 | 2.70657... | 4 | 9.18820... | 7 | 7.60361... |
| 2 | 5.32524... | 5 | 9.75972... | 8 | 5.32524... |
| 3 | 7.60361... | 6 | 9.18820... | 9 | 2.70657... |

**Fig. 8.13** Approximations to a solution of Eq. (8.39) by Jarratt's method

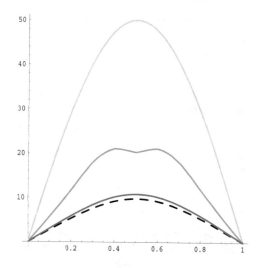

**Table 8.8** The error estimates $\|\bar{u}_n^{(i)} - x^*\|_\infty$ and the computational order of convergence

| $n$ | Newton's method | Jarratt's method | $\rho_J$ |
|---|---|---|---|
| 0 | | | |
| 1 | | | |
| 2 | | $1.14055\ldots$ | $2.90570\ldots$ |
| 3 | | $4.20955\ldots \times 10^{-2}$ | $3.95012\ldots$ |
| 4 | $1.86971\ldots$ | $2.68433\ldots \times 10^{-7}$ | $4.00007\ldots$ |
| 5 | $4.81043\ldots \times 10^{-1}$ | $1.30434\ldots \times 10^{-27}$ | $4.00000\ldots$ |
| 6 | $6.36842\ldots \times 10^{-2}$ | | |
| 7 | $2.23232\ldots \times 10^{-3}$ | | |
| 8 | $5.48580\ldots \times 10^{-6}$ | | |
| 9 | $6.62575\ldots \times 10^{-11}$ | | |
| 10 | $1.93310\ldots \times 10^{-20}$ | | |
| 11 | $3.29096\ldots \times 10^{-39}$ | | |

our a priori error estimates are competitive. Note also that computational order of convergence (8.32) for Jarratt's method (see the last column of Table 8.8), where the max-norm is used, the $R$-order of convergence at least four obtained in Theorem 8 is computationally reached.

An important feature of the two steps of Newton's method is that we can guarantee its semilocal convergence under the same convergence conditions as for Newton's method, for example under mild Newton-Kantorovich conditions, that is, let $x_0 \in \Omega$, suppose $\Gamma_0 = [F'(x_0)]^{-1} \in \mathscr{L}(Y, \Omega)$ exists at $x_0$, $F'$ is continuous on $\Omega$ and

(c$_1$) $\|\Gamma_0\| \leq \beta$.
(c$_2$) $\|y_0 - x_0\| = \|\Gamma_0 F(x_0)\| \leq \eta$.
(c$_3$) $\|F'(x) - F'(y)\| \leq K\|x - y\|^p$,   $x, y \in \Omega$,   $p \in [0, 1]$.

We observe that the conditions (i)–(v) are overly restrictive. In addition, the family of iterative processes (8.33) only use F' in its algorithm. So, we now prove that iterations given in (8.33) are convergent under mild Newton-Kantorovich conditions (c$_1$)–(c$_3$) as Newton's method.

**Theorem 9** *Let X and Y be two Banach spaces and $F : \Omega \subseteq X \to Y$ a Fréchet differentiable operator on a non-empty open convex subset $\Omega$. Let $x_0 \in \Omega$ and suppose that all conditions (c$_1$)–(c$_3$) hold. If $\tilde{a}_0 < \left(\frac{3}{2(1-3\theta)}\right)^p r$, $\tilde{a}_0 g(\tilde{a}_0)^p < 1$ and $h(\tilde{a}_0)^{1+p}\tilde{\ell}(\tilde{a}_0)^p < 1$, where $\tilde{a}_0 = K\beta\eta^p$ and $\tilde{\ell}(x) = x\left(f(x) + g(x)^{1+p}/2\right)$, and $B(x_0, \tilde{R}\eta) \subset \Omega$, where $\tilde{R} = \frac{g(\tilde{a}_0)}{1-\tilde{d}_0}$ and $\tilde{d}_0 = h(\tilde{a}_0)\tilde{\ell}(\tilde{a}_0)$, then the sequence $\{x_n\}$, given by (8.33) and starting at $x_0$, converges to a solution $x^*$ of Eq. (8.1), the solution $x^*$ and the iterates $x_n$, $y_n$ belong to $\overline{B(x_0, \tilde{R}\eta)}$.*

We provide a numerical test where we apply the convergence result previously obtained. We now apply Theorem 9.

If we choose $p = 1/2$ and $\lambda = \mu = 1$ in (8.37), then boundary value problem (8.37) is reduced to the following:

$$\begin{cases} \dfrac{d^2x(t)}{dt^2} + x(t)^{3/2} + x(t) = 0, \\ x(0) = x(1) = 0. \end{cases} \tag{8.40}$$

To prove the convergence of iterations (8.33) to a solution of (8.40), we discretize the problem as in the previous boundary value problem. Observe that, in this case, we cannot apply Theorem 8 to approximate a solution of problem (8.40), since the second derivative of the operator $F(\bar{x})$ does not exist in the origin, but we can Theorem 9. We take again $n = 10$ and $\bar{u}_0 = 50\sin(\pi\bar{x})$ as initial approximation. Then

$$\bar{u}_0 = \begin{pmatrix} 15.4508\ldots \\ 29.3892\ldots \\ 40.4508\ldots \\ 47.5528\ldots \\ 50 \\ 47.5528\ldots \\ 40.4508\ldots \\ 29.3892\ldots \\ 15.4508\ldots \end{pmatrix}$$

and after two iterations by Jarratt's method, we have

$$\bar{u}_2^{(2)} = \begin{pmatrix} 27.2444\ldots \\ 52.7953\ldots \\ 73.9844\ldots \\ 88.0739\ldots \\ 93.0222\ldots \\ 88.0739\ldots \\ 73.9844\ldots \\ 52.7953\ldots \\ 27.2444\ldots \end{pmatrix}.$$

We now choose $\bar{u}_2^{(2)}$ as the initial iteration $x_0$ in Theorem 9, so that the hypotheses of Theorem 9 are then satisfied. If 50 significant digits are used, we obtain the following values of the parameters:

$$\beta = 29.4665\ldots, \qquad \eta = 0.114027\ldots, \qquad K = 0.015.$$

**Table 8.9** Numerical
solution of (8.40)

| $i$ | $x_i^*$ | $i$ | $x_i^*$ | $i$ | $x_i^*$ |
|---|---|---|---|---|---|
| 1 | 27.2110... | 4 | 87.9658... | 7 | 73.8937... |
| 2 | 52.7305... | 5 | 92.9080... | 8 | 52.7305... |
| 3 | 73.8937... | 6 | 87.9658... | 9 | 27.2110... |

**Table 8.10** The error
$\|\bar{u}_n^{(i)} - x^*\|_\infty$ and the
computational order of
convergence

| $n$ | Newton's method | Jarratt's method | $\rho_J$ |
|---|---|---|---|
| 0 | 42.9080... | 42.9080... | |
| 1 | 82.1136... | 25.7101... | |
| 2 | 20.5670... | $1.14132... \times 10^{-1}$ | 3.83272... |
| 3 | 2.42426... | $1.09691... \times 10^{-10}$ | 3.99977... |
| 4 | $4.52774... \times 10^{-2}$ | $9.40227... \times 10^{-47}$ | 4.00000... |
| 5 | $1.65341... \times 10^{-5}$ | | |
| 6 | $2.20683... \times 10^{-12}$ | | |
| 7 | $3.93139... \times 10^{-26}$ | | |

**Fig. 8.14** Approximations to
a solution of Eq. (8.40)

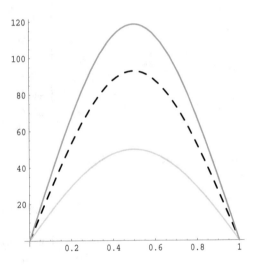

The domain of existence of solutions of equation $F(\bar{x}) = 0$ is therefore $\overline{B(x_0, 0.156166...)}$. The numerical solution of (8.40) is given in Table 8.9 and obtained after four iterations by Jarratt's method.

After that, in Table 8.10, the errors $\|\bar{u}_n^{(i)} - x^*\|_\infty$ ($i = 1, 2$) obtained for Newton's method ($\{\bar{u}_n^{(1)}\}$) and Jarratt's method ($\{\bar{u}_n^{(2)}\}$) are shown with the stopping criterion $\|\bar{u}_n^{(i)} - x^*\|_\infty < C \times 10^{-50}$. We have also added the computational order of convergence $\rho_J$ for Jarratt's method, which is closed to four, so that it has been also reached under mild differentiability convergence conditions.

Finally, the approximations and the numerical solution of (8.40) are interpolated (continuous lines and discontinuous line, respectively) to obtain the approximated solution of (8.40), see Fig. 8.14.

**Acknowledgements** This scientific work has been supported by the 'Proyecto MTM2011-28636-C02-01' of the Ministry of science and innovation of Spain.

# References

1. Amat, S., Busquier, S.: Geometry and convergence of some third-order methods. Southwest J. Pure Appl. Math. **2**, 61–72 (2001)
2. Argyros, I.K., Szidarovszky, F.: The Theory and Applications of Iteration Methods. CRC Press, Boca Raton, FL (1993)
3. Argyros, I.K., Chen D., Qian, Q.: The Jarratt method in Banach space setting. J. Comput. Appl. Math. **51**, 103–106 (1994)
4. Bruns, D.D., Bailey, J.E.: Nonlinear feedback control for operating a nonisothermal CSTR near an unstable steady state. Chem. Eng. Sci. **32**, 257–264 (1977)
5. Conway, J.B.: A Course in Functional Analysis. Springer, New York (1990)
6. Davis, H.T.: Introduction to Nonlinear Differential and Integral Equations. Dover, New York (1962)
7. Ezquerro, J.A., Hernández, M.A.: A uniparametric Halley-type iteration with free second derivative. Int. J. Pure Appl. Math. **6**, 103–114 (2003)
8. Ezquerro, J.A., Hernández, M.A., Romero, N.: A modification of Cauchy's method for quadratic equations. J. Math. Anal. Appl. **339**(2), 954–969 (2008)
9. Ezquerro, J.A., Hernández, M.A., Romero, N.: On some one-point hybrid iterative methods. Nonlinear Anal. Ser. A Theory Methods Appl. **72**, 587–601 (2010)
10. Ezquerro, J.A., Hernández, M.A., Romero, N.: Newton-like methods for operators with bounded second Fréchet derivative. Monografías del Seminario Matemático García Galdeano. **35**, 137–144 (2010)
11. Ezquerro, J.A., Hernández, M.A., Romero, N.: Solving nonlinear integral equations of Fredholm type with high order iterative methods. J. Comput. Appl. Math. **236**(6), 1449–1463 (2011)
12. Gander, W.: On Halley's iteration method. Am. Math. Mon. **92**, 131–134 (1985)
13. Ganesh, M., Joshi, M.C.: Numerical solvability of Hammerstein integral equations of mixed type. IMA J. Numer. Anal. **11**, 21–31 (1991)
14. Hairer, E., Wanner, G.: Solving Ordinary Differential Equations II: Stiff and Differential-Algebraic Problems. Springer, Berlin (1991)
15. Hernández, M.A.: The Newton method for operators with Hölder continuous first derivative. J. Optim. Theory Appl. **109**, 631–648 (2001)
16. Hernández, M.A.: Chebyshev's approximation algorithms and applications. Comput. Math. Appl. **41**, 433–445 (2001)
17. Hernández, M.A., Romero, N.: On a new multiparametric family of Newton-like methods. Appl. Numer. Anal. Comput. Math. **2**, 78–88 (2005)
18. Hernández, M.A., Romero, N.: On a characterization of some Newton-like methods of $R$-order at least three. J. Comput. Appl. Math. **183**(1), 53–66 (2005)
19. Hernández, M.A., Romero, N.: General study of iterative processes of R-order at least three under weak convergence conditions. J. Optim. Theory Appl. **133**, 163–177 (2007)
20. Hernández, M.A., Romero, N.: Application of iterative processes of R-order at least three to operators with unbounded second derivative. Appl. Math. Comput. **185**, 737–747 (2007)
21. Hernández, M.A., Romero, N.: Toward a unified theory for third $R$-order iterative methods for operators with unbounded second derivative. Appl. Math. Comput. **215**(6), 2248–2261 (2009)
22. Hernández, M.A., Salanova, M.A.: Index of convexity and concavity: application to Halley method. Appl. Math. Comput. **103**, 27–49 (1999)

23. Jerome, J.W., Varga, R.S.: Generalizations of Spline Functions and Applications to Nonlinear Boundary Value and Eigenvalue Problems, Theory and Applications of Spline Functions. Academic, New York (1969)
24. Kantorovich, L.V., Akilov, G.P.: Functional Analysis. Pergamon Press, Oxford (1982)
25. Keller, H.B.: Numerical Methods for Two-Point Boundary Value Problems. Dover Publications, New York (1992)
26. Kneisl, K.: Julia sets for the super-Newton method, Cauchy's method, and Halley's method. Chaos **11**(2), 359–370 (2001)
27. Macnamee, J.M.: Numerical Methods for Roots of Polynomials - Part I. Studies in Computational Mathematics, vol. 14. Elsevier, Amsterdam (2007)
28. Potra, F.A., Ptak, V.: Nondiscrete Induction and Iterative Processes. Pitman Advanced Publishing Program, London (1984)
29. Safiev, R.A.: On some iterative processes, Ž. Vyčcisl. Mat. Fiz. **4**, 139–143 (1964). Translated into English by L.B. Rall as MRC Technical Summary Report, vol. 649, University of Wisconsin-Madison (1966)
30. Schröeder, E.: Über unendlich viele Algotithmen zur Auflösung der Gleichugen. Math. Ann. **2**, 317–365 (1870)
31. Traub, J.F.: Iterative Methods for the Solution of Equations. Prentice Hall, Englewood Cliffs (1964)
32. Varona, J.L.: Graphic and numerical comparison between iterative methods. Math. Intell. **24**, 37–46 (2002)
33. Weerakoon, S., Fernando, T.G.I.: A variant of Newton's method with accelerated third-order convergence. Appl. Math. Lett. **13**, 87–93 (2000)

# Chapter 9
# Measures of the Basins of Attracting $n$-Cycles for the Relaxed Newton's Method

**J.M. Gutiérrez, L.J. Hernández, Á.A. Magreñán, and M.T. Rivas**

**Abstract** The relaxed Newton's method modifies the classical Newton's method with a parameter $h$ in such a way that when it is applied to a polynomial with multiple roots and we take as parameter one of these multiplicities, it is increased the order of convergence to the related multiple root.

For polynomials of degree three or higher, the relaxed Newton's method may possess extraneous attracting (even super-attracting) cycles. The existence of such cycles is an obstacle for using the relaxed Newton's method to find the roots of the polynomial. Actually, the basins of these attracting cycles are open subsets of $\mathbb{C}$.

The authors have developed some algorithms and implementations that allow to compute the measure (area or probability) of the basin of a $p$-cycle when it is taken in the Riemann sphere. In this work, given a non negative integer $n$, we use our implementations to study the basins of non-repelling $p$-cycles, for $1 \leq p \leq n$, when we perturb the relaxing parameter $h$. As a consequence, we quantify the efficiency of the relaxed Newton's method by computing, up to a given precision, the measure of the different attracting basins of non-repelling cycles. In this way, we can compare the measure of the basins of the ordinary fixed points (corresponding to the polynomial roots) with the measure of the basins of the point at infinity and the basins of other non-repelling $p$-cyclic points for $p > 1$.

J.M. Gutiérrez (✉) • L.J. Hernández • M.T. Rivas
Departamento de Matemáticas y Computación, Universidad de La Rioja, Logroño, Spain
e-mail: jmguti@unirioja.es; luis-javier.hernandez@unirioja.es; maria-teresa.rivas@unirioja.es

Á.A. Magreñán
Departamento de TFG/TFM, Universidad Internacional de La Rioja, Logroño, Spain
e-mail: angel.magrenan@unir.net

© Springer International Publishing Switzerland 2016
S. Amat. S. Busquier (eds.), *Advances in Iterative Methods for Nonlinear Equations*, SEMA SIMAI Springer Series 10,
DOI 10.1007/978-3-319-39228-8_9

## 9.1  Introduction

Given a complex polynomial $q$, the associated Newton's iteration map is the rational map

$$N_q : \widehat{\mathbb{C}} \to \widehat{\mathbb{C}}, \quad N_q(z) = z - \frac{q(z)}{q'(z)}$$

and the iteration map associated to the relaxed Newton's method is given by

$$N_{q,h} : \widehat{\mathbb{C}} \to \widehat{\mathbb{C}}, \quad N_{q,h}(z) = z - h \cdot \frac{q(z)}{q'(z)} \ ,$$

where $h$ is a complex parameter and $\widehat{\mathbb{C}} = \mathbb{C} \cup \{\infty\}$ where $\mathbb{C}$ is the field of complex numbers. *Newton's method* and the *the relaxed Newton's method* are two well-known iterative processes for solving nonlinear equations. Starting at an initial seed $z_0 \in \mathbb{C}$, both methods define sequences that, under appropriate conditions, converge to one of the roots of the polynomial $q$. For Newton's method all the roots are super-attracting fixed points of $N_q$. However, for the relaxed Newton's method, the character of the roots as fixed points of $N_{q,h}$ depends on $h$ and could be even repelling. In addition, $\infty$ is always a repelling fixed point for Newton's method whereas it could become an attracting fixed point for the relaxed Newton's method depending on the values of $h$.

For quadratic polynomials, the dynamical behavior of Newton's method is completely understood since the studies of Cayley and Schröder at the end of the nineteenth century. In fact, it is equivalent to the dynamics of the map $z \mapsto z^2$ in the Riemann sphere. For polynomials of degree three or higher, the behavior of Newton's method is more complicated because it may possess extraneous attracting (even super-attracting) cycles. The study of Newton's method applied to cubic polynomials can be reduced, via conjugation with an adequate Möbius transformation, to the study of a one-parameter family of polynomials. For instance, in [5, 10] the one-parameter family of cubics $q_\lambda(z) = z^3 + (\lambda - 1)z - \lambda$, $\lambda \in \mathbb{C}$ is considered. It can be shown that there exist values of the parameter $\lambda$ such that the Newton's map for $q_\lambda$, denoted by $N_{q_\lambda}$, has super-attracting cycles. The existence of such cycles, of course, forms a barrier to using Newton's method to find the roots of the polynomial, as their basins will be open subsets of $\mathbb{C}$.

There are also some results, see [3], that prove that given $r$ distinct points in $\mathbb{C}$, there exists a polynomial $q(z)$ of degree at most $r + 1$ so that the corresponding Newton's map, or even the relaxed Newton's map, for $q(z)$ has the given points as a super-attracting cycle. This improves the result in [13], which shows how to find such a polynomial of degree $2r$.

The basin of attraction is a classical topic in dynamical systems (see [1] or [2] for instance). Roughly speaking, the basin of attraction related to an attractor is the set of initial points whose orbits converge to such attractor under the action of a given iteration function $f$. The basin of attraction related to a $p$-cycle $\{z_0, \ldots, z_{p-1}\}$ is the union of the basins of each $z_i$, $i = 0, \ldots, p - 1$ obtained under the action of $f^p$.

In this work, given a non negative integer $n$ and a rational map $f$, we have developed some algorithms and implementations to study the basins of $p$-cyclic points for $1 \leq p \leq n$. In these algorithms, we consider the spherical multiplier instead of the absolute value of the standard multiplier. For a given rational map $f$, the spherical multiplier $sm(f)$ has the advantage that is a real bounded function and, moreover, the spherical multiplier of a $p$-cyclic point agrees with the absolute value of the standard multiplier. Therefore we can divide the $p$-cyclic points $C_p(f)$ of a rational function $f$ as the disjoint union

$$C_p(f) = C_p^0(f) \cup C_p^{(0,1)}(f) \cup C_p^1(f) \cup C_p^{(1,\infty)}(f)$$

of super-attracting points, attracting and non-super-attracting points, indifferent points and repelling points depending if $sm(f)(x) = 0$, $0 < sm(f)(x) < 1$, $sm(f)(x) = 1$ or $sm(f)(x) > 1$, $x \in C_p(f)$. Note that the attracting $p$-cyclic points are given by $sm(f)(x) < 1, x \in C_p(f)$ and the non-repelling $p$-cyclic points by $sm(f)(x) \leq 1, x \in C_p(f)$. We use different type of algorithms:

1. We give an spherical plot with the different basins of non-repelling $p$-cyclic points for $1 \leq p \leq n$.
2. Our study also contains an interesting implementation which gives the measure (up to a given precision) of the different basins of non-repelling $p$-cyclic points for $1 \leq p \leq n$.
3. We also give a plot with the different repelling $p$-cycles, $1 \leq p \leq n$, which approaches the Julia set of a rational function and a different algorithm is used to obtain neighborhoods at the Julia set by taking the spherical multiplier of iterations of the rational map at a point of the sphere.
4. For a given uniparametric family of rational maps, the parameter planes are divided in regions having attracting $p$-cycles with the same $p$. This method also gives the different values of the non negative integer $p$, $1 \leq p \leq n$, such that the corresponding rational map has an attracting $p$-cycle.

One of the objectives of this work is to present a brief description of these algorithms and implementations and to give some applications of these tools for the study of the dynamic of the relaxed Newton's method when it is applied to cubical polynomials and the relaxed parameter is perturbed.

In order to study the applications of these algorithms we consider the following notation:

Let $C_p^{\leq 1}(N_{q,h})$ be the set of non-repelling $p$-cycles of the rational function obtained when relaxed Newton's method with complex parameter $h$ is applied to a fixed polynomial $q(z)$.

Let $B_p^{\leq 1}(N_{q,h})$ be the union of the attracting basins of non-repelling $p$-cycles and $A_p^{\leq 1}(N_{q,h})$ is the area of $B_p^{\leq 1}(N_{q,h})$. When we divide the area by $4\pi$ (the area of the unit 2-sphere), we have a measure that can be interpreted as the probability of a point to converge to an attracting $p$-cycle. In this way, we also consider the probability $P_p^{\leq 1}(N_{q,h}) = A_p^{\leq 1}(N_{q,h})/4\pi$.

Since the point at infinity, $\infty$, is also a fixed point for Newton's method, for $p = 1$ we denote $C_{1(\neq\infty)}^{\leq 1}(N_{q,h}) = C_1^{\leq 1}(N_{q,h}) \setminus \{\infty\}$, $B_{1(\neq\infty)}^{\leq 1}(N_{q,h})$ the union of the basins of the attracting finite fixed points. We shall use the following notation $P_{1(\neq\infty)}^{\leq 1}(N_{q,h}) = A_{1(\neq\infty)}^{\leq 1}(N_{q,h})/4\pi$, where $A_{1(\neq\infty)}^{\leq 1}(N_{q,h})$ is the area of $B_{1(\neq\infty)}^{\leq 1}(N_{q,h})$.

In this study, the probability $P_{1(\neq\infty)}^{\leq 1}(N_{q,h})$ is called the *initial efficiency* of the method $N_{q,h}$. The more favorable behavior for a numerical method as a root-finder is attained when the initial efficiency is equal to 1. But when this coefficient is strictly less than 1, it is also interesting to know what is the dynamical reason of this phenomenon. In fact, we distinguish two situations:

1. For values $h$ such that $P_{1(\neq\infty)}^{\leq 1}(N_{q,h}) < P_1^{\leq 1}(N_{q,h})$, the area of the basin of the point at infinity is not zero.
2. If $P_1^{\leq 1}(N_{q,h}) < \sum_{p=1}^{\infty} P_p^{\leq 1}(N_{q,h})$, the area of the basins of some non-repelling $p$-cycles for $p > 1$ is not zero.

Note that for both Newton's and the relaxed Newton's methods, the only extraneous fixed point (a fixed point that is not a root of the polynomial) is $\infty$. This situation changes if we consider other iterative algorithm, as Chebyshev's method, that could introduce extraneous fixed points. In this case, the notion of initial efficiency must consider also this situation.

For a given polynomial $q(z)$ and the corresponding uniparametric family of iteration functions $N_{q,h}(z)$, derived from the relaxed Newton's method, we are interested in the following questions:

1. To analyze the existence of attracting $p$-cycles. We can do it, by drawing the parameter plane for $h \in \mathbb{C}$. For each critical point, the plane is divided into different disjoint regions associated to each value of the non-negative integer $p$.
2. To give a graphic study of the evolution of the basins and its bifurcations when the parameter $h$ runs on an interval of real numbers containing the value 1, which correspond to the classical Newton's method,
3. To compute the measure of the basins of the fixed points and the non-repelling $p$-cyclic points, for different values of $h$.
4. To analyse the inequalities of the initial efficiency

$$P_{1(\neq\infty)}^{\leq 1}(N_{q,h}) \leq P_1^{\leq 1}(N_{q,h}) \leq \sum_{p=1}^{\infty} P_p^{\leq 1}(N_{q,h}) \leq 1$$

as a function of the parameter $h$.

## 9.2    Mathematical Framework

We refer the reader to [8] for a more detailed description of the mathematical framework that we have used to develop and to implement the algorithms employed in this work. Nevertheless, in order to make the study as more self-contained as possible, we include a brief summary with some definitions, notations, basic tools and known results.

### 9.2.1    Discrete Semi-flows on Metric Spaces and Basins

Let $(X, d)$ be a metric space with metric $d$. Given a discrete semi-flow induced by a continuous map $f: X \rightarrow X$, the triple $(X, d, f)$ will be called *metric discrete semi-flow*. Given an integer $n \geq 0$, $f^n$ denotes the $n$th composition $f \circ \cdots \circ f$ and $f^0 = id_X$.

Let $X = (X, f)$ be a discrete semi-flow. A point $x \in X$ is said to be a *fixed point* if, for all $n \in \mathbb{N}$, $f^n(x) = x$; $x$ is said to be a *periodic point* if there exists $n \in \mathbb{N}$, $n \neq 0$, such that $f^n(x) = x$ and $x$ is said to be a *p-cyclic point* if $f^p(x) = x$ and $f^{p-1}(x) \neq x$. The subset of fixed points of a discrete semi-flow $(X, f)$ will be denoted by $\mathrm{Fix}(f)$, the subset of periodic points by $P(f)$ and the subset of $p$-cyclic points by $C_p(f)$.

Next we introduce a notion of end point based on the existence of the metric $d$; for other notions and properties of end points of a dynamical system, we refer the reader to [6, 7].

**Definition 1**    Given a metric discrete semi-flow $X = (X, d, f)$, the *end point space* of $X$ is defined as the quotient set

$$\Pi(X) = \frac{\{(f^n(x))_{n \in \mathbb{N}} \mid x \in X\}}{\sim},$$

where, given $x, y \in X$, $(f^n(x)) \sim (f^n(y))$ if and only if

$$(d(f^n(x), f^n(y))) \xrightarrow[n \to +\infty]{} 0.$$

An element $a = [(f^n(x))] \in \Pi(X)$ is called an *end point* of the metric discrete semi-flow $X$.

Note that, if $a \in C_p(f)$, we can interpret that $a$ is an end point of the form $\bar{a} = [(a, f(a), f^2(a), \cdots, f^p(a) = a, f^{p+1}(a) = f(a), \cdots)] \in \Pi(X)$.

We can define the natural map

$$\omega: X \to \Pi(X)$$

given by $\omega(x) = [(f^n(x))] = [(x, f(x), f^2(x), \dots)]$.

The map $\omega$ allows us to decompose any metric discrete semi-flow in the way shown below.

**Definition 2** Let $X$ be a metric discrete semi-flow. The subset denoted by

$$B(a) = \omega^{-1}(a), \quad a \in \Pi(X)$$

is called the *basin of the end point a*.

There is an induced partition of $X$ given by

$$X = \bigsqcup_{a \in \Pi(X,d)} B(a),$$

which will be called $\omega$-*decomposition* of the metric discrete semi-flow $X$.

## 9.2.2 Basins of Rational Functions on the Riemann Sphere

Let $S^2 = \{(r_1, r_2, r_3) \in \mathbb{R}^3 \mid r_1^2 + r_2^2 + r_3^2 = 1\}$ be the unit 2-sphere, $\hat{\mathbb{C}} = \mathbb{C} \cup \{\infty\}$ the Alexandroff compactification of $\mathbb{C}$ and $\mathbf{P}^1(\mathbb{C})$ the complex projective line. The stereographic projection and the change from homogenous to absolute coordinates (explicit formulas can be seen in [8]) give the canonical bijections:

$$S^2 \xrightarrow{\bar{\theta}} \hat{\mathbb{C}} \xleftarrow{\theta} \mathbf{P}^1(\mathbb{C}) .$$

We recall that a surface with a 1-dimensional complex structure is said to be a *Riemann surface* and a Riemann surface of genus 0 is said to be a *Riemann sphere*. Since $\mathbf{P}^1(\mathbb{C})$ has a canonical structure of 1-dimensional complex manifold, we can use the bijections above, to give to $S^2$ and $\hat{\mathbb{C}}$ the structure of a Riemann sphere.

Let $\varphi: \mathbb{C} \to \mathbb{C}$ be a rational function of the form $\varphi(u) = a \dfrac{F(u)}{G(u)}$, where $a \in \mathbb{C}$, $a \neq 0$, $F(u), G(u) \in \mathbb{C}[u]$ are monic polynomials and $F$ and $G$ have not a common root. The complex function $\varphi(u) = \frac{F(u)}{G(u)}$ has a canonical extension $\varphi^+: \hat{\mathbb{C}} \to \hat{\mathbb{C}}$ applying the roots of $G(u) = 0$ to $\infty$ and $\varphi^+(\infty) \in \{\infty, 0, a\}$. The value $\varphi^+(\infty)$ depends on the degrees of $F$ and $G$.

The bijection $\theta: \mathbf{P}^1(\mathbb{C}) \to \hat{\mathbb{C}}$ induces the map $g: \mathbf{P}^1(\mathbb{C}) \to \mathbf{P}^1(\mathbb{C})$ defined by $g = \theta^{-1} \varphi^+ \theta$, which is expressed in homogeneous coordinates as follows:

$$g([z, t]) = [F_1(z, t), G_1(z, t)],$$

where $F_1, G_1 \in \mathbb{C}[z, t]$ are homogeneous polynomials, with the same degree of homogeneity, such that $F_1(z, 1) = F(z)$ and $G_1(z, 1) = G(z)$, $z \in \mathbb{C}$.

In a similar way, we can consider the bijection $\tilde{\theta}$ and the map $f = \tilde{\theta}^{-1} \varphi^+ \tilde{\theta}$ to obtain a discrete dynamical system on the 2-sphere $(S^2, f)$. In all cases, it is said that the maps $\varphi, \varphi^+, g, f$ are rational maps.

We also recall that one has two natural metrics on $S^2$, on the one hand, since $S^2$ is a subspace of $\mathbb{R}^3$, the usual Euclidean metric of $\mathbb{R}^3$ induces the Euclidean metric $d^E$ on $S^2$, which is called the *chordal metric*; on the other hand, we have as well that $S^2$ inheres a Riemannian metric $d^R$ from the canonical Riemannian structure of $S^2 \subset \mathbb{R}^3$. The connection between Riemannian metric $d^R$ and Euclidean metric $d^E$ on $S^2$ is given by the expression:

$$d^E(x, y) = 2 \operatorname{sen} \left( \frac{d^R(x, y)}{2} \right), \quad x, y \in S^2.$$

Using the bijections $\tilde{\theta}, \theta$, one can translate these metric structures from $S^2$ to $\hat{\mathbb{C}}$ and $\mathbf{P}^1(\mathbb{C})$ to obtain metric spaces $(S^2, d)$, $(\hat{\mathbb{C}}, d_1)$ and $(\mathbf{P}^1(\mathbb{C}), d_2)$, where $d$ can be taken either the chordal metric $d^E$ or the Riemannian metric $d^R$. Note that $\tilde{\theta}, \theta$ are isometries and the discrete dynamical systems $(S^2, d, f)$, $(\hat{\mathbb{C}}, d_1, \varphi^+)$ and $(\mathbf{P}^1(\mathbb{C}), d_2, g)$ are isomorphic.

In this work, we consider the metric discrete semi-flow $(S^2, d^E, f)$ induced by a rational map $f$, the map $\omega: S^2 \to \Pi(S^2, d^E)$ given by $\omega(x) = [(x, f(x), f^2(x), \dots)]$, $x \in S^2$, and the corresponding $\omega$-decomposition of the 2-sphere as a disjoint union of basins of end points.

### 9.2.3  Spherical Multipliers

Recall that if $(V, <, >_V)$ and $(W, <, >_W)$ are Euclidean vectorial spaces provided with an escalar product and the corresponding norm $\|v\|_V = < v, v >^{\frac{1}{2}}$, then the norm of a linear transformation $T : V \to W$ is defined by

$$\|T\| = \sup\{\|T(v)\|_W \mid \|v\|_V \leq 1\}.$$

We note that for the real vectorial space $V = \mathbb{C}$ and the usual escalar product, a complex linear map $T : \mathbb{C} \to \mathbb{C}$ of the form $T(z) = \lambda z$ can also be taken as a real linear endomorphism of a 2-dimensional real vectorial space. In this case is clear that $\|T\|$ is the absolute values of $\lambda$.

Since $S^2$ has a canonical Riemannian structure, if $f : S^2 \to S^2$ is a rational function, one has that for a given point $x \in S^2$, there is an induced linear transformation $T_x(f) : T_x(S^2) \to T_{f(x)}(S^2)$ on the Euclidean tangent spaces at $x$ and $f(x)$ of the Riemannian manifold $S^2$.

Then, the *spherical multiplier* of a rational function $f : S^2 \to S^2$ at a point $x \in S^2$ is given by

$$sm(f)(x) = \|T_x(f)\| \, .$$

We remark the following facts: (1) if $x$ is a $p$-cyclic point, the spherical multiplier of $f^p$ at $x$ agrees with the absolute value of the standard multiplier of $(\varphi^+)^p$ at $\tilde{\theta}(x) \in \hat{\mathbb{C}}$, $\varphi^+ = \tilde{\theta}f(\tilde{\theta})^{-1}$; (2) the spherical multiplier is a bounded function from the 2-sphere to $\mathbb{R}$ (notice that the standard multiplier in general is not a bounded function).

If $sm(f)(x) = 0$ and $x$ is a $p$-cyclic point, it is said that $x$ is a *super-attracting $p$-cyclic point*; if $0 < sm(f)(x) < 1$, $x$ is said to be an *attracting $p$-cyclic point*; if $sm(f)(x) = 1$, $x$ is an *indifferent $p$-cyclic point* and when $sm(f)(x) > 1$, $x$ is said to be a *repelling $p$-cyclic point*. In this work, for $X \in \{S^2, \hat{\mathbb{C}}, \mathbf{P}^1(\mathbb{C})\}$, we focus on the study of the subset $C_p^{\leq 1}(f) \subset C_p(f)$ of non-repelling $p$-cyclic points and its corresponding attraction basins.

### 9.2.4 Lebesgue Measures on the 2-Sphere

Recall that a spherical triangle $ABC$ is formed by connecting three points on the surface of a 2-sphere with great arcs, so that these three points do not lie on a great circle of the sphere. The angle $\angle A$ at the vertex $A$ is measured as the angle between the tangents to the incident sides in the vertex tangent plane. Note that a pair of unitary tangent vectors at a vertex determines a canonical arc in the unit 1-sphere $S^1$ contained in the tangent plane to the 2-sphere at this vertex and we can find the measure (angle) of this arc. If we put the condition that each angle of the triangle is smaller than $\pi$, we can avoid a possible ambiguity between the triangle or its complement on the 2-sphere.

Let $ABC$ be a spherical triangle and let $ABCD$ be a spherical quadrilateral on a 2-sphere of radius $R$ with angles at vertexes smaller than $\pi$. Then, the non-negative real number $(\angle A + \angle B + \angle C - \pi)$ is called the excess of the spherical triangle. Similarly, the excess of a spherical quadrilateral $ABCD$ is the non-negative real number $(\angle A + \angle B + \angle C + \angle D - 2\pi)$.

On the 2-sphere one can introduce a Lebesgue measure using for example a 2-volumen form induced by its Riemannian structure and, in particular, we have an induced area for spherical triangles and quadrilaterals.

There is a result known as Girard's theorem that asserts that the area of an spherical triangle (or a quadrilateral) is equal to the excess multiplied by $R^2$ and in the case that $R = 1$, the area is equal to the excess. We recall that a similar formula is used in the hyperbolic plane to give the area of a triangle, but in this case one has to take the defect $\pi - (\angle A + \angle B + \angle C)$.

In [8], using consecutive subdivisions of spherical quadrilaterals, an algorithm have been developed to approach the area (or probability) of any spherical region that can be approached by spherical quadrilaterals of the subdivisions.

The measure of many regions whose boundary is given by a smooth curve can be computed in many cases using suitable coordinates and the usual integration formulas. However, many problems appear when one wants to compute the area of a region whose frontier is not a smooth curve (for instance, a Julia set). In these cases, the techniques introduced in [8] are more appropriated for developing some computational algorithms to give the area or probability of these "more complicated regions" of the 2-sphere.

## 9.3   Algorithms for Computing Basins of Non-repelling Cyclic Points, Measures and Initial Efficiency, Julia Sets and Attracting Cyclic Points

Along the previous section, we have introduced some mathematical techniques and basic theoretical aspects necessary to build computer programs with the ability of representing attraction basins of end points associated to a given rational function. In this section, we present the algorithms that we have developed to study the basins induced by a rational function $f$ on the Riemann 2-sphere and other algorithms to complement our dynamical study. We can complement these algorithms with the ones given by Chicharro–Cordero–Torregrosa [4], Magreñán [11] and Varona [15] for plotting the basins of attraction of different iterative processes in the complex plane.

### 9.3.1   Algorithm 1: Spherical Plots of Basins of Fixed and Non-repelling Cyclic Points

With the target of finding an end point associated to a point $x \in S^2$, the rational map $f$ must be iterated to obtain a finite sequence

$$(x, f(x), f^2(x), f^3(x), \ldots, f^{k-1}(x), f^k(x)).$$

In this context, we remind that a maximum number of iterations $l$ must be considered and a certain precision $c_1$ must be prefixed to determine when to stop the iterative process while programming the function which returns such sequence. That is why we shall always work with sequences in which $k \leq l$. Now we explain how our first algorithm works. It is divided into several sub-algorithms in the following way.

One of the basic sub-algorithms is based on the notion of Cauchy sequence in order to stop the iterative process. After each iteration, there will be two possible cases:

1. If the chordal distance from $f^{k-1}(x)$ to $f^k(x)$ is lower than $10^{-c_1}$, then take as output the list $[f^k(x), k]$; otherwise, case 2) is applied.
2. If $k < l$, a new iteration is done and case 1) is applied again; otherwise (if $k = l$), then the output $[f^l(x), l]$ is taken.

Let us suppose that we know some fixed points $\{x_1, x_2, \ldots, x_{m+1}\}$ of a rational function $f$.

A second basic sub-algorithm of our program is devoted to decide if a point $x$ is on the basin of a fixed point $x_i \in \{x_1, x_2, \ldots, x_{m+1}\}$. For a point $x \in S^2$, consider the iteration sequence

$$(x, f(x), f^2(x), f^3(x), \ldots, f^{k-1}(x), f^k(x)).$$

If there exists a value $i \in \{1, \ldots, m + 1\}$ such that the chordal distance from $f^k(x)$ to the fixed point $x_i$ is lower than $10^{-c_2}$ ($c_2$ is non negative integer), then the sub-algorithm returns $(i, k)$. Otherwise, $k = l$ and the output must be $(0, l)$, where $l$ is the maximum number of iterations which was prefixed beforehand.

It is very important to remark that if we start an iteration using $f^p$ instead $f$, the fixed points of $f^p$ are $p$-periodic points of $f$. We use this fact to compute the basins of a $p$-cyclic point using all the algorithms developed for fixed points, but using $f^p$ instead $f$.

We remark that in our algorithms we are working with two tolerances: $10^{-c_1}$ and $10^{-c_2}$. This fact is related with the order of convergence. When the multiplicity of a fixed point is greater than 1 and the iteration process has finished close to this multiple fixed point, the distance between the last iterations could be less than the distance from the last iterated point to the fixed point. These two precision parameters will be used in order to take into account this possibility.

A basic sub-algorithm associated to a rational function $f$ and an integer $n$ gives a list of non-repelling 1-cycles, 2-cycles, ..., and $n$-cycles. The output for each non-repelling $p$-cyclic point $a$ is represented in homogeneous coordinates $\{z, t\}$ of $a$ (using the brace notation of Mathematica). The sub-algorithm also gives for each non-repelling $p$-cyclic point $a$ its spherical multiplier $sm(a)$, which verifies that $0 \leq sm(a) \leq 1$.

Combining the above sub-algorithms, the authors have implemented in *Mathematica* the function:

```
SphericalPlotNonRepellingCyclesBasins[{P, Q}, untiln, subdivision]
```

where $P, Q$ are the numerator and denominator of a rational function. The argument *untiln* is a positive integer. For instance, if *untiln*=3, we can obtain a spherical plot with the basin of non-repelling 1-cyclic points (fixed points), 2-cyclic points and 3-cyclic points. The argument *subdivision* is again a non negative integer which denotes the number of consecutive subdivisions of the standard cubic subdivision of the sphere. For instance, for *subdivision*=2, this second subdivision has $6 \times 4^2 = 96$ spherical quadrilaterals.

Let us illustrate this algorithm with an example. Take the iteration function

$$\varphi(z) = \frac{P(z)}{Q(z)}, \ P(z) = (2i/3)(i + \sqrt{3})z + ((1 - i\sqrt{3})/6)z^4, \ Q(z) = 1, \quad (9.1)$$

and the following parameters *untiln*=3 and *subdivision*=8. We obtain an output (see Table 9.1) consisting in an spherical picture of the basins of non-repelling $p$-cyclic points for $p = 1, p = 2, p = 3$ and a list with information associated to 1-cyclic, 2-cyclic and 3-cyclic points. We have distributed all the output given by function *SphericalPlotNonRepellingCyclesBasins* in Table 9.1 as follows: In the first column, we have an spherical picture with basins of different colors associated to non-repelling $p$-cyclic points ($p \in \{1, 2, 3\}$); in the second column, $p$ runs from 1 to *untiln*=3; in the third column, we have some color palettes and in the last column we can find the different non-repelling cyclic points for $p = 1, p = 2$ and $p = 3$. In this case, we obtain for $p = 1$ one non-repelling 1-cyclic point, for $p = 2$ an empty list of non-repelling 2-cyclic points and for $p = 3$ three non-repelling 3-cyclic points.

In this example, the point at infinity $a_1^1 = \{1, 0\}$ has spherical multiplier $sm(f)(a_1^1) = 0.$, where $f = \tilde{\theta}^{-1}\varphi + \tilde{\theta}$. The fixed point $a_1^1$ is a super-attractor and the grey is its associated color. In the spherical picture, all the grey points are in the basin of the non-repelling 1-cyclic point $a_1^1$. We remark that in the first palette of the third column it also appears the black color. In our algorithm, the black color is associated either to points that do not belong to the union of the basins of non-repelling 1-cycles, 2-cycles and 3-cycles or to points where the iteration process

**Table 9.1** Output coming from our first algorithm

| Basins | | Colors | Non-repelling cyclic points |
|---|---|---|---|
| | Complement | ⬛ | |
| | $p = 1$ | | $\{0., \{1, 0\}\}$ |
| | $p = 2$ | | $\{ \}$ |
| | $p = 3$ | | $\{9.46228 * 10^{-15}, \{1., 1\}\}$ |
| | | | $\{5.74839 * 10^{-14}, \{-0.5 + 0.866025 \cdot i, 1\}\}$ |
| | | | $\{5.74839 * 10^{-14}, \{-0.5 - 0.866025 \cdot i, 1\}\}$ |

We can see the basins of non-repelling $p$-cycles for $1 \leq p \leq 3$ for the iteration function given in (9.1)

has reached the maximum limit previously prefixed and the distance between the last two iterated points is greater than the precision $10^{-c_1}$. In this example the black color does not appear in the spherical picture. This means that the unique points that could have black color are in the Julia set (the boundary of one of the basins), but in this case the area of this Julia set is zero; that is, they probability to appear in the spherical picture is zero. In the second row appears "Null" and an empty list. This means that for this rational function there are no repelling 2-cycles. In the third row, we can see a palette with three colors and three points $a_1^3 = \{1., 1\}$, $a_2^3 = \{-0.5 + 0.866025i, 1\}$, $a_3^3 = \{-0.5 - 0.866025i, 1\}$ whose spherical multiplier is zero (up to precision $10^{-14}$). These 3-cyclic points are super-attractors and the corresponding three colors have been used to give color to their corresponding spherical basins.

Each $p$-cyclic point $a$ generates an end point $\tilde{a} = [(f^k(a))_{k \in \mathbb{N}}]$. Then, for a given integer $n$, one has the induced decomposition of the 2-sphere $S^2 = (S^2 \setminus D_n) \cup D_n$, where $D_n$ is the union of basins of non-repelling cyclic points:

$$D_n = \left( \bigsqcup_{a \in C_1^{\leq 1}(f)} \omega^{-1}(\tilde{a}) \right) \cup \cdots \cup \left( \bigsqcup_{a \in C_n^{\leq 1}(f)} \omega^{-1}(\tilde{a}) \right).$$

For a more detailed description of the initial version of these graphic algorithms, see [9]. Here we can find some implementations in *Sage* and in *Mathematica* of these algorithms. The main difference of the present algorithms and implementations with the previous versions is that now the repelling $p$-cyclic points are removed. Then only the calculuses to compute the basins of non-repelling $p$-cyclic points are considered. This fact produces faster algorithms. Moreover, the present version gives, in the same spherical plot, the basins of $p$-cyclic points for all different values of $p$ between 1 and a chosen $n$.

### 9.3.2 Algorithm 2: Areas of the Basins of End Points Associated with Non-repelling Cyclic Points

In order to compute the measure of a region contained in the 2-sphere, we have used a cubical structure of the 2-sphere and a procedure to construct consecutive subdivisions of the initial cubic structure. For each $p$, $1 \leq p \leq n$, and for a given point $a$ in the interior of a 2-cube of the cubical subdivided structure of the sphere, the algorithm iterates $f^p$ to give the sequence $f^p(a), (f^p)^2(a), (f^p)^3(a), \cdots$. This process stops when either the distance of the last two points is less than a prefixed precision $10^{-c_1}$ or when the prefixed superior limit of iterations has been reached. For each non-repelling $p$-cyclic point these processes are used to construct a cubic sub-complex that approaches its basin, and the area of this cubic sub-complex is

**Table 9.2**  Output of our second algorithm: areas of basins of non-repelling $p$-cycles for $1 \le p \le 3$ for the iteration function given in (9.1)

| | Colors | Non-repellling cyclic points | Areas |
|---|---|---|---|
| Complement | | | $2.86349 * 10^{-12}$ |
| $p = 1$ | | $\{0., \{1, 0\}\}$ | $7.92117$ |
| $p = 2$ | | $\{ \}$ | |
| $p = 3$ | | $\{9.46228 * 10^{-15}, \{1., 1\}\}$ | $1.5462$ |
| | | $\{5.74839 * 10^{-14}, \{-0.5 + 0.866025i, 1\}\}$ | $1.5495$ |
| | | $\{5.74839 * 10^{-14}, \{-0.5 - 0.866025i, 1\}\}$ | $1.5495$ |

computed. For a higher subdivided cubic structure, the area of this sub-complex is a good approximation of the area of the basin of the given non-repelling $p$-cyclic point. Moreover, the area of the complement of all these cubic sub-complexes approaches the area of the complement of the union of these basins in the 2-sphere.

The authors have implemented in *Mathematica* the function:

```
AreaComplementInftyNonRepellingCyclicPoints[{P, Q}, untiln,
iter, precpoints, precroots, subdivision]
```

For the same example (9.1) and the following parameters

```
AreaComplementInftyNonRepellingCyclicPoints[{P, Q}, 3, 50, 3, 3, 6]
```

the following output is obtained (see Table 9.2):

```
{2.86349*10^-12, {{7.92117}, {}, {1.5495, 1.5495, 1.5462}},
{{{0., {1,  0}}}, {}, {{5.74839*10^-14, {-0.5 - 0.866025 I, 1}},
{5.74839*10^-14, {-0.5 + 0.866025 I, 1}}, {9.46228*10^-15, {1., 1}}}}}}
```

## 9.3.3   Algorithm 3: The Julia Set of a Rational Function in the Riemann Sphere

Our approach to the Julia set of a rational function is based in two different algorithms: the first method compute repelling $p$-cyclic points, $1 \le p \le n$ (the Julia set is the closure of the set of repelling cyclic points); the second procedure looks for points $x \in S^2$ such that $sm(f^p)(x) > 1$.

The first algorithm has been implemented by the *Mathematica* function

```
SphericalPlotRepellingCycles[{P, Q}, n]
```

that can been executed using two arguments: the numerator and denominator of the rational function $\{P, Q\}$ and the upper bound $n$. For instance, the rational function $P/Q = 2z/-(1 + 3z^2)$ is obtained when the relaxed Newton's method is applied to the polynomial $z^3 + z$ for the relaxing parameter $h = 3$. Taking the first argument

**Table 9.3** Spherical plots of repelling $p$-cyclic points for $1 \le p \le n$, the relaxed Newton's method with $h = 3$, $n = 3, 4, 5, 6$ and polynomial $z^3 + z$

| $n = 3$ | $n = 4$ | $n = 5$ | $n = 6$ |
|---|---|---|---|
| | | | |

the pair of polynomials $P(z) = 2z$, $Q(z) = -(1 + 3z^2)$, we can see in Table 9.3 the graphic output corresponding to four executions of the *Mathematica* function related to the values of the second argument $n \in \{3, 4, 5, 6\}$. In the corresponding plot each $p$ has an associated color; for example, the green color corresponds to repelling 6-cyclic points, the red color is for 5-cyclic points, et cetera. In this case, one can see that the repelling $p$-cyclic points are contained in a great circle of the unit sphere and therefore this suggests that the corresponding Julia set is this great circle.

The algorithm SphericalPlotRepellingCycles has to deal with rational functions of high degree and this increases the execution time. To avoid this difficulties we also have developed the implemented *Mathematica* function

```
SphericalPlotNeigborhoodsRepellingCycles[{P, Q}, ntimes, subdivision]
```

Given a point $x \in S^2$ and a rational function $f : S^2 \to S^2$, we can consider the finite sequence

$$(x, f(x), f^2(x), f^3(x), \ldots, f^{p-1}(x))$$

and we can take the spherical multiplier:

$$\|T_x(f^p)\| = \|T_x(f)\| \, \|T_{f(x)}(f)\| \cdots \|T_{f^{p-1}(x)}(f)\|$$

Note that if $x$ is a repelling $p$-cyclic point, one has that $\|T_x(f^p)\| > 1$ and this inequality holds in a small neighborhood at $x$. Using this property of the spherical multiplier and subdivisions of the canonical cubic structure of the 2-sphere we can find neighborhoods of repelling $p$-cyclic points; that is, neighborhoods at the Julia set contained in $S^2$. Using the formula above, we can compute the spherical

**Table 9.4** Small neighborhoods (non black color) of the Julia set of the relaxed Newton's method applied to $z^3 + z$ for $h = 1$ and $h = 3$

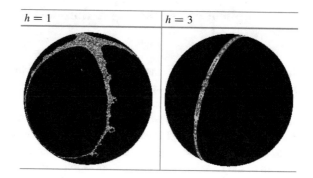

multiplier of $f^p$ at a point $x \in S^2$ without computing rational functions of higher degree.

We can see in Table 9.4 graphical outputs which correspond to Julia sets of rational functions obtained when the relaxed Newton's method is applied to the polynomial $z^3 + z$ for the values $h = 1, 3$, respectively; we have taken *ntimes* $= 8$ and *subdivision* $= 8$. The different colors are assigned depending on the values of the spherical multiplier; the black color corresponds to the region where the spherical multiplier is $\leq 1$.

Compare the plots given on the right in Tables 9.3 and 9.4. In the first Table, we have points contained in a great circle and in the second Table we have a small neighborhood of this great circle.

### 9.3.4 Algorithm 4: The Parameter Plane

The parameter plane is a well-known technique to graphically and numerically understand the dynamical behavior of an iteration function [12–14]. It is based on the classical Fatou's theorem (see [2], for instance) that ensures that the basin of each attracting cycle contains at least one critical point. In fact, if the iteration map is related to a method for finding the roots of a polynomial $q(z)$, then the study of the orbits of the free critical points (critical points that are not roots of the polynomial) plays an important role in the dynamical behavior of such an iteration map. The parameter plane is used in the following section to study the dynamical behavior of the relaxed Newton's method.

## 9.4 Applications of the Algorithms to the Relaxed Newton's Method

In this section we apply the algorithms presented in the previous section to the relaxed Newton's method applied to a couple of cubic polynomials. In fact, we have chosen the generic polynomial $q_1(z) = z^3 + z$, with three distinct complex roots, and the polynomial $q_2(z) = (z-1)^2(z+1)$, with a double root and a simple root, as representative examples of cubic polynomials in order to study the dynamical behavior of the relaxed Newton's method. We can see other numerical experiments related to the relaxed Newton's method applied to cubical polynomial in [11].

### 9.4.1 Attracting p-Cyclic Points for the Relaxed Newton's Method

The study of the critical points of a rational point and the convergence of the corresponding postcritical point is applied to find the attracting $p$-cyclic points.

#### 9.4.1.1 Dynamic of the Relaxed Newton's Method Applied to $q_1(z) = z^3 + z$

Let us consider the iteration map obtained by applying the relaxed Newton's method to the polynomial $q_1(z) = z^3 + z$:

$$N_{q_1,h}(z) = \frac{z\left((3-h)z^2 - h + 1\right)}{3z^2 + 1}. \tag{9.2}$$

The critical points of $N_{q_1,h}(z)$ are the solutions of the equation

$$\left(3z^2 + 1\right)^2 - h\left(3z^4 + 1\right) = 0. \tag{9.3}$$

In general, for $h \neq 3$, there are four complex solutions of (9.3):

$$cp_1(h) = -\frac{\sqrt{-\frac{3}{-3+h} - \frac{\sqrt{3}\sqrt{4h-h^2}}{-3+h}}}{\sqrt{3}},$$

$$cp_2(h) = \frac{\sqrt{-\frac{3}{-3+h} - \frac{\sqrt{3}\sqrt{4h-h^2}}{-3+h}}}{\sqrt{3}},$$

$$cp_3(h) = -\frac{\sqrt{-\frac{3}{-3+h} + \frac{\sqrt{3}\sqrt{4h-h^2}}{-3+h}}}{\sqrt{3}}$$

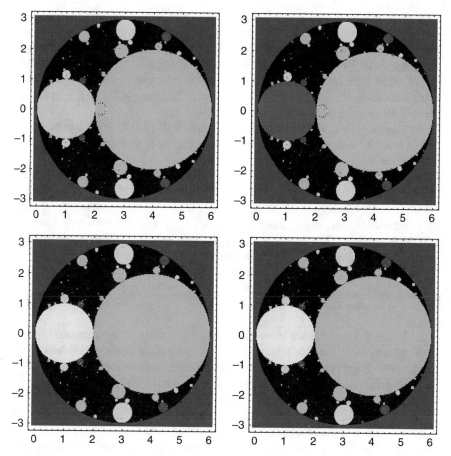

**Fig. 9.1** Parameter plane associated to each of the four critical points $cp_i(h)$, $i = 1, 2, 3, 4$ of the iteration map (9.2)

and

$$cp_4(h) = \frac{\sqrt{-\frac{3}{-3+h} + \frac{\sqrt{3}\sqrt{4h-h^2}}{-3+\lambda}}}{\sqrt{3}}.$$

If $h = 3$, these four solutions collapse in two: $\pm 1/\sqrt{3}$.

As we have said, the study of the orbits of the critical points is crucial in the dynamical behavior of an iterative method. In concrete, to determinate if there exists any attracting periodic orbit different to the roots of the polynomial $q_1(z)$, we can draw the parameter space in terms of $h \in \mathbb{C}$. In Fig. 9.1, we show the four parameter planes related to each of the critical points $cp_i(h)$, $i = 1, 2, 3, 4$.

A point is painted in cyan if the iteration of the method starting in any of the critical points converges to the root $-i$, in magenta if it converges to the root $i$ and in yellow if the iteration converges to the root 0. Moreover, it appears in green the convergence to 2-cycles, in red to 3-cycles, in dark blue to fixed point $\infty$ and the rest of colors are related to the convergence of different $p$-cycles with $p > 3$. If after 2000 iterations with a tolerance of $10^{-6}$ there is no convergence to a $p$-cycle, the pixel of the relaxing parameter is painted in black color.

As a consequence, every point $h$ of the plane with $|h-1| > 1$ is not a good choice of $h$ in terms of numerical behavior when the objective is to reach the three roots of the polynomial.

In Fig. 9.1 we can observe that for every value of $h$ with $|h - 1| < 1$ the iteration of every point converges to one of the roots. Moreover, it is clear that for values of $h$ with $|h - 4| < 2$ there exists attracting 2-cycles. Inside the circle $|h - 3| < 3$ but outside the circles $|h - 1| < 1$ and $|h - 4| < 2$ there are other attracting cycles of length bigger than 2. For $|h - 3| > 3$ the point at infinity is an attractor and its basin (taking higher enough number of iterations) cover the 2-sphere.

By using the algorithms described in the previous section, and according to Fig. 9.1, we can make now a graphical and numerical approach of the areas of the basins of attraction of the non-repelling cyclic points for the cubical polynomial $z^3 + z$.

In Tables 9.5 and 9.6 we show the results for a range of representative values of the parameter $h$. For this first approach, we have chosen only real values of the parameter $h$. For them, we have derived the following conclusions.

If the parameter $h$ is in the circle $|h - 1| < 1$, the roots of the polynomial $q_1(z)$ are the only attracting fixed points, the basins of the roots cover the 2-sphere and there are not attracting cycles of length greater than one. For this range of values we have chosen $h = 1$ (the classical Newton's method) and $h = 1.5$ as representative ones for the tables, but other values of $h$ have been also considered. In fact, we have checked that when $|h - 1|$ tends to 1, the spherical multipliers of the fixed points tend to 1.

From a dynamical point of view, $h = 2$ is an interesting value, because a bifurcation happens in it. When $h = 2$, all of the three roots are indifferent and there are not non-repelling cyclic points for $p > 1$. The graphic approach for $h = 2$ has a very slow convergence to the indifferent fixed points and a higher number of iterations of the associated rational function is needed (note that in this case the output of the program gives us the fixed points with a different layout. Now the yellow corresponds to the root 0 and the cyan is related to the basin of $-i$.)

If the parameter $h$ is in the interval $(2, 6)$, the three roots are repelling fixed points and some attracting 2-cycles appear. The basins of these non-repelling 2-cyclic points cover the 2-sphere. In addition, when $h$ tends to 2 and $h$ tends to 6, the spherical multipliers of the 2-cyclic points tend to 1. We have chosen $h \in \{2.5, 3, 3.5, 4, 4.5, 5, 5.5\}$ as representative values in this range.

Another bifurcation value occurs for $h = 6$. The point at infinity is indifferent and for $p > 1$ there are not non-repelling $p$-cyclic points. With the parameters and tolerances used in this graphical approach, we see that almost all the 2-sphere is the

**Table 9.5** Spherical plots and areas of basins of non-repelling *p*-cycles for $1 \leq p \leq 2$ and $h \in \{1, 1.5, 2, 2.5, 3, 3.5\}$ for $z^3 + z$

| Basins | | | Non-repelling cyclic points | Areas |
|---|---|---|---|---|
| $h = 1$ | | | | |
| | Complement | ■ | | $4.3201 * 10^{-12}$ |
| | $p = 1$ | | $\{0., \{0, 1\}\}$ | 8.25884 |
| | | | $\{8.88178 * 10^{-16}, \{-i, 1\}\}$ | 2.15376 |
| | | | $\{8.88178 * 10^{-16}, \{i, 1\}\}$ | 2.15376 |
| | $p = 2$ | | { } | |
| $h = 1.5$ | | | | |
| | Complement | ■ | | $4.3201 * 10^{-12}$ |
| | $p = 1$ | | $\{0.5, \{0, 1\}\}$ | 8.76401 |
| | | | $\{0.5, \{-i, 1\}\}$ | 1.90118 |
| | | | $\{0.5, \{i, 1\}\}$ | 1.90118 |
| | $p = 2$ | | { } | |
| $h = 2$ | | | | |
| | Complement | ■ | | 0.0183957 |
| | $p = 1$ | | $\{1., \{i, 1\}\}$ | 1.30531 |
| | | | $\{1., \{-i, 1\}\}$ | 1.30531 |
| | | | $\{1., \{0, 1\}\}$ | 9.93735 |
| | $p = 2$ | | { } | |
| $h = 2.5$ | | | | |
| | Complement | ■ | | $4.3201 * 10^{-12}$ |
| | $p = 1$ | | { } | |
| | $p = 2$ | | $\{0.09, \{-0.377964, 1\}\}$ | 6.28319 |
| | | | $\{0.09, \{0.377964, 1\}\}$ | 6.28319 |

(continued)

**Table 9.5** (continued)

| Basins | | | Non-repelling cyclic points | Areas |
|---|---|---|---|---|
| $h = 3$ | | | | |
| | Complement | ◼ | | $4.3201 * 10^{-12}$ |
| | $p = 1$ | | { } | |
| | $p = 2$ | ◼ | $\{2.22045 * 10^{-16}, \{-0.57735, 1\}\}$ | 6.28319 |
| | | ◼ | $\{0., \{0.57735, 1\}\}$ | 6.28319 |
| $h = 3.5$ | | | | |
| | Complement | ◼ | | $4.3201 * 10^{-12}$ |
| | $p = 1$ | | { } | |
| | $p = 2$ | ◼ | $\{0.00510204, \{-0.774597, 1\}\}$ | 6.28319 |
| | | ◼ | $\{0.00510204, \{0.774597, 1\}\}$ | 6.28319 |

basin of the point at infinity. There appear some areas in black that could disappear if we change our parameters, allowing a higher number of iterations.

For $h > 6$ the point at infinity is an attractor and its basin (taking an enough number of iterations) cover the 2-sphere.

To end this section and to complement the graphical information shown in Tables 9.5 and 9.6, we show in Table 9.7 the Julia sets related to the relaxed Newton's method applied to the polynomial $q_1(z) = z^3 + z$ for the values of $h$ considered in the aforementioned tables. We have done this, by using the third algorithm introduced in the previous section.

### 9.4.1.2 Dynamic of the Relaxed Newton's Method Applied to $q_2(z) = (z-1)^2(z+1)$

Let us consider the iteration map obtained by applying the relaxed Newton's method to the polynomial $q_2(z) = (z-1)^2(z+1)$:

$$N_{q_2,h}(z) = \frac{(3-h)z^2 + h + z}{3z + 1}. \tag{9.4}$$

**Table 9.6** Spherical plots and areas of basins of non-repelling *p*-cycles for $1 \leq p \leq 2$ and $h \in \{4, 4.5, 5, 5.5, 6, 6.5\}$ for $z^3 + z$

| Basins | | | Non-repelling cyclic points | Areas |
|---|---|---|---|---|
| *h* = 4 | | | | |
| | Complement | ■ | | $4.3201 * 10^{-12}$ |
| | *p* = 1 | | { } | |
| | *p* = 2 | ■ | $\{6.66134 * 10^{-16}, \{1, 1\}\}$ | 6.28319 |
| | | | $\{6.66134 * 10^{-16}, \{-1, 1\}\}$ | 6.28319 |
| *h* = 4.5 | | | | |
| | Complement | ■ | | $4.3201 * 10^{-12}$ |
| | *p* = 1 | | { } | |
| | *p* = 2 | ■ | $\{0.0277778, \{1, 0.774597\}\}$ | 6.28319 |
| | | | $\{0.0277778, \{1, -0.774597\}\}$ | 6.28319 |
| *h* = 5 | | | | |
| | Complement | ■ | | $4.3201 * 10^{-12}$ |
| | *p* = 1 | | { } | |
| | *p* = 2 | ■ | $\{0.16, \{1, -0.57735\}\}$ | 6.28319 |
| | | | $\{0.16, \{1, 0.57735\}\}$ | 6.28319 |
| *h* = 5.5 | | | | |
| | Complement | ■ | | $4.3201 * 10^{-12}$ |
| | *p* = 1 | | { } | |
| | *p* = 2 | ■ | $\{0.464876, \{-0.377964, 1\}\}$ | 6.28319 |
| | | | $\{0.464876, \{0.377964, 1\}\}$ | 6.28319 |

(continued)

**Table 9.6** (continued)

| Basins | | | Non-repelling cyclic points | Areas |
|---|---|---|---|---|
| $h = 6$ | | | | |
| | Complement | ■ | | 0.028371 |
| | $p = 1$ | ■ | $\{1, \{1, 0\}\}$ | 12.538 |
| | $p = 2$ | | $\{\ \}$ | |
| $h = 6.5$ | | | | |
| | Complement | ■ | | $4.3201 * 10^{-12}$ |
| | $p = 1$ | ■ | $\{0.857143, \{1, 0\}\}$ | 12.5664 |
| | $p = 2$ | | $\{\ \}$ | |

The critical points of $N_{q2,h}$ are the (complex) solutions of the equation

$$(3z + 1)^2 - h \left(3z^2 + 2z + 3\right) = 0. \tag{9.5}$$

Note that for $h = 3$ the equation has no solution, whereas for $h \neq 3$ there are two critical points, namely

$$cp_1(h) = -\frac{h + 2\sqrt{2}\sqrt{(3-h)h} - 3}{3(h-3)}, cp_2(h) = \frac{-h + 2\sqrt{2}\sqrt{(3-h)h} + 3}{3(h-3)}.$$

Figure 9.2 shows the parameter planes associated to each of the two critical-point functions $cp_i(h)$, $i = 1, 2$ of the iteration map (9.4). In fact, the parameter plane related to $cp_1(h)$ is shown in the left-side figure and the parameter plane related to $cp_2(h)$ appears in the right figure. In both cases, for each $h \in \mathbb{C}$, a point is painted in cyan if the orbit of a critical point converges to the simple root $-1$, in magenta if it converges to the double root 1 and in yellow if the iteration diverges to $\infty$. It appears in dark blue if it converges to a 2-cycle and the rest of colors are related to the convergence of different $p$-cycles with $p > 2$. Moreover, points in black mean that there is no convergence to a $p$-cycle or to any fixed point, after 2000 iterations with a tolerance of $10^{-6}$.

**Table 9.7** Small neighborhoods (non black color) at the Julia set of the relaxed Newton's method applied to $z^3 + z$ for $h \in \{1, 1.5, 2, 2.5, 3, 3.5, 4, 4.5, 5, 5.5, 6, 6.5\}$

| $h$ | Basins | Nbhs at Julia sets | $h$ | Basins | Nbhs at Julia sets |
|-----|--------|--------------------|-----|--------|--------------------|
| 1 | | | 4 | | |
| 1.5 | | | 4.5 | | |
| 2 | | | 5 | | |
| 2.5 | | | 5.5 | | |
| 3 | | | 6 | | |
| 3.5 | | | 6.5 | | |

We can see, as in the case of simple roots, that every $h$ in the parameter plane with $|h - 1| < 1$ is a good choice in terms of its numerical behavior. The orbit of the first critical point converges to the simple root whereas the orbit of the second one converges to the double root. Consequently, there are not other attracting behaviors. But outside the ball $|h - 1| < 1$, the situation is completely different. There are situations of no convergence to the roots, as we can see in Fig. 9.2 or in its magnification Fig. 9.3. So these values of $h$ are not a good choice when the objective is to reach the two roots of the polynomial. However, from a dynamical point of view, the problem is very attracting. For instance, when $|h - 2| < 2$ the orbit of the second critical point converges again to the double root, but there are values of $h$ outside the ball $|h - 1| < 1$ and inside the ball $|h - 2| < 2$ where the orbit of the first critical point goes to a $p$-cycle. In Fig. 9.3 we can see a dark blue

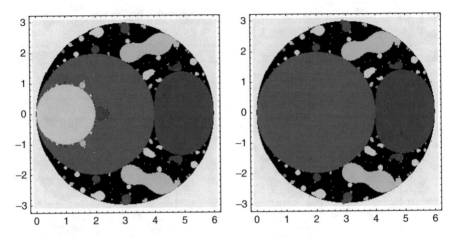

**Fig. 9.2** Parameter planes associated to each of the two critical points $cp_i(h)$, $i = 1, 2$ of the iteration map (9.4)

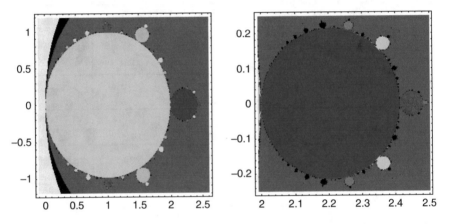

**Fig. 9.3** Magnification of the parameter plane associated to the critical point $cp_1(h)$ of the iteration map (9.4) (left-side figure in Fig. 9.2)

ball (approximately centered at 2.2 and with radius 0.2) corresponding with values of $h$ such the orbit of the first critical point converges to a 2-cycle.

Outside the ball $|h-2| < 2$ but inside the ball $|h-3| < 3$, the dynamical situation is complicated, with the presence of cycles of different length and a region in dark that must be explored in a more detailed way (we have plotted cycles just until order 4). Note that there is a kind of ellipse, having the minor edge on the real interval $(4, 6)$, where both critical points converge to a 2-cycle. So in this case, this 2-cycle presents the only attracting behavior.

In Tables 9.8, 9.9, and 9.10, we use the algorithms introduced in this paper for doing a graphical and numerical approach of the areas of the basins of attraction of the non-repelling cyclic points for the relaxed Newton's method applied to the

**Table 9.8** Spherical plots and areas of basins of non-repelling $p$-cycles for $1 \leq p \leq 2$ and $h \in \{1, 1.5, 2, 2.5, 3, 3.5\}$ for $(z-1)^2(z+1)$

| Basins | | | Non-repelling cyclic points | Areas |
|---|---|---|---|---|
| $h = 1$ | | | | |
| | Complement | ⬛ | | $4.3201 * 10^{-12}$ |
| | $p = 1$ | 🟦 🟫 | $\{0., \{-1, 1\}\}$ $\{0.5, \{1, 1\}\}$ | 2.93386 9.63251 |
| | $p = 2$ | | $\{\ \}$ | |
| $h = 1.5$ | | | | |
| | Complement | ⬛ | | $4.3201 * 10^{-12}$ |
| | $p = 1$ | 🟦 🟫 | $\{0.5, \{-1, 1\}\}$ $\{0.25, \{1, 1\}\}$ | 2.47313 10.0932 |
| | $p = 2$ | | $\{\ \}$ | |
| $h = 2$ | | | | |
| | Complement | ⬛ | | 0.00654705 |
| | $p = 1$ | 🟦 🟫 | $\{1, \{-1., 1\}\}$ $\{0, \{1, 1\}\}$ | 1.58614 10.9737 |
| | $p = 2$ | | $\{\ \}$ | |
| $h = 2.5$ | | | | |
| | Complement | ⬛ | | 0.102641 |
| | $p = 1$ | 🟫 | $\{0.25, \{1, 1\}\}$ | 12.4637 |
| | $p = 2$ | | $\{\ \}$ | |

(continued)

**Table 9.8** (continued)

| Basins | | | Non-repelling cyclic points | Areas |
|---|---|---|---|---|
| $h = 3$ | | | | |
| | Complement | | | $4.3201 * 10^{-12}$ |
| | $p = 1$ | | $\{0.5, \{1, 1\}\}$ | 12.5664 |
| | $p = 2$ | | $\{\ \}$ | |
| $h = 3.5$ | | | | |
| | Complement | | | 2.52286 |
| | $p = 1$ | | $\{0.75, \{1, 1\}\}$ | 10.0435 |
| | $p = 2$ | | $\{\ \}$ | |

cubical polynomial $q_2(z) = (z - 1)^2(z + 1)$. In Tables 9.8 and 9.9 we have chosen, according to Fig. 9.2, some significant values of the relaxing parameter $h$ in the real case: $h \in \{1, 1.5, 2, 2.5, 3, 3.5, 4, 4.5, 5, 5.5, 6, 6.5\}$. In addition, in Table 9.10 we consider the value $h = 2.02$. Of course, complex values of $h$ could be also considered. From a detailed study of the results of the simulations given in these tables, we obtain some interesting conclusions:

If the parameter $h$ is in the interval $(0, 2)$, the basins of the two roots cover the 2-sphere. In addition, when $h$ tends to 2, the spherical multiplier of the double root decreases to 0 and the spherical multiplier of the simple root increases until 1.

For $h = 2$, the double root is a super-attractor, the simple root is indifferent and there is not non-repelling $p$-cyclic points for $p > 1$. In this case, we have to take a higher maximum of the number of iterations because there is a slow convergence to the indifferent point. For this reason, one could have a less precision in the calculus of areas.

When the real parameter $h$ surpass the value $h = 2$, the dynamical situation becomes very complicated; for instance, until the value 2.4 (approximately) there are attracting 2-cycles (dark blue color in the parameter plane). In Table 9.10 we show the situation for $h = 2.02$. In this case, the cyan region obtained for $h = 2$ is divided in two basins corresponding to the two cyclic points (in green and orange). In the interval $(2.4, 2.5)$ (approximately) we can find more non-repelling $p$-cyclic points for $p > 2$. Next, for $h \in (2.5, 4)$ (the left value is again given approximately), the situation becomes easier and the double root is the only attracting point.

**Table 9.9** Spherical plots and areas of basins of non-repelling *p*-cycles for $1 \leq p \leq 2$ and $h \in \{4, 4.5, 5, 5.56, 6.5\}$ for $(z-1)^2(z+1)$

| Basins | | | Non-repelling cyclic points | Areas |
|---|---|---|---|---|
| $h = 4$ | | | | |
| | Complement | ■ | | 1.53814 |
| | $p = 1$ | | $\{1, \{1, 1\}\}$ | 11.0382 |
| | $p = 2$ | | $\{\ \}$ | |
| $h = 4.5$ | | | | |
| | Complement | ■ | | $4.3201 * 10^{-12}$ |
| | $p = 1$ | | $\{\ \}$ | |
| | $p = 2$ | | $\{0.375, \{1, 0.315789 - 0.611524 \cdot i\}\}$ | 6.28319 |
| | | | $\{0.375, \{1, 0.315789 + 0.611524 \cdot i\}\}$ | 6.28319 |
| $h = 5$ | | | | |
| | Complement | ■ | | $4.3201 * 10^{-12}$ |
| | $p = 1$ | | $\{\ \}$ | |
| | $p = 2$ | | $\{0.25, \{1, 0.125 - 0.484123 \cdot i\}\}$ | 6.28319 |
| | | | $\{0.25, \{1, 0.125 + 0.484123 \cdot i\}\}$ | 6.28319 |
| $h = 5.5$ | | | | |
| | Complement | ■ | | $4.3201 * 10^{-12}$ |
| | $p = 1$ | | $\{\ \}$ | |
| | $p = 2$ | | $\{0.475, \{1, 0.0425532 - 0.323376 \cdot i\}\}$ | 6.28319 |
| | | | $\{0.475, \{1, 0.0425532 + 0.323376 \cdot i\}\}$ | 6.28319 |

(continued)

**Table 9.9** (continued)

| Basins | | | Non-repelling cyclic points | Areas |
|---|---|---|---|---|
| $h = 6$ | | | | |
| | Complement | ■ | | 0.0127437 |
| | $p = 1$ | ■ | $\{1, \{1, 0\}\}$ | 12.5536 |
| | $p = 2$ | | $\{\ \}$ | |
| $h = 6.5$ | | | | |
| | Complement | ■ | | 0.00890963 |
| | $p = 1$ | ■ | $\{0.857143, \{1, 0\}\}$ | 12.5575 |
| | $p = 2$ | | $\{\ \}$ | |

If the parameter $h$ is in the interval $(2, 4)$, the spherical multiplier of the double root increases its value from 0 to 1. It is also interesting to note that for $h = 3$ the point at infinity is not a fixed point, and $z = 1$ is the only attracting fixed point ($z = -1$ is repelling).

For $h = 4$, the double root is indifferent and the simple root is repelling. In this case, the orbits of the two critical points converge to the double root $z = 1$ and then there is not other attracting behavior. All the 2-sphere appears with a cyan color which in this case correspond to the double root. We remark that for $h = 4$ the convergence is very slow and we have had to increment the number of iterations. We have also modified the tolerances $10^{-c_1}$, $10^{-c_2}$ taking $c_1$ higher than $c_2$ and a small resolution has been used. It is interesting to compare the left side Fig. 9.6 and the plot given in Table 9.9 for $h = 4$. Note that the region having a higher spherical multiplier (left side Fig. 9.6) corresponds with the black region (Table 9.9 for $h = 4$) of points that will need a higher number of iterations to change a black spherical cubic pixel into a cyan pixel.

If the parameter $h$ is in the interval $(4, 6)$, the basins of two 2-cyclic non-repelling points cover the 2-sphere.

For $h = 6$, the point at infinity is indifferent and its basin cover all the sphere, but a higher number of iterations are needed. For $h > 6$, the point at infinity is attracting and its basin cover all the sphere. Therefore, there are not non-repelling $p$-cyclic points for $p > 1$.

**Table 9.10** Spherical plots and areas of basins of non-repelling *p*-cycles for $1 \leq p \leq 2$ and $h = 2.02$ for $(z - 1)^2(z + 1)$

| $h = 2.02$ | Basins | | |
|---|---|---|---|

| | Complement | $p = 1$ | $p = 2$ |
|---|---|---|---|
| Colors | ■ | | |
| Non-repelling cyclic points | | $sm = 0.01$ $\{1., 1\}$ | $sm = 0.919588$ $\{1, -0.858317\}, \{-0.875746, 1\}$ |
| Areas | 1.52926 | 11.0371 | 0.949088, 0.56299 |

The cyan region is the basin of the double root $z = 1$. Note that there are regions in green, brown and black colors and the areas of the basins of the two 2-cyclic points is non zero

Note that a graphical study showing the Julia sets related to the relaxed Newton's method applied to the cubical polynomial $q_2(z) = (z - 1)^2(z + 1)$ could be also done.

## 9.4.2 Initial Efficiency and Areas of Basins of the Relaxed Newton's Method

For a fixed polynomial $q(z)$, one of our objectives is the analysis of the in-equation of initial efficiency

$$P_{1(\neq\infty)}^{\leq 1}(N_{q,h}) \leq P_1^{\leq 1}(N_{q,h}) \leq P_1^{\leq 1}(N_{q,h}) + P_2^{\leq 1}(N_{q,h}) \leq \cdots \leq \sum_{p=1}^{\infty} P_p^{\leq 1}(N_{q,h}) \leq 1$$

**Table 9.11** Obstruction to the initial efficiency associated to non-repelling $p$-cycles of the relaxed Newton's method applied to $(z^3 + z)$ for $1 \leq p \leq 4$

| $z^3 + z$ | $P^{\leq 1}_{1(\neq \infty)}$ Initial efficiency | $P^{\leq 1}_1$ | $P^{\leq 1}_2$ | $P^{\leq 1}_3$ | $P^{\leq 1}_4$ | $\sum_{p=1}^4 P^{\leq 1}_p$ |
|---|---|---|---|---|---|---|
| $h = 1$ | 1 | 1 | 0 | 0 | 0 | 1 |
| $h = 1.5$ | 1 | 1 | 0 | 0 | 0 | 1 |
| $h = 2$ | 0.998536 | 0.998536 | 0 | 0 | 0 | 0.998536 |
| $h = 2.5$ | 0 | 0 | 1 | 0 | 0 | 1 |
| $h = 3$ | 0 | 0 | 1 | 0 | 0 | 1 |
| $h = 3.5$ | 0 | 0 | 1 | 0 | 0 | 1 |
| $h = 4$ | 0 | 0 | 1 | 0 | 0 | 1 |
| $h = 4.5$ | 0 | 0 | 1 | 0 | 0 | 1 |
| $h = 5$ | 0 | 0 | 1 | 0 | 0 | 1 |
| $h = 5.5$ | 0 | 0 | 1 | 0 | 0 | 1 |
| $h = 6$ | 0 | 0.997742 | 0 | 0 | 0 | 0.997742 |
| $h = 6.5$ | 0 | 1 | 0 | 0 | 0 | 1 |
| $h = 7$ | 0 | 1 | 0 | 0 | 0 | 1 |

**Table 9.12** Obstruction to the initial efficiency associated to non-repelling $p$-cycles of the relaxed Newton's method applied to $(z - 1)^2(z + 1)$ for $1 \leq p \leq 4$

| $(z - 1)^2(z + 1)$ | $P^{\leq 1}_{1(\neq \infty)}$ Init. ef. | $P^{\leq 1}_1$ | $P^{\leq 1}_2$ | $P^{\leq 1}_3$ | $P^{\leq 1}_4$ | $\sum_{p=1}^4 P^{\leq 1}_p$ |
|---|---|---|---|---|---|---|
| $h = 1$ | 1 | 1 | 0 | 0 | 0 | 1 |
| $h = 1.5$ | 1 | 1 | 0 | 0 | 0 | 1 |
| $h = 2$ | 0.999479 | 0.999479 | 0 | 0 | 0 | 0.999479 |
| $h = 2.02$ | 0.872566 | 0.872566 | 0.120327 | 0 | 0 | 0.998632 |
| $h = 2.5$ | 0.991832 | 0.991832 | 0 | 0 | 0 | 0.991832 |
| $h = 3$ | 1 | 1 | 0 | 0 | 0 | 1 |
| $h = 3.5$ | 0.799237 | 0.799237 | 0 | 0 | 0 | 0.799237 |
| $h = 4$ | 0.878395 | 0.878395 | 0 | 0 | 0 | 0.878395 |
| $h = 4.5$ | 0 | 0 | 1 | 0 | 0 | 1 |
| $h = 5$ | 0 | 0 | 1 | 0 | 0 | 1 |
| $h = 5.5$ | 0 | 0 | 1 | 0 | 0 | 1 |
| $h = 6$ | 0 | 0.998986 | 0 | 0 | 0 | 0.998986 |
| $h = 6.5$ | 0 | 1 | 0 | 0 | 0 | 1 |
| $h = 7$ | 0 | 1 | 0 | 0 | 0 | 1 |

depending on the real parameter $h$.

In this work, we analyze two polynomials $q_1(z) = z^3 + z$ and $q_2(z) = (z - 1)^2(z + 1)$ and we apply the implementations given in Sect. 9.3.2 to compute areas and probabilities.

Approximations to the initial efficiency for these polynomials depending on the real values of the parameter $h$ are given in Tables 9.11 and 9.12. In the following we describe some properties of the initial efficiency in term of the relaxed real parameter $h$. In general, the calculus of areas, probabilities, initial efficiency have less accuracy for values of $h$ such that the induced rational maps have indifferent $p$-cyclic points. In these cases, we have incremented the number of iterations in order to have better results.

### 9.4.2.1   A Study of Initial Efficiency and Areas Associated to $z^3 + z$

For this representative polynomial we have analyzed the situation for some real values of the relaxing parameter $h$, actually for $h \in (0, 7)$, to obtain the following conclusions:

(i) When the parameter $h$ is in $(0, 2)$, the sum of probabilities of the three roots is equal to 1 and the initial efficiency of the relaxed Newton's method is equal to 1.

(ii) When $h = 2$, an approximation of the initial efficiency of the relaxed Newton's method is equal to 0.998536. In this case, we have incremented the maximum of the number of iterations and the result suggests that using a large number of iterations and adequate tolerances the initial efficiency tends to 1.

(iii) When the parameter $h$ is in $(2, 6)$, the sum of probabilities of the fixed points is equal to 0 and the initial efficiency of the relaxed Newton's method is equal to 0. The obstruction of this initial efficiency is due to the basins of two non-repelling 2-cyclic points. This fact is also confirmed by the plot of a neighborhood of the Julia set given on the left in Fig. 9.4 and by the plot of repelling point given in Table 9.3. The graphic plots given in Tables 9.5 and 9.6 also confirm this result.

(iv) When $h = 6$, the initial efficiency of the relaxed Newton's method is equal to 0. An approximation of probability associated to the point at infinity is 0.997742 and taking a larger number of iterations this probability tends to 1. We have a plot of a neighborhood of the corresponding Julia set (in the right in Fig. 9.4).

(v) When the parameter $h$ belongs to the interval $(6, 7)$, the probability at the point at infinity for the relaxed Newton's method is equal to 1, but the initial efficiency of the relaxed Newton's method is equal to 0. The obstruction is due to the basin of the point at infinity.

**Fig. 9.4** Neighborhoods (*non black color*) of the Julia set of the relaxed Newton's method for $h = 3$ (in the *left*) and for $h = 6$ (in the *right*) applied to $z^3 + z$

A quantification of the areas (probabilities) for each root and for each non-repelling $p$-cyclic point is detailed in Tables 9.5 and 9.6.

### 9.4.2.2  A Study of Initial Efficiency and Areas Associated to $(z-1)^2(z+1)$

For this polynomial with a double root and a simple root and for the same range of real values of the relaxing parameter $h$, that is $h \in (0, 7)$, we have:

(i) When the parameter is in $(0, 2)$, the sum of probabilities of the two roots is equal to 1 and the initial efficiency of the relaxed Newton's method is equal to 1. A quantification of the areas (probabilities) for each root is obtained in Tables 9.8, 9.9, and 9.10. The probability of convergence to the double root is greater than the probability of convergence to the simple root. When $h$ tends to 2, we also note that the area of the basin of the double root increases and the area of the simple root decreases.

(ii) When $h = 2$, the sum of probabilities of the basins of the double root and the simple root is equal to 0.999479 and the initial efficiency of the relaxed Newton's method is equal to 0.999479. We also have a plot of a neighborhood of the corresponding Julia set, see Fig. 9.5. In Table 9.8 we note that for $h = 2$ the cyan color corresponds to the indifferent simple root. Since for this value of $h$ we have a very slow convergence we have increased the number of iterations in the corresponding calculus of graphics, areas and initial efficiency.

(iii) When the real value of the parameter $h$ is in $(2, 4)$, the probability corresponding to the simple root is zero. In the sub-interval $(2, 2.5)$, the initial efficiency the relaxed Newton's method is less than 1. According to Fig. 9.2 the dynamics for $h$ in the subinterval $(2, 2.5)$ is more complicated because there are points which converges to 2-cyclic points. In Fig. 9.5, we can see that in both cases ($h = 2$ and $h = 2.02$) the neighborhoods of the corresponding Julia set are similar. We also note that for $h = 2.02$ the graphics given in Fig. 9.2 and the calculus of the area given in Table 9.10 confirms that the basin of the indifferent simple root for $h = 2$ corresponds to the basin of an attracting 2-cycle for $h = 2.02$.

**Fig. 9.5** Two sights of two neighborhood (*non black color*) of the Julia sets of the relaxed Newton's method for $h = 2$ and for $h = 2.02$ applied to $(z-1)^2(z+1)$

**Fig. 9.6** Neighborhoods (*non black color*) of the Julia set of the relaxed Newton's method for $h = 4$ (*left*) and for $h = 6$ (*right*) applied to $(z - 1)^2(z + 1)$

(iv) When $h = 4$, the actual approximation of the initial efficiency of the relaxed Newton's method is equal to 0.878395. This is related to the plot of a neighborhood of the corresponding Julia set given in the left side Fig. 9.6. We think that taking a larger number of iterations and a better election of tolerances for $h = 4$ the initial efficiency will tend to 1.

(v) When the parameter $h$ is in $(4, 6)$, the initial efficiency of the relaxed Newton's method is equal to 0. The obstruction is due to basins of 2-cyclic points.

(vi) When $h = 6$, the initial efficiency of the relaxed Newton's method is equal to 0. We also have a plot of a neighborhood of the corresponding Julia set, see right side in Fig. 9.6. In these pictures we note that there is a black part that corresponds to the basin of the point at infinity.

(vii) When the parameter $h > 6$, the initial efficiency of the relaxed Newton's method is equal to 0. The obstruction is due to the basin of the point at infinity.

## 9.5   Conclusion

The implementations presented in this work and the results given by the computer simulations allow us to study the existence of non-repelling $p$-cyclic points and to give the measure (area or probability) of the different basins. Moreover, with these implementations we can draw neighborhoods of the corresponding Julia sets.

The graphical approach and the calculus of measures of basins of non-repelling cyclic points permit us to find out some analogies and differences when the relaxed Newton's method (in the case of real values of the relaxing parameter) is applied either to a cubical polynomial with three simple roots or to a polynomial with a double root and a simple root. The following remarks correspond with simulations effectuated with a chosen maximum number of iterations except when we have been working with singular values of the real relaxing parameter associated to indifferent $p$-cyclic points. For these singular (bifurcation) values of the real relaxing parameter we have incremented the number of iterations in order to obtain a better graphical approximation.

In general, when we are working with non-repelling cyclic points whose spherical parameter is less than 1, the values obtained for the measures of basins and for the initial efficiency are quite reliable, but when some non-repelling cyclic point is indifferent (or its spherical parameter is close to 1) we have less accuracy on the results.

For the two chosen representative polynomials, we can highlight the following analogies:

1. When the real parameter $h$ increases, there is a first interval such that the basins of the roots cover almost all 2-sphere. This finishes with a singular value ($h = 2$) of the relaxed real parameter where some indifferent non-repelling fixed points appear.
2. After this singular value, it starts a new sub-interval where there is a non-repelling 2-cycle and this interval finishes with a singular value of the relaxing real parameter ($h = 6$), where there is a unique non-repelling fixed point (the point at infinity) which is indifferent.
3. After this second singular value of the parameter, the point at infinity becomes an attracting fixed point and its basin covers all the sphere.
4. In both cases the initial efficiency is 1 for $h < 2$ and 0 for $h > 4$ (due to the presence of attracting 2-cycles for $2 < h < 4$ and to the attraction character of the point at infinity for $h > 6$).

Amongst the differences, we have:

1. In the subinterval $(2,6)$ there are important differences between the behavior of the two polynomials. At the beginning the polynomial with three simple roots has only attracting 2-cycles while the polynomial with a double root and a simple root in some cases it combines an attracting 2-cycle and an attracting fixed point. In the last part of the subinterval there is only an attracting 2-cycle.
2. In the subinterval $(2,4)$ the initial efficiency associated to the polynomial with three simple roots is zero. However, for the polynomial with a double and simple root, the initial efficiency is greater that zero and it is less than 1 in $(2,4)$ except for $h = 3$ which it is equal to 1 and it coincides with the probability associated to the double root.

It would be interesting to develop a deep study for the cases when the sequence of postcritical points converges to an indifferent $p$-cyclic point. The numerical simulations of this work show a nice connection between the bifurcation values of the real relaxing parameter and these indifferent $p$-cyclic points and a further work will be necessary to know if these indifferent points are or not in their associated Julia sets.

**Acknowledgements** This scientific work has been supported by the project PROFAI13/15 of the Universidad of La Rioja and the project MTM2011-28636-C02-01.

# References

1. Amat, S., Busquier, S., Plaza, S.: Review of some iterative root-finding methods from a dynamical point of view. Sci. Ser. A Math. Sci. **10**, 3–35 (2004)
2. Beardon, A.F.: Iteration of Rational Functions. Springer, New York (2000)
3. Campbell, J.T., Collins, J.T.: Specifying attracting cycles for Newton maps of polynomials. J. Differ. Equ. Appl. **19**, 1361–1379 (2013)
4. Chicharro, F., Cordero, A., Torregrosa, J.R.: Drawing dynamical and parameters planes of iterative families and methods. Sci. World J. **2013**, Article ID 780153 (2013)
5. Curry, J.H., Garnett, L., Sullivan, D.: On the iteration of a rational function: computer experiments with Newton's method. Commun. Math. Phys. **91**, 267–277 (1983)
6. García-Calcines, J.M., Hernández, L.J., Rivas, M.T.: Limit and end functors of dynamical systems via exterior spaces. Bull. Belg. Math. Soc. Simon Stevin **20**, 937–959 (2013)
7. García-Calcines, J.M., Hernández, L.J., Rivas, M.T.: A completion construction for continuous dynamical systems. Topol. Methods Nonlinear Anal. **44**, 497–526 (2014)
8. Gutiérrez, J.M., Hernández, L.J., Marañón, M., Rivas, M.T.: Influence of the multiplicity of the roots on the basins of attraction of Newton's method. Numer. Algorithm **66**(3), 431–455 (2014)
9. Hernández, L.J., Marañón, M., Rivas, M.T.: Plotting basins of end points of rational maps with Sage. Tbil. Math. J. **5**(2), 71–99 (2012)
10. Kriete, H.: Holomorphic motions in the parameter space for relaxed Newton's method. Kodai Math. J. **25**, 89–107 (2002)
11. Magreñán, Á.A.: Estudio de la dinámica del método de Newton amortiguado. Ph.D. thesis, Serv. de Publ. Univ. La Rioja (2013)
12. Magreñán, Á.A.: Different anomalies in a Jarratt family of iterative root-finding methods. Appl. Math. Comput. **233**, 29–38 (2014)
13. Plaza, S., Romero, N.: Attracting cycles for relaxed Newton's method. J. Comput. Appl. Math. **235**, 3238–3244 (2011)
14. Roberts, G.E., Horgan-Kobelski, J.: Newton's versus Halley's method: a dynamical systems approach. Int. J. Bifurcat. Chaos Appl. Sci. Eng. **14**(10), 3459–3475 (2004)
15. Varona, J.L.: Graphic and numerical comparison between iterative methods. Math. Intell. **24**(1), 37–46 (2002)

# Chapter 10
# On Convergence and Efficiency in the Resolution of Systems of Nonlinear Equations from a Local Analysis

Miquel Grau-Sánchez and Miquel Noguera

**Abstract** The aim of this chapter is to provide an overview of theoretical results and numerical tools in some iterative schemes to approximate solutions of nonlinear equations. Namely, we examine the concept of iterative methods and their local order of convergence, numerical parameters that allow us to assess the order, and the development of inverse operators (derivative and divided differences). We also provide a detailed study of a new computational technique to analyze efficiency. Finally, we end the chapter with a consideration of adaptive arithmetic to accelerate computations.

## 10.1 Iteration Functions

Many problems in computational sciences and other disciplines can be formulated by means of an equation like the following

$$\mathscr{F}(x) = 0, \tag{10.1}$$

where $\mathscr{F} : D \subset \mathbb{X} \longrightarrow \mathbb{Y}$ is a continuous operator defined on a nonempty convex subset $D$ of a Banach space $\mathbb{X}$ with values in a Banach space $\mathbb{Y}$. We face the problem of approximating a local unique solution $\alpha \in \mathbb{X}$ of Eq. (10.1). Since the exact solution of this equation can rarely be found, then we need to use iterative techniques

M. Grau-Sánchez (✉)
Department of Mathematics, Technical University of Catalonia · BarcelonaTech,
Campus Nord, Edifici Omega, 08034 Barcelona, Spain
e-mail: miquel.grau@upc.edu

M. Noguera
Department of Mathematics, Technical University of Catalonia · BarcelonaTech,
Edifici TR5 (ESEIAAT), 08222 Terrassa, Spain
e-mail: miquel.noguera@upc.edu

© Springer International Publishing Switzerland 2016
S. Amat, S. Busquier (eds.), *Advances in Iterative Methods for Nonlinear Equations*, SEMA SIMAI Springer Series 10, DOI 10.1007/978-3-319-39228-8_10

to approximate it to the desired precision from one or several initial approximations. This procedure generates a sequence of approximations of the solution.

Traub [41] includes a classification of iteration functions, according to the information that is required to carry them out. We build up a sequence $\{x_n\}_{n\geq1}$ in a Banach space $\mathbb{X}$ using the initial conditions $x_{-k}, \ldots, x_{-1}, x_0, 0 \leq k \leq j-1$. Traub's classification of iteration functions is the following.

**Type I.**  Term $x_{n+1}$ is obtained using only the information at $x_n$ and no other information. That is, given $x_0 \in D$ we have

$$x_{n+1} = \Phi(x_n), \quad n \geq 0.  \tag{10.2}$$

The function $\Phi$ is called a *one-point iteration function* and Eq. (10.2) is called the one-point iterative method without memory.

**Type II.**   Term $x_{n+1}$ is obtained using the information at $x_n$ and previous information at $x_{n-1}, \ldots, x_{n-j} \in D$. Namely,

$$x_{n+1} = \Phi(x_n ; x_{n-1}, \ldots, x_{n-j}), \quad n \geq 0, \quad j \geq 1.  \tag{10.3}$$

Function $\Phi$ is called a *one-point iteration function with memory* and Eq. (10.3) is called a one-point iterative method with memory ($j$ points). The semicolon in (10.3) is written to distinguish the information provided by the new data from the information that was previously used.

**Type III.**   Term $x_{n+1}$ is determined using new information at $x_n$ and previous information at $\varphi_1 = \varphi_1(x_n)$, $\varphi_2 = \varphi_2(\varphi_1, x_n), \ldots, \varphi_r = \varphi_r(\varphi_{r-1}, \ldots, \varphi_1, x_n) \in D$, $r \geq 1$. That is,

$$x_{n+1} = \Phi(x_n, \varphi_1, \ldots, \varphi_r), \quad n \geq 0, \quad r \geq 1.  \tag{10.4}$$

Here, function $\Phi$ is called a *multipoint iteration function* without memory and Eq. (10.4) is called a multipoint iterative method without memory ($r$ steps).

**Type IV.**   Term $x_{n+1}$ is obtained from new information at $x_n$ and previous information at

$$\varphi_1 = \varphi_1(x_n ; x_{n-1}, \ldots, x_{n-j}),$$

$$\vdots$$

$$\varphi_r = \varphi_r(x_n, \varphi_1, \ldots, \varphi_{r-1} ; x_{n-1}, \ldots, x_{n-j}).$$

Namely,

$$x_{n+1} = \Phi(x_n, \varphi_1, \ldots, \varphi_r; x_{n-1}, \ldots, x_{n-j}), \quad n \geq 0, \quad r \geq 1, \quad j \geq 1.  \tag{10.5}$$

Function $\Phi$ is called a *multipoint iteration function with memory* and (10.5) is called a multipoint iteration method with memory ($r$ steps and $j$ points).

## 10.1.1   One-Dimensional Case

In particular, when the Banach spaces $\mathbb{X} = \mathbb{Y} = \mathbf{R}$, we have to solve the most simple, classical nonlinear problem. Namely, let $f : I \subseteq \mathbf{R} \to \mathbf{R}$ be a nonlinear function. We have to approximate a simple root $\alpha$ of the equation

$$f(x) = 0, \tag{10.6}$$

where $I$ is a neighborhood of $\alpha$. An approximation of $\alpha$ is usually obtained by means of an iterative function of type **I, II, III** or **IV**, defined in (10.2), (10.3), (10.4) or (10.5) whereby a sequence $\{x_n\}_{n \geq 1}$ is considered that converging converges to $\alpha$.

**Definition 1** The sequence $\{x_n\}$ is said to converge to $\alpha$ with order of convergence $\rho \in \mathbf{R}$, $\rho \geq 1$, if there exists a positive real constant $C \neq 0$ and $C \neq \infty$ such that

$$\lim_{n \to \infty} \frac{|e_{n+1}|}{|e_n|^\rho} = C, \tag{10.7}$$

where $e_n = x_n - \alpha$ is the *error* in the $n$th iterate, and the constant $C$ is called the *asymptotic error constant* (see [41]).

The local order of convergence of an iterative method in a neighborhood of a root is the order of its corresponding sequence generated by the iterative function and the corresponding initial approximations. For iterative methods without memory, the local order is a positive integer. The convergence is said to be linear if $\rho = 1$, quadratic if $\rho = 2$, cubic if $\rho = 3$, and, in general, superlinear if $\rho > 1$, superquadratic if $\rho > 2$, and so on.

The one-point iterative method without memory (10.7) can be written as

$$e_{n+1} = C e_n^\rho + O(e_n^{\rho+1}), \ n \geq n_0. \tag{10.8}$$

The expression (10.8) is called the *error difference equation* for the one-point iterative method. Note that the higher order terms in (10.8) are powers of $\rho + 1$.

For the one-point iterative method without memory, an approximation of the number of correct decimal places in the $n$th iterate, $d_n$, is given by

$$d_n = -\log_{10} |x_n - \alpha|. \tag{10.9}$$

From (10.8), for $n$ large enough we have $e_{n+1} \approx C e_n^\rho$, which using logarithms yields

$$d_{n+1} \approx -\log_{10} C + \rho \cdot d_n, \tag{10.10}$$

from which follows

$$d_{n+1} \approx \rho \cdot d_n. \tag{10.11}$$

This means that, in each iteration, the number of correct decimal places is approximately the number of correct decimals in the previous iteration multiplied by the local error.

This is in agreement with Wall's definition [42]. That is, the local order of convergence of a one-point iteration function indicates the rate of convergence of the iteration method. Then, Wall defines the order $\rho$ of the iteration formula by

$$\rho = \lim_{n\to\infty} \frac{\log|e_{n+1}|}{\log|e_n|} = \lim_{n\to\infty} \frac{d_{n+1}}{d_n}. \tag{10.12}$$

This expression will be used later on when we define some parameters employed in the computation of the local order of convergence of an iterative method.

For the one-point iterative method with memory (10.3) the error difference equation is

$$e_{n+1} = Ce_n^{a_1} e_{n-1}^{a_2} \cdots e_{n-j+1}^{a_j} + o(e_n^{a_1} e_{n-1}^{a_2} \cdots e_{n-j+1}^{a_j}), \tag{10.13}$$

where $a_k$ are nonnegative integers for $1 \le k \le j$ and $o(e_n^{a_1} e_{n-1}^{a_2} \cdots e_{n-j+1}^{a_j})$ represents terms with high order than the term $e_n^{a_1} e_{n-1}^{a_2} \cdots e_{n-j+1}^{a_j}$. In this case, the order of convergence $\rho$ is the unique real positive root of the indicial polynomial (see [27, 28, 40, 41]) of the error difference equation (10.13) given by

$$p_j(t) = t^j - a_1 t^{j-1} - \cdots - a_{j-1}t - a_j. \tag{10.14}$$

Notice that $p_j(t)$ in (10.14) has a unique real positive root $\rho$ on account of Descartes's rule of signs. Moreover, we can write $e_{n+1} = Ce_n^\rho + o(e_n^\rho)$.

### 10.1.2 Multidimensional Case

When the Banach spaces $\mathbb{X} = \mathbb{Y} = \mathbf{R}^m$ we have to solve a system of nonlinear equations. Namely, let $F : D \subset \mathbf{R}^m \longrightarrow \mathbf{R}^m$ be a nonlinear function and $F \equiv (F_1, F_2, \ldots, F_m)$ with $F_i : D \subseteq \mathbf{R}^m \to \mathbf{R}$, $i = 1, 2, \ldots, m$, where $D$ is an open convex domain in $\mathbf{R}^m$, so that we have to approximate a solution $\alpha \in D$ of the equation $F(x) = 0$.

Starting with a given set of initial approximations of the root $\alpha$, the iteration function $\Phi : D \longrightarrow D$ of type **I**, **II**, **III** or **IV** is defined by (10.2), (10.3), (10.4) or (10.5), whereby a sequence $\{x_n\}_{n\le 1}$ XXX is considered to converge to $\alpha$.

**Definition 2** The sequence $\{x_n\}$ converges to $\alpha$ with *an order of convergence* of at least $\rho \in \mathbf{R}$, $\rho \ge 1$, if there is a positive real constant $0 < C < \infty$ such that

$$\|e_{n+1}\| \le C\|e_n\|^\rho, \tag{10.15}$$

where $e_n = x_n - \alpha$ is the error in the $n$th iterate, and the constant $C$ is called the asymptotic error constant (see [41]). Here the norm used is the maximum norm.

The local order of convergence of an iterative method in a neighborhood of a root is the order of the corresponding sequence generated (in $\mathbf{R}^m$) by the iterative function $\Phi$ and the corresponding initial approximations.

Without using norms, a definition of the local order of convergence for the one-step iterative method without memory can be considered as follows. The local order of convergence is $\rho \in \mathbb{N}$ if there is an $\rho$–linear function $C \in \mathscr{L}\left(\mathbf{R}^m \times \overset{\rho}{\cdots} \times \mathbf{R}^m, \mathbf{R}^m\right) \equiv \mathscr{L}_\rho\left(\mathbf{R}^m, \mathbf{R}^m\right)$ such that

$$e_{n+1} = C e_n^\rho + O\left(e_n^{\rho+1}\right), \ n \geq n_0 \tag{10.16}$$

where $e_n^\rho$ is $(e_n, \overset{\rho}{\cdots}, e_n) \in \mathbf{R}^m \times \overset{\rho}{\cdots} \times \mathbf{R}^m$. When $0 < C < \infty$ exists for some $\rho \in [1, \infty)$ from (10.15), then $\rho$ is the R-order of convergence of the iterative method defined by Ortega and Rheinboldt [27]. Moreover, the local order $\rho$ of (10.16) is also the R-order of convergence of the method.

For the one-point iterative method with memory, the error difference equation can be expressed by

$$e_{n+1} = C e_n^{a_1} e_{n-1}^{a_2} \cdots e_{n-j+1}^{a_j} + o(e_n^{a_1} e_{n-1}^{a_2} \cdots e_{n-j+1}^{a_j}), \tag{10.17}$$

where $C \in \mathscr{L}_{a_1 + \cdots + a_j}\left(\mathbf{R}^m, \mathbf{R}^m\right)$ and $a_k$ are nonnegative integers for $1 \leq k \leq j$.

As in the one-dimensional case, we can write the equation associated with (10.17), $p_j(t) = t^j - a_1 t^{j-1} - \cdots - a_{j-1} t - a_j = 0$. If we apply Descartes's rule to the previous polynomial, there is a unique real positive root $\rho$ that coincides with the local order of convergence (see [27, 40]).

## 10.2  Computational Estimations of the Order

After testing the new iterative methods, we need to check the theoretical local order of convergence. The parameter Computational Order of Convergence (COC) is used in most studies published after Weerakoon and Fernando [43]. This parameter can only be used when the root $\alpha$ is known. To overcome this problem, the following parameters have been introduced:

Approximated Computational Order of Convergence (ACOC) by Hueso et al. [22],

Extrapolated Computational Order of Convergence (ECOC) by Grau et al. [12], and Pétkovic Computational Order of Convergence (PCOC) by Petković [29].

The paper by Grau et al. [14] examines the relations between the parameters COC, ACOC and ECOC and the theoretical convergence order of iterative methods without memory.

Subsequently, using Wall's definition of the order (10.12), four new parameters (CLOC, ACLOC, ECLOC and PCLOC) were given in [19] to check this order. Note that the last three parameters do not require knowledge of the root.

Generalizations of COC, ACOC and ECOC from the one-dimensional case to the multi-dimensional one can be found in [15]. They will be presented in detail in the sequel.

### 10.2.1 Computational Order of Convergence and Its Variants

Let $\{x_n\}_{n\geq 1}$ be a sequence of real numbers converging to $\alpha$. It is obtained by carrying out an iteration function in $\mathbf{R}$, starting with an initial approximation $x_0$, or $x_{-j+1}, \ldots x_0$, of the root $\alpha$ of (10.6). Let $\{e_n\}_{n\geq 1}$ be the sequence of errors given by $e_n = x_n - \alpha$. If functions (10.2)–(10.5) have local order of convergence $\rho$, then from (10.10) we have

$$\log |e_n| \approx \rho \cdot \log |e_{n-1}| + \log C,$$

$$\log |e_{n-1}| \approx \rho \cdot \log |e_{n-2}| + \log C.$$

By subtracting the second expression from the first one we get

$$\rho \approx \frac{\log |e_n \, / \, e_{n-1}|}{\log |e_{n-1} \, / \, e_{n-2}|} . \tag{10.18}$$

This expression is the same as that described in papers by Weerakoon and Fernando [43], and Jay [23].

**Definition 3** The values $\bar{\rho}_n$ (COC), $\widehat{\rho}_n$ (ACOC), $\widetilde{\rho}_n$ (ECOC) and $\check{\rho}_n$ (PCOC) are defined by

$$\bar{\rho}_n = \frac{\log |e_n \, / e_{n-1}|}{\log |e_{n-1} \, / e_{n-2}|}, \quad e_n = x_n - \alpha, \quad n \geq 3, \tag{10.19}$$

$$\widehat{\rho}_n = \frac{\log |\hat{e}_n \, / \hat{e}_{n-1}|}{\log |\hat{e}_{n-1} \, / \hat{e}_{n-2}|}, \quad \hat{e}_n = x_n - x_{n-1}, \quad n \geq 4, \tag{10.20}$$

$$\widetilde{\rho}_n = \frac{\log |\tilde{e}_n \, / \, \tilde{e}_{n-1}|}{\log |\tilde{e}_{n-1} \, / \tilde{e}_{n-2}|}, \quad \tilde{e}_n = x_n - \tilde{\alpha}_n, \quad n \geq 5, \tag{10.21}$$

$$\widetilde{\alpha}_n = x_n - \frac{(\delta x_{n-1})^2}{\delta^2 x_{n-2}}, \quad \delta x_n = x_{n+1} - x_n,$$

$$\check{\rho}_n = \frac{\log |\check{e}_n|}{\log |\check{e}_{n-1}|}, \quad \check{e}_n = \frac{f(x_n)}{f(x_{n-1})}, \quad n \geq 2. \tag{10.22}$$

Note that the first variant of COC, ACOC, involves the parameter $\hat{e}_n = x_n - x_{n-1}$ and the second variant ECOC is obtained using Aitken's extrapolation procedure [1]. That is, from the iterates $x_{n-2}, x_{n-1}, x_n$ we can obtain the approximation $\widetilde{\alpha}_n$ of the root $\alpha$.

Sequences $\{\widetilde{\rho}_n\}_{n\geq5}$ and $\{\hat{\rho}_n\}_{n\geq4}$ converge to $\rho$. The details of the preceding claim can be found in [14], where the relations between the error $e_n$ and $\check{e}_n$ and $\hat{e}_n$ are also described.

From a computational viewpoint, ACOC has the least computational cost, followed by PCOC. Inspired by (10.12) given in [42], in our study [19] we present four new parameters that will be described in the following section.

## 10.2.2   New Parameters to Compute the Local Order of Convergence

**A. Definitions**   Given the sequence $\{x_n\}_{n\geq1}$ of iterates converging to $\alpha$ with order $\rho$, we consider the sequences of errors $e_n = x_n - \alpha$ and error parameters $\hat{e}_n = x_n - x_{n-1}$, $\widetilde{e}_n = x_n - \widetilde{\alpha}_n$ and $\check{e}_n = \frac{f(x_n)}{f(x_{n-1})}$ defined previously in (10.20), (10.21), (10.22). From the preceding, we define the following sequences $\{\overline{\lambda}_n\}_{n\geq2}$ (CLOC), $\{\hat{\lambda}_n\}_{n\geq3}$ (ACLOC), $\{\widetilde{\lambda}_n\}_{n\geq4}$ (ECLOC) and $\{\check{\lambda}_n\}_{n\geq2}$ (PCLOC):

$$\overline{\lambda}_n = \frac{\log |e_n|}{\log |e_{n-1}|}, \quad \hat{\lambda}_n = \frac{\log |\hat{e}_n|}{\log |\hat{e}_{n-1}|}, \quad \widetilde{\lambda}_n = \frac{\log |\widetilde{e}_n|}{\log |\widetilde{e}_{n-1}|}, \quad \check{\lambda}_n = \frac{\log |f(x_n)|}{\log |f(x_{n-1})|}. \tag{10.23}$$

Note the analogy between $\overline{\lambda}_n$ and the definitions given by Wall in [42] and by Tornheim in [40]. To obtain $\overline{\lambda}_n$, we need knowledge of $\alpha$; while to obtain $\hat{\lambda}_n, \widetilde{\lambda}_n$ and $\check{\lambda}_n$ we do not. The new parameters CLOC, ACLOC, ECLOC and PCLOC have a lower computational cost than their predecessors. A detailed description of their convergence can be found in our studies [19] and [20].

**B. Relations Between Error and Error Parameters**   In the case of iterative methods to obtain approximates of the root $\alpha$ of $f(x) = 0$, where $f : I \subset \mathbf{R} \to \mathbf{R}$, the error difference equation is given by

$$e_{n+1} = C e_n^\rho \left(1 + O(e_n^\sigma)\right), \quad 0 < \sigma \leq 1, \tag{10.24}$$

where $C$ is the asymptotic error constant. With the additional hypothesis on the order, say $\rho \geq (1 + \sqrt{5})/2$, in [19] the relations between $\rho$ and the parameters $\overline{\lambda}_n$, $\widehat{\lambda}_n$, $\widetilde{\lambda}_n$ and $\check{\lambda}_n$ are presented.

Using (10.24) and the definitions of $\hat{e}_n$, $\tilde{e}_n$ and $\check{e}_n$, we obtain the following theoretical approximations of $e_n$. Namely,

$$e_n \approx C^{\frac{1}{1-\rho}} \left( \frac{\hat{e}_n}{\hat{e}_{n-1}} \right)^{\rho^2/(\rho-1)} \qquad n \geq 3, \tag{10.25a}$$

$$e_n \approx C^{\frac{\rho-1}{2\rho-1}} \left( \tilde{e}_n \right)^{\rho^2/(2\rho-1)} \qquad n \geq 3, \tag{10.25b}$$

$$e_n \approx C^{\frac{1}{1-\rho}} \left( \check{e}_n \right)^{\rho/\rho-1} \qquad n \geq 2. \tag{10.25c}$$

From the preceding (10.25a), (10.25b) and (10.25c), we can obtain bounds of the error to predict the number of correct figures and establish a stopping criterion, all without knowledge of the root $\alpha$.

**C. Numerical Test** The convergence of the new parameters has been tested in six iterative schemes with local convergence order equal to 2, 3, 4, $(1 + \sqrt{5})/2$, $1 + \sqrt{2}$ and $1 + \sqrt{3}$ respectively, in a set of seven real functions shown in Table 10.1. The first three methods are one-point iterative methods without memory, known as the Newton method, the Chebyshev method [11] and the Schröder method [35]. The other three are iterative methods with memory, namely the Secant method and two of its variants (see [13]).

They are defined by

$$\phi_1(x_n) = x_n - u(x_n), \tag{10.26}$$

$$\phi_2(x_n) = \phi_1(x_n) - \frac{1}{2} L(x_n) u(x_n), \tag{10.27}$$

**Table 10.1** Test functions, their roots and the initial points considered

| $f(x)$ | $\alpha$ | $x_0$ | $\{x_{-1}, x_0\}$ |
|---|---|---|---|
| $f_1(x) = x^3 - 3x^2 + x - 2$ | 2.89328919630449778906356 | 2.50 | $\{2.25, 2.60\}$ |
| $f_2(x) = x^3 + \cos x - 2$ | 1.17257796475397000126733333 | 1.50 | $\{1.50, 2.50\}$ |
| $f_3(x) = 2\sin x + 1 - x$ | 2.38006127313933901721254 | 2.50 | $\{1.00, 2.00\}$ |
| $f_4(x) = (x + 1) e^{x-1} - 1$ | 0.557145598997611416858672 | 1.00 | $\{0.00, 0.75\}$ |
| $f_5(x) = e^{x^2+7x-30} - 1$ | 3.0 | 2.94 | $\{2.90, 3.10\}$ |
| $f_6(x) = e^{-x} + \cos x.$ | 1.746139530408012417650703 | 1.50 | $\{1.60, 1.90\}$ |
| $f_7(x) = x - 3\ln x$ | 1.85718386020783533645698 | 2.00 | $\{1.00, 2.00\}$ |

$$\phi_3(x_n) = \phi_2(x_n) - \left(\frac{1}{2}L(x_n)^2 - M(x_n)\right)u(x_n), \tag{10.28}$$

$$\phi_4(x_n) = x_n - [x_{n-1}, x_n]_f^{-1} f(x_n), \tag{10.29}$$

$$\phi_5(x_n) = \phi_4(x_n) - [x_n, \phi_4(x_n)]_f^{-1} f(\phi_4(x_n)), \tag{10.30}$$

$$\phi_6(x_n) = \phi_4(x_n) - [x_n, 2\phi_4(x_n) - x_n]_f^{-1} f(\phi_4(x_n)), \tag{10.31}$$

where

$$u(x) = \frac{f(x)}{f'(x)}, \ L(x) = \frac{f''(x)}{f'(x)}u(x), \ M(x) = \frac{f'''(x)}{3!f'(x)}u(x)^2, \ [x, y]_f^{-1} = \frac{y - x}{f(y) - f(x)}.$$

The numerical results can be found in [20]. For each method from (10.26) to (10.31) and each function in Table 10.1, we have applied the four techniques with adaptive multi-precision arithmetic (see below) derived from relations (10.25a), (10.25b) and (10.25c) and the desired precision that for this study is $10^{-2200}$. The number of necessary iterations to obtain the desired precision and the values of iterated points $x_1, \ldots, x_I$ are the same.

From the results of [20], we can conclude that CLOC gives the best approximation of the theoretical order of convergence of an iterative method. However, knowledge of the root is required. Conversely, as we can see in the definitions (10.23) of ACLOC, ECLOC and PCLOC, these parameters do not involve the expression of the root $\alpha$. Actually, in real problems we want to approximate the root that is not known in advance. For practical purposes, we recommend ECLOC because it presents the best approximation of the local order (see [20]). Nevertheless, PCLOC is a good practical parameter in many cases because it requires fewer computations.

## 10.2.3   Multidimensional Case

A generalization to several variables of some parameters is carried out to approximate the local convergence order of an iterative method presented in the previous sections. In order to define the new parameters, we substitute the absolute value by the maximum norm, and all computations are done using the components of the vectors. Let $\{x_n\}_{n \in \mathbb{N}}$ be a convergence sequence of $\mathbf{R}^m$ towards $\alpha \in \mathbf{R}^m$, where $x_n = (x_n^{(1)}, x_n^{(2)}, \ldots, x_n^{(m)})^t$ and $\alpha = (\alpha^{(1)}, \alpha^{(2)}, \ldots, \alpha^{(m)})^t$. We consider the vectorial sequence of the error $e_n = x_n - \alpha$ and the following vectorial sequences of parameters:

$$\hat{e}_n = x_n - x_{n-1}, \qquad \tilde{e}_n = \max_{1 \le r \le m} \left| \frac{\left(\delta x_{n-1}^{(r)}\right)^2}{\delta^2 x_{n-2}^{(r)}} \right| \tag{10.32}$$

where $\delta x_n = x_{n+1} - x_n$. Notice that $\tilde{e}_n$ is the $\delta^2$-Aitken procedure applied to the components of $x_{n-1}, x_n$ and $x_{n+1}$, and all parameters are independent of knowledge of the root.

**Definitions** Let $\{\overline{\rho}_n\}_{n\geq 3}$, $\{\widehat{\rho}_n\}_{n\geq 4}$, $\{\widetilde{\rho}_n\}_{n\geq 5}$, $\{\check{\rho}_n\}_{n\geq 3}$, $\{\overline{\lambda}_n\}_{n\geq 2}$, $\{\widehat{\lambda}_n\}_{n\geq 3}$, $\{\widetilde{\lambda}_n\}_{n\geq 4}$ y $\{\check{\lambda}_n\}_{n\geq 2}$ be the following real sequences:

- Parameters COC, $\{\overline{\rho}_n\}_{n\geq 3}$ and CLOC, $\{\overline{\lambda}_n\}_{n\geq 2}$:

$$\overline{\rho}_n = \frac{\log(\|e_n\|/\|e_{n-1}\|)}{\log(\|e_{n-1}\|/\|e_{n-2}\|)}, \ n \geq 3, \quad \overline{\lambda}_n = \frac{\log\|e_n\|}{\log\|e_{n-1}\|}, \ n \geq 2. \quad (10.33a)$$

- Parameters ACOC, $\{\widehat{\rho}_n\}_{n\geq 4}$ and ACLOC $\{\widehat{\lambda}_n\}_{n\geq 3}$:

$$\widehat{\rho}_n = \frac{\log(\|\hat{e}_n\|/\|\hat{e}_{n-1}\|)}{\log(\|\hat{e}_{n-1}\|/\|\hat{e}_{n-2}\|)}, \ n \geq 4, \quad \widehat{\lambda}_n = \frac{\log\|\hat{e}_n\|}{\log\|\hat{e}_{n-1}\|}, \ n \geq 3. \quad (10.33b)$$

- Parameters ECOC $\{\widetilde{\rho}_n\}_{n\geq 5}$ and ECLOC $\{\widetilde{\lambda}_n\}_{n\geq 4}$:

$$\widetilde{\rho}_n = \frac{\log(\|\tilde{e}_n\|/\|\tilde{e}_{n-1}\|)}{\log(\|\tilde{e}_{n-1}\|/\|\tilde{e}_{n-2}\|)}, \ n \geq 5, \quad \widetilde{\lambda}_n = \frac{\log\|\tilde{e}_n\|}{\log\|\tilde{e}_{n-1}\|}, \ n \geq 4. \quad (10.33c)$$

- Parameters PCOC $\{\check{\rho}_n\}_{n\geq 3}$ and PCLOC, $\{\check{\lambda}_n\}_{n\geq 2}$:

$$\check{\rho}_n = \frac{\|F(x_n)\|/\|F(x_{n-1})\|}{\|F(x_{n-1})\|/\|F(x_{n-2})\|}, \ n \geq 3, \quad \check{\lambda}_n = \frac{\log\|F(x_n)\|}{\log\|F(x_{n-1})\|}, \ n \geq 2. \quad (10.33d)$$

Approximations COC, ACOC and ECOC have been used in Grau et al. [15]. A complete study of these parameters has been carried out to compute the local convergence order for four iterative methods and seven systems of nonlinear equations.

## 10.3 The Vectorial Error Difference Equation

Here we present a generalization to several variables of a technique used to compute analytically the error equation of iterative methods without memory for one variable. We consider iterative methods to find a simple root of a system of non-linear equations

$$F(x) = 0,$$

where $F : D \subseteq \mathbf{R}^m \longrightarrow \mathbf{R}^m$ is sufficiently differentiable and $D$ is an open convex domain in $\mathbf{R}^m$. Assume that the solution of $F(x) = 0$ is $\alpha \in D$, at which $F'(\alpha)$ is nonsingular.

The key idea is to use formal power series. The vectorial expression of the error equation obtained by carrying out this procedure, is

$$e_{n+1} = G\left(F'(\alpha), F''(\alpha), \ldots\right) e_n^\rho + O\left(e_n^{\rho+1}\right),$$

where $\rho$ is a nonnegative integer. If the iterative scheme is with memory we obtain [see (10.13)]

$$e_{n+1} = H\left(F'(\alpha), F''(\alpha), \ldots\right) e_n^{a_1} e_{n-1}^{a_2} \cdots e_{n-j+1}^{a_j} + o\left(e_n^{a_1} e_{n-1}^{a_2} \cdots e_{n-j+1}^{a_j}\right),$$

where $a_k$ are nonnegative integers for $1 \le k \le j$.

### 10.3.1   Notation

To obtain the vectorial equation of the error, we need some known results that, for ease of reference, are included in the following. Let $F : D \subseteq \mathbf{R}^m \longrightarrow \mathbf{R}^m$ be sufficiently differentiable (Fréchet-differentiable) in $D$, and therefore with continuous differentials. If we consider the $k$th derivative of $F$ at $a \in \mathbf{R}^m$, we have the $k$-linear function

$$F^{(k)}(a) : \mathbf{R}^m \times \overset{k}{\cdots} \times \mathbf{R}^m \longrightarrow \mathbf{R}^m$$

$$(h_1, \ldots, h_k) \longmapsto F^{(k)}(a)(h_1, \ldots, h_k).$$

That is, $F^{(k)}(a) \in \mathscr{L}\left(\mathbf{R}^m \times \overset{k}{\cdots} \times \mathbf{R}^m, \mathbf{R}^m\right) \equiv \mathscr{L}_k(\mathbf{R}^m, \mathbf{R}^m)$. It has the following properties:

P1.   $F^{(k)}(a)(h_1, \ldots, h_{k-1}, \cdot) \in \mathscr{L}(\mathbf{R}^m, \mathbf{R}^m) \equiv \mathscr{L}(\mathbf{R}^m)$.
P2.   $F^{(k)}(a)(h_{\sigma(1)}, \ldots, h_{\sigma(k)}) = F^{(k)}(a)(h_1, \ldots, h_k)$, where $\sigma$ is any permutation of the set $\{1, 2, \ldots, k\}$.

Notice that from P1 and P2 we can use the following notation:

N1.   $F^{(k)}(a)(h_1, \ldots, h_k) = F^{(k)}(a) h_1 \cdots h_k$. For $h_j = h$, $1 \le j \le k$, we write $F^{(k)}(a) h^k$.
N2.   $F^{(k)}(a) h^{k-1} F^{(l)}(a) h^l = F^{(k)}(a) F^{(l)}(a) h^{k+l-1}$.

Hence, we can also express $F^{(k)}(a) (h_1, \ldots, h_k)$ as

$$F^{(k)}(a) (h_1, \ldots, h_{k-1}) h_k = F^{(k)}(a) (h_1, \ldots, h_{k-2}) h_{k-1} h_k$$

$$\vdots$$

$$= F^{(k)}(a) h_1 \cdots h_k .$$

For any $q = a + h \in \mathbf{R}^m$ lying in a neighborhood of $a \in \mathbf{R}^m$, assuming that $[F'(a)]^{-1}$ exists, and taking into account this notation, we write Taylor's formulae in the following way:

$$F(a + h) = F(a) + F'(a) h + \frac{1}{2!} F^{(2)}(a) h^2 + \cdots + \frac{1}{p!} F^{(p)}(a) h^p + O_{p+1} ,$$

$$= F(a) + F'(a) \left( h + \sum_{k=2}^{p} A_k(a) h^k + O_{p+1} \right) , \qquad (10.34)$$

where $A_k(a) = \dfrac{1}{k!} [F'(a)]^{-1} F^{(k)}(a) \in \mathcal{L}_k (\mathbf{R}^m, \mathbf{R}^m) , \ 2 \le k \le p,$

and $O_{p+1} = O(h^{p+1}) .$

## 10.3.2 Symbolic Computation of the Inverse of a Function of Several Variables

We assume that $F : D \subseteq \mathbf{R}^m \longrightarrow \mathbf{R}^m$ has at least $p$-order derivatives with continuity on $D$ for any $x \in \mathbf{R}^m$ lying in a neighborhood of a simple zero, $\alpha \in D$, of the system $F(x) = 0$. We can apply Taylor's formulae to $F(x)$. By setting $e = x - \alpha$, the local order, and assuming that $[F'(\alpha)]^{-1}$ exists, we have

$$F(x) = F(\alpha + e) = \Gamma \left( e + \sum_{k=2}^{p-1} A_k e^k \right) + O_p , \qquad (10.35)$$

where

$$A_k = A_k(\alpha) , \ k \ge 2, \ \text{with} \ \Gamma = F'(\alpha) ,$$

$$e^k = (e, \overset{k}{\cdots}, e) \in \mathbf{R}^m \times \overset{k}{\cdots} \times \mathbf{R}^m \ \text{and} \ O_p = O(e^p).$$

Moreover, from (10.35) noting the identity by $I$, the derivatives of $F(x)$ can be written as

$$F'(x) = \Gamma \left( I + \sum_{k=2}^{p-1} k A_k e^{k-1} \right) + O_p, \tag{10.36}$$

$$F''(x) = \Gamma \left( \sum_{k=2}^{p-2} k(k-1) A_k e^{k-2} \right) + O_{p-1}, \tag{10.37}$$

$$F'''(x) = \Gamma \left( \sum_{k=3}^{p-3} \frac{k!}{(k-3)!} A_k e^{k-3} \right) + O_{p-2}, \tag{10.38}$$

and so forth up to order $p$.

By developing a formal series expansion of $e$, the inverse of $F'(x)$ is

$$[F'(x)]^{-1} = \left( I + \sum_{j=1}^{4} K_j e^j + O_5 \right) \Gamma^{-1}, \tag{10.39}$$

where

$$K_1 = -2 A_2,$$

$$K_2 = 4 A_2^2 - 3 A_3,$$

$$K_3 = -8 A_2^3 + 6 A_2 A_3 + 6 A_3 A_2 - 4 A_4,$$

$$K_4 = 16 A_2^4 - 12 A_2^2 A_3 - 12 A_2 A_3 A_2 - 12 A_3 A_2^2 + 8 A_2 A_4 + 8 A_4 A_2.$$

*Example  (Newton Method)* We consider the Newton's method that we can write as

$$X = x - F'(x)^{-1} F(x). \tag{10.40}$$

The expression of the error $E = X - \alpha$ in terms of $e$ is built up by subtracting $\alpha$ from both sides of (10.40) and taking into account (10.35) and (10.39). Namely,

$$E = e - \left( I + \sum_{j=1}^{3} K_j e^j + O_4 \right) \Gamma^{-1} \Gamma \left( e + \sum_{k=2}^{4} A_k e^k + O_5 \right)$$

$$= A_2 e^2 + 2(A_3 - A_2^2) e^3 + (3 A_4 - 4 A_2 A_3 - 3 A_3 A_2 + 4 A_2^3) e^4 + O_5. \tag{10.41}$$

The result (10.41) agrees with the classical asymptotic constant in the one-dimensional case and states that Newton's method has at least local order 2. Note that the terms $A_2A_3$ and $A_3A_2$ are noncommutative.

### 10.3.3  A Development of the Inverse of the First Order Divided Differences of a Function of Several Variables

We assume that $F : D \subseteq \mathbf{R}^m \longrightarrow \mathbf{R}^m$ has, at least, fifth-order derivatives with continuity on $D$. We consider the first divided difference operator of $F$ in $\mathbf{R}^m$ as a mapping

$$[-, -; F] : D \times D \longrightarrow \mathscr{L}(\mathbf{R}^m, \mathbf{R}^m)$$
$$(x + h, x) \longrightarrow [x + h, x; F] ,$$

which, for all $x, x + h \in D$, is defined by

$$[x + h, x; F] h = F(x + h) - F(x) , \tag{10.42}$$

where $\mathscr{L}(\mathbf{R}^m, \mathbf{R}^m)$ denotes the set of bounded linear functions (see [27, 32] and references therein). For $F$ sufficiently differentiable in $D$, we can write:

$$F(x + h) - F(x) = \int_x^{x+h} F'(z)\, dz = \int_0^1 F'(x + th)\, dt. \tag{10.43}$$

By developing $F'(x + th)$ in Taylor's series at the point $x \in \mathbf{R}^m$ and integrating, we obtain

$$[x + h, x; F] = F'(x) + \frac{1}{2}F''(x) h + \cdots + \frac{1}{p!}F^{(p)}(x) h^{p-1} + O_p. \tag{10.44}$$

By developing $F(x)$ and its derivatives in Taylor's series at the point $x = \alpha + e$ lying in a neighborhood of a simple zero, $\alpha \in D$, of the system $F(x) = 0$, and assuming that $[F'(\alpha)]^{-1}$ exists, we obtain the expressions (10.35) and (10.38). Next, by replacing these expressions in (10.44), we obtain:

$$[x + h, x; F] = \Gamma \left( I + A_2(2e + h) + A_3(3\,e^2 + 3\,e\,h + h^2) + \ldots \right), \tag{10.45}$$

or more precisely

$$[x + h, x; F] = \Gamma \left( I + \sum_{k=1}^{p-1} S_k(h, e) + O_p(\varepsilon, e) \right), \tag{10.46}$$

where $S_k(h, e) = A_{k+1} \sum_{j=1}^{k+1} \binom{k+1}{j} e^{k-j+1} h^{j-1}, \quad k \geq 1.$

We say that a function depending on $\varepsilon$ and $e$ is an $O_p(\varepsilon, e)$ if it is an $O(\varepsilon^{q_0} e^{q_1})$ with $q_0 + q_1 = p$, $q_i \geq 0$, $i = 0, 1$.

Setting $y = x + h$, $\varepsilon = y - \alpha$ and $h = \varepsilon - e$ in (10.45) and (10.46) we obtain

$$[y, x; F] = \Gamma \left( I + A_2(\varepsilon + e) + A_3(\varepsilon^2 + \varepsilon e + e^2) + \dots \right), \tag{10.47}$$

or more precisely

$$[y, x; F] = \Gamma \left( I + \sum_{k=1}^{p-1} T_k(\varepsilon, e) + O_p(\varepsilon, e) \right), \tag{10.48}$$

where $T_k(\varepsilon, e) = A_{k+1} \sum_{j=0}^{k} \varepsilon^{k-j} e^j.$

If we expand in formal power series of $e$ and $\epsilon$, the inverse of the divided difference given in (10.47) or in (10.48) is:

$$[y, x; F]^{-1} = \left( I - A_2(\epsilon + e) - A_3(\varepsilon^2 + \varepsilon e + e^2) + \left( A_2(\epsilon + e) \right)^2 + O_3(\varepsilon, e) \right) \Gamma^{-1}. \tag{10.49}$$

Notice that Eq. (10.49) is written explicitly until the 2nd-degree in $\varepsilon$ and $e$, while in each specific circumstance it will be adapted and reduced to the necessary terms, with an effective contribution to the computation of the local order of convergence.

These developments of the divided difference operator (10.49) were first used in our study Grau et al. [18].

*Example (Secant Method)* The generic case, (10.47), (10.48) or (10.49) can be adapted to different cases. For example, the well-known iterative method called the Secant method [27, 32] is defined by the algorithm:

$$x_{n+1} = x_n - [x_{n-1}, x_n; F]^{-1} F(x_n), \quad x_0, x_1 \in D. \tag{10.50}$$

If $y = x_{n-1}$ and $x = x_n$ in (10.47) then we obtain an expression of the operator $[x_{n-1}, x_n; F]$ in terms of $e_{n-1} = x_{n-1} - \alpha$ and $e_n = x_n - \alpha$. If we expand in formal power series of $e_{n-1}$ and $e_n$ the inverse of the divided difference operator in the

Secant method we obtain

$$[x_{n-1}, x_n; F]^{-1} = (I - A_2 (e_{n-1} + e_n) + (A_2^2 - A_3) e_{n-1}^2 + o(e_{n-1}^2)) \Gamma^{-1},$$
(10.51)

where $A_2^2 e_{n-1}^2 = (A_2 e_{n-1})^2$. The expression of the error $e_{n+1} = x_{n+1} - \alpha$ in terms of $e_n$ and $e_{n-1}$ for the Secant method is built up by subtracting $\alpha$ from both sides of (10.50). Taking into account (10.35) and (10.51), we have

$$e_{n+1} = e_n - (I - A_2(e_{n-1} + e_n) + (A_2^2 - A_3)e_{n-1}^2 + o(e_{n-1}^2)) \cdot$$
$$\cdot (e_n + A_2 e_n^2 + O(e_n^3))$$
$$= A_2 e_{n-1} e_n + (A_3 - A_2^2) e_{n-1}^2 e_n + o(e_{n-1}^2 e_n),$$
(10.52)

where the indicial polynomial [see (10.17)] of the error difference equation (10.52) is $t^2 - t - 1 = 0$, with only one positive real root, which is the R-order of convergence of the Secant method, $\phi = (1 + \sqrt{5})/2$. The second term of the right side of (10.52) would give order 2, since its associated polynomial equation is $t^2 - t - 2 = 0$. This result agrees with the classical asymptotic constant in the one-dimensional case and states that the Secant method has at least local order $\phi$.

A more complete expression of the error expression for the Secant method can be found in our studies Grau et al. [13] and Ezquerro et al. [7]. A generalization of Ostrowski method for several variables can be found in [17].

## 10.4 Efficiency Indices

We are interested in comparing iterative processes to approximate a solution $\alpha$ of a system of nonlinear equations. In the scalar case, the parameters of the efficiency index (*EI*) and computational efficiency (*CE*) are possible indicators of the efficiency of the scheme. Then, we consider the computational efficiency index (*CEI*) as a generalization to the multi-dimensional case. We show the power of this parameter by applying it to some numerical examples.

### 10.4.1 Efficiency Index and Computational Efficiency

To compare different iterative methods for solving scalar nonlinear equations, the efficiency index suggested by Ostrowski [28] is widely used,

$$EI = \rho^{1/a},$$
(10.53)

where $\rho$ is the local order of convergence of the method and $a$ represents the number of evaluations of functions required to carry out the method per iteration.

Another classical measure of efficiency for iterative methods applied to scalar nonlinear equations is the computational efficiency proposed by Traub [41],

$$CE = \rho^{1/\omega}, \tag{10.54}$$

where $\omega$ is the number of operations, expressed in product units, that are needed to compute each iteration without considering the evaluations of the functions. In general, if we are interested in knowing the efficiency of a scalar scheme, the most frequently used parameter is $EI$, instead of any combination of this parameter with $CE$.

The efficiency index for Newton's method is $2^{1/2} \approx 1.414$ because an iterative step of Newton requires the computation of $f(x)$ and $f'(x)$ then $a = 2$, and the local order is $\rho = 2$. Note that the parameter $EI$ is independent of the expression of $f$ and its derivative, while the parameter $CE$ does not consider the evaluation of the computational cost of the functions of the algorithm.

More precisely, note that an iteration step requires two actions: first the calculation of new functions values; and then the combination of data to calculate the next iterate. The evaluation of functions requires the invocation of subroutines, whereas the calculation of the next iterate requires only a few arithmetic operations. In general, these few arithmetic operations are not considered in the scalar case.

## 10.4.2   Computational Efficiency Index

The traditional way to present the computational efficiency index of iterative methods (see [28, 41]) is adapted for systems of nonlinear equations. When we deal with a system of nonlinear equations, the total operational cost is the sum of the evaluations of functions (the function and the derivatives involved) and the operational cost of doing a step of the iterative method.

In $m$-dimensional case, the choice of the most suitable iterative method, $x_{n+1} = \Phi(x_n)$, depends mainly on its efficiency which also depends on the convergence order and the computational cost. The number of operations per iteration increases the computational cost in such a way that some algorithms will not be used because they are not efficient. In general, we have a scheme such as the following

$$\Phi(x_n) = x_n - \Theta_n^{-1} F(x_n),$$

where instead of computing the inverse of the operator $\Theta_n$, we solve the following linear system

$$\Theta_n y_n = -F(x_n),$$
$$x_{n+1} = x_n + y_n.$$

Therefore, we choose the *LU*-decomposition plus the resolution of two linear triangular systems in the computation of the inverse operator that appears. In other words, in the multi-dimensional case we have to perform a great number of operations, while in the scalar case the number of operations is reduced to a very few products.

Let $\ell$ be the conversion factor of quotients into products (the time needed to perform a quotient, in time of product units). Recall that the number of products and quotients that we need to solve an *m*-dimensional linear system, using the *LU*-decomposition is

$$\omega_1 = \frac{m}{6}\left(2\,m^2 - 3\,m + 1\right) + \ell\,\frac{m}{2}\,(m-1),$$

and to solve the two triangular linear systems with ones in the main diagonal of matrix $L$ we have $\omega_2 = m\,(m-1) + \ell\,m$ products. Finally, the total number of products is

$$\frac{m}{6}\left(2\,m^2 + 3\,(1+\ell)\,m + 3\,\ell - 5\right).$$

**Definition 4** The computational efficiency index (*CEI*) and the computational cost per iteration ($\mathscr{C}$) are defined by (see [13, 16, 18, 32])

$$CEI(\mu_0, \mu_1, m, \ell) = \rho^{\dfrac{1}{\mathscr{C}(\mu_0, \mu_1, m, \ell)}}, \qquad (10.55)$$

where $\mathscr{C}(\mu_0, \mu_1, m, \ell)$ is the computational cost given by

$$\mathscr{C}(\mu_0, \mu_1, m) = a_0(m)\,\mu_0 + a_1(m)\,\mu_1 + \omega(m, \ell), \qquad (10.56)$$

$a_0(m)$  represents the number of evaluations of the scalar functions $(F_1, \ldots, F_m)$ used in one step of the iterative method.

$a_1(m)$  is the number of evaluations of scalar functions of $F'$, say $\dfrac{\partial F_i}{\partial x_j}$, $1 \le i, j \le m$.

$\omega(m, \ell)$  represents the number of products needed per iteration.

The constants $\mu_0$ and $\mu_1$ are the ratios between products and evaluations required to express the value of $\mathscr{C}(\mu_0, \mu_1, m)$ in terms of products, and $\ell$ is the cost of one quotient in products.

Note that:

$$CEI(\mu_0, \mu_1, m, \ell) > 1, \qquad \lim_{m \to \infty} CEI(\mu_0, \mu_1, m, \ell) = 1.$$

Notice that for $\mu_0 = \mu_1 = 1$ and $\omega(m) = 0$, (10.55) is reduced to (10.53), that is the classic efficiency index of an iterative method, say $EI = \rho^{1/a}$ in the scalar case. Also observe that, if $a_0 = a_1 = 0$, (10.55) is written in the scalar case as (10.54); namely, $CE = \rho^{1/\omega}$.

According to (10.56), an estimation of the factors $\mu_0, \mu_1$ is required. To do this, we express the cost of the evaluation of the elementary functions in terms of products, which depend on the machine, the software and the arithmetic used. In [10, 38], we find comparison studies between a multi-precision library (MPFR) and other computing libraries. In Tables 10.2 and 10.3 our own estimation of the cost of the elementary functions is shown in product units, where the running time of one product is measured in milliseconds.

The values presented in Table 10.2 have been rounded to 5 unities because of the huge variability in the different repetitions that were carried out. In contrast, the averages are shown in Table 10.3, since the variability was very low. In addition, the compilator of C++ that was used ensures that the function `clock()` gives exactly the CPU time invested by the program. Table 10.3 shows that some relative values for the product are lower in multiple precision than in double precision, although the absolute time spent on a product is much higher in multiple precision.

This measure of computing efficiency is clearly more satisfactory than considering only the number of iterations or only the number of evaluated functions, which are used widely by others authors. Any change of software or hardware requires us to recompute the elapsed time of elemental functions, quotients and products.

In this section we compare some free derivative iterative methods that use the divided difference operator (see [13]). Firstly, we recall the classical Secant method (10.50) and we study a few two-step algorithms with memory.

**Table 10.2** Computational cost of elementary functions computed with Matlab 2009b and Maple 13 on an Intel®Core(TM)2 Duo CPU P8800 (32-bit machine) processor with Microsoft Windows 7 Professional, where $x = \sqrt{3} - 1$ and $y = \sqrt{5}$

| Software | $x * y$ | $x/y$ | $\sqrt{x}$ | $\exp(x)$ | $\ln(x)$ | $\sin(x)$ | $\cos(x)$ |
|---|---|---|---|---|---|---|---|
| Matlab 2009b | 4.5E−7 ms | 10 | 55 | 80 | 145 | 35 | 50 |
| Maple13 16 digits | 1.2E−3 ms | 1 | 10 | 25 | 45 | 25 | 20 |
| Maple13 1024 digits | 4.0E−2 ms | 1 | 5 | 45 | 10 | 90 | 90 |
| Maple13 4096 digits | 3.5E−1 ms | 1 | 5 | 50 | 10 | 105 | 105 |

**Table 10.3** Computational cost of elementary functions computed with a program written in C++, compiled by `gcc(4.3.3)` for i486-linux-gnu with `libgmp` (v.4.2.4) and `libmpfr` (v.2.4.0) libraries on an Intel®Xeon E5420, 2.5 GHz, 6 MB cache processor where $x = \sqrt{3} - 1$ and $y = \sqrt{5}$

| Arithmetics | $x * y$ | $x/y$ | $\sqrt{x}$ | $\exp(x)$ | $\ln(x)$ | $\sin(x)$ | $\cos(x)$ |
|---|---|---|---|---|---|---|---|
| C++ double | 2.3E−7 ms | 29 | 29 | 299 | 180 | 181 | 192 |
| C++ MPFR 1024 digits | 1.16E−2 ms | 2.4 | 1.7 | 62 | 57 | 69 | 65 |
| C++ MPFR 4096 digits | 1.04E−1 ms | 2.5 | 1.7 | 88 | 66 | 116 | 113 |

### 10.4.3   Examples of Iterative Methods

**Secant Method**  We call $\Phi_0$ to the well-known iterative secant method (10.50). That is, by setting $x_{-1}, x_0 \in D$

$$x_{n+1} \,=\, \Phi_0(x_{n-1}, x_n) \,=\, x_n - [x_{n-1}, x_n; F]^{-1} \, F(x_n), \qquad (10.57)$$

with the local order of convergence $\phi = (1 + \sqrt{5})/2 = 1.618\ldots$

**Frozen Divided Difference Method**  We consider two steps of the Secant method and by setting $x_{-1}, x_0$ given in $D$,

$$\begin{cases} y_n = \Phi_0(x_{n-1}, x_n), \\ x_{n+1} = \Phi_1(x_{n-1}, x_n) \,=\, y_n - [x_{n-1}, x_n; F]^{-1} \, F(y_n), \quad n \geq 0. \end{cases} \qquad (10.58)$$

In this case, the local order is at least 2. Therefore, this method is a quadratic method where the divided difference operator is only computed once, which is why the reason why it is called the frozen divided difference method (for more details see [18]).

The next two iterative methods are pseudo-compositions of two known schemes, which are two-step algorithms.

**First Superquadratic Method**  We take the Secant method twice. That is, by putting $x_{-1}, x_0$ given in $D$,

$$\begin{cases} y_n = \Phi_0(x_{n-1}, x_n), \\ x_{n+1} = \Phi_2(x_{n-1}, x_n) \,=\, y_n - [x_n, y_n; F]^{-1} \, F(y_n), \quad n \geq 0. \end{cases} \qquad (10.59)$$

The order of the two-step iterative method with memory defined in (10.59) is $1 + \sqrt{2} = 2.414\ldots$

**Second Superquadratic Method**  We define the pseudo-composition of the Secant method with the Kurchatov method [9, 26]. The result is the following scheme: Given $x_{-1}, x_0$ in $D$,

$$\begin{cases} y_n = \Phi_0(x_{n-1}, x_n), \\ x_{n+1} = \Phi_3(x_{n-1}, x_n) \,=\, y_n - [x_n, 2y_n - x_n; F]^{-1} \, F(y_n), \quad n \geq 0. \end{cases} \qquad (10.60)$$

This two-step scheme with memory has a local order of convergence equal to $1 + \sqrt{3} = 1.732\ldots$

Finally, we observe that we have moved from a superlinear method such as the Secant method with local order equal to $(1 + \sqrt{5})/2$ to a superquadratic method with local order equal to $1 + \sqrt{3}$.

### 10.4.4   *Comparisons Between These Methods*

We study the efficiency index of the four iterative methods $\Phi_j$, $0 \le j \le 3$, given by (10.57), (10.58), (10.59) and (10.60) respectively. The computational efficiency index ($CEI_j$) of each iterative method and the computational cost per iteration ($\mathscr{C}_j$) are defined in (10.55) as

$$CEI_j(\mu, m, \ell) = \rho^{\dfrac{1}{\mathscr{C}_j(\mu, m, \ell)}},$$

where

$$\mathscr{C}_j(\mu, m, \ell) = a_j(m)\mu + \omega_j(m, \ell).$$

Note that we denote $\mu_0$ by $\mu$ in these examples. For each method $\Phi_j$ for $j = 0, 1, 2, 3$, Table 10.4 shows: the local order of convergence $\rho_j$; the number of evaluations of $F$ (NEF); the number of computations of the divided differences (DD); the value of $a_j(m)$; and $\omega_j(m, \ell)$. In order to obtain these results, we consider the following computational divided difference operator (10.61). To compute the $[x_{n-1}, x_n; F]$ operator, we need $m^2$ divisions and $m(m - 1)$ scalar evaluations. Note that, for the $[x_n, y_n; F]$ or $[x_n, 2y_n - x_n; F]$ operators we need $m^2$ scalar evaluations.

$$[y, x; F]_{ij}^{(1)} = \big(F_i(y_1, \ldots, y_{j-1}, y_j, x_{j+1}, \ldots, x_m) - \tag{10.61}$$

$$F_i(y_1, \ldots, y_{j-1}, x_j, x_{j+1}, \ldots, x_m)\big) / (y_j - x_j), \quad 1 \le i, j \le m.$$

Summarizing the results of Table 10.4 we have

$$\mathscr{C}_0(\mu, m, \ell) = \frac{m}{6}(2m^2 + 6m\mu + 3m + 9\ell m + 3\ell - 5), \qquad \rho_0 = \tfrac{1+\sqrt{5}}{2};$$

$$\mathscr{C}_1(\mu, m, \ell) = \frac{m}{6}(2m^2 + 6m\mu + 9m + 9\ell m + 6\mu + 9\ell - 11), \quad \rho_1 = 2;$$

$$\mathscr{C}_2(\mu, m, \ell) = \frac{m}{3}(2m^2 + 6m\mu + 3m + 9\ell m + 3\ell - 5), \qquad \rho_2 = 1 + \sqrt{2};$$

$$\mathscr{C}_3(\mu, m, \ell) = \frac{m}{3}(2m^2 + 6m\mu + 3m + 9\ell m + 3\mu + 3\ell - 2), \quad \rho_3 = 1 + \sqrt{3}.$$

In order to compare the corresponding $CEI$s we use the following quotient

$$R_{ij} = \frac{\log CEI_i}{\log CEI_j} = \frac{\log \rho_i}{\log \rho_j} \frac{\mathscr{C}_j}{\mathscr{C}_i},$$

and we have the following theorem [13]. In Table 10.5, we present the different situations of this theorem.

**Table 10.4** Local convergence order and computational cost of methods $\Phi_j$ for $0 \leq j \leq 3$

| | $\rho_j$ | NEF | DD | $a_j(m)$ | $\omega_j(m, \ell)$ |
|---|---|---|---|---|---|
| $\Phi_0$ | $(1 + \sqrt{5})/2$ | 1 | 1 | $m(m-1) + m = m^2$ | $m(2m^2 + 3m + 9\ell m + 3\ell - 5)/6$ |
| $\Phi_1$ | 2 | 2 | 1 | $a_0(m) + m = m^2 + m$ | $p_0(m) + m(m-1) + m\ell =$ $m(2m^2 + 9m + 9\ell m + 9\ell - 11)/6$ |
| $\Phi_2$ | $1 + \sqrt{2}$ | 1 | 2 | $a_0(m) + m^2 = 2m^2$ | $2p_0(m) =$ $m(2m^2 + 3m + 9\ell m + 3\ell - 5)/3$ |
| $\Phi_3$ | $1 + \sqrt{3}$ | 2 | 2 | $a_1(m) + m = 2m^2 + m$ | $2p_0(m) + 1 =$ $m(2m^2 + 3m + 9\ell m + 3\ell - 2)/3$ |

**Table 10.5** The four situations of Theorem 1

| $m = 2$ | $m = 3$ | $4 \leq m \leq 11$ | $m \geq 12$ |
|---|---|---|---|
| $CEI_0 > CEI_2$ | | | |
| $CEI_1 > CEI_2$ | $CEI_1 > CEI_0 > CEI_2$ | $CEI_1 > CEI_0 > CEI_2$ | |
| $CEI_1 > CEI_3$ | $CEI_1 > CEI_3$ | $CEI_1 > CEI_3 > CEI_2$ | $CEI_1 > CEI_0 > CEI_3 > CEI_2$ |

**Theorem 1** *For all $\ell \geq 1$ we have:*

1. *$CEI_1 > CEI_2$ and $CEI_3$, for all $m \geq 2$.*
2. *$CEI_0 > CEI_2$ for all $m \geq 2$.*
3. *$CEI_1 > CEI_0$, for all $m \geq 3$.*
4. *$CEI_3 > CEI_2$ for all $m \geq 4$.*
5. *$CEI_3 > CEI_0$, for all $m \geq 12$,*

## 10.4.5 Numerical Results

The numerical computations listed in Tables 10.7, 10.8, 10.9, 10.10 10.11, 10.12 and 10.13 were performed on an MPFR library of C++ multiprecision arithmetics [39] with 4096 digits of mantissa. All programs were compiled by `gcc(4.3.3)` for i486-linux-gnu with `libgmp` (v.4. 2.4) and `libmpfr` (v.2.4.0) libraries on an Intel®Xeon E5420, 2.5 GHz and 6 MB cache processor. For this hardware and software, the computational cost of the quotient with respect to the product is $\ell = 2.5$ (see Table 10.3). Within each example the starting point is the same for all methods tested. The classical stopping criterion $\|e_I\| = \|x_I - \alpha\| > 0.5 \cdot 10^{-\varepsilon}$ and $\|e_{I+1}\| \leq 0.5 \cdot 10^{-\varepsilon}$, with $\varepsilon = 4096$, is replaced by

$$E_I = \frac{\|\hat{e}_I\|}{\|\hat{e}_{I-1}\|} < 0.5 \cdot 10^{-\eta},$$

where $\hat{e}_I = x_I - x_{I-1}$ and $\eta = \frac{\rho-1}{\rho^2} \varepsilon$ [see (10.25a)]. Notice that this criterium is independent of the knowledge of the root. Furthermore, in all computations we have substituted the computational order of convergence (COC) [43] by the approximation ACOC, $\hat{\rho}_I$ (10.33b).

*Examples* We present the system defined by

$$F_i(x_1, \ldots, x_m) = \sum_{\substack{j=1 \\ j \neq i}}^{m} x_j - \exp(-x_i) = 0, \quad 1 \leq i \leq m, \tag{10.62}$$

where $m = 3, 5, 13$ and $\mu = 87.8$ for arithmetics of 4096 digits, since in (10.62) $\mu$ is independent from $m$. The three values of $m$ correspond to three situations of the Theorem 1 (see Table 10.6). Tables 10.7, 10.8 and 10.9 show the results obtained for the iterative methods $\Phi_0$, $\Phi_1$, $\Phi_2$ and $\Phi_3$ respectively.

**Table 10.6**  The three cases of Theorem 1 for $\mu = 87.8$ and $\ell = 2.5$

| Case 1. $m = 3$ | Case 2. $m = 5$ | Case 3. $m = 13$ |
|---|---|---|
| $CEI_1 > CEI_0 > CEI_2 > CEI_4$ | $CEI_1 > CEI_0 > CEI_4 > CEI_2$ | $CEI_1 > CEI_4 > CEI_0 > CEI_2$ |

**Table 10.7**  Numerical results for case 1, where $m = 3$ and $t_p = 0.1039$

| $\Phi_i$ | $I$ | $T$ | $D_I$ | $CEI$ | $TF$ | $|\Delta\hat{\rho}_I|$ |
|---|---|---|---|---|---|---|
| $\Phi_0$ | 17 | 1540 | 3130 | 1.000573924 | 4013.15 | 5.18E−6 |
| $\Phi_1$ | 10 | 1340 | 3310 | 1.000621515 | 3705.95 | 7.79E−3 |
| $\Phi_2$ | 9 | 1620 | 3960 | 1.000525578 | 4382.20 | 1.76E−5 |
| $\Phi_3$ | 7 | 1480 | 2140 | 1.000517189 | 4453.27 | 2.05E−3 |

**Table 10.8**  Numerical results for case 2, where $m = 5$ and $t_p = 0.1039$

| $\Phi_i$ | $I$ | $T$ | $D_I$ | $CEI$ | $TF$ | $|\Delta\hat{\rho}_I|$ |
|---|---|---|---|---|---|---|
| $\Phi_0$ | 17 | 4290 | 3470 | 1.000205229 | 11220.74 | 2.65E−6 |
| $\Phi_1$ | 10 | 3040 | 2180 | 1.000246133 | 9356.20 | 8.25E−3 |
| $\Phi_2$ | 8 | 4030 | 1880 | 1.000187944 | 12252.59 | 1.12E−5 |
| $\Phi_3$ | 7 | 3850 | 2140 | 1.000195783 | 11762.06 | 1.28E−3 |

**Table 10.9**  Numerical results for case 3, where $m = 13$ and $t_p = 0.1039$

| $\Phi_i$ | $I$ | $T$ | $D_I$ | $CEI$ | $TF$ | $|\Delta\hat{\rho}_I|$ |
|---|---|---|---|---|---|---|
| $\Phi_0$ | 17 | 29070 | 2630 | 1.000029533 | 77967.67 | 3.99E−6 |
| $\Phi_1$ | 11 | 20460 | 2950 | 1.000039330 | 58546.41 | 7.58E−3 |
| $\Phi_2$ | 9 | 30950 | 3380 | 1.000027046 | 85137.03 | 1.45E−5 |
| $\Phi_3$ | 7 | 24810 | 1650 | 1.000029786 | 77305.42 | 1.58E−3 |

For each case, we present one table where we read the method $\Phi_i$, the number of iterations needed $I$ to reach the maximum precision requested, the computational elapsed time $T$ in milliseconds of the C++ execution for these iterations, the correct decimals reached in $D_I$ approximately, the computational efficiency index $CEI$, the time factor $\mathrm{TF} = 1/\log CEI$, an error's higher bound of the ACOC computation $\Delta\hat{\rho}_I$ where $\rho = \hat{\rho}_I \pm \Delta\hat{\rho}_I$.

**Case 1** We begin with the system (10.62) for $m = 3$ where $CEI_1 > CEI_0 > CEI_2 = CEI_3 > CEI_4$. The root $\alpha = (\alpha_i)$, $1 \leq i \leq m$, and the two initial points $x_{-1}$, $x_0$ are

$$\alpha_1 = -0.8320250398, \quad \alpha_{2,3} = 1.148983754,$$
$$x_{-1} = (-0.8, 1.1, 1.1)^t \quad x_0 = (-0.9, 1.2, 1.2)^t.$$

The numerical results of this case are shown in Table 10.7.

**Case 2** The second case is the system (10.62) for $m = 5$ where $CEI_1 > CEI_0 > CEI_4 > CEI_2 = CEI_3$. The numerical results of this case are shown in Table 10.8. The root $\alpha$ and the two initial points $x_{-1}$, $x_0$ are

$$\alpha_{1,2,5} = -2.153967996, \quad \alpha_{3,4} = 6.463463374,$$
$$x_{-1} = (-2.1, -2.1, 6.4, 6.4, -2.1)^t \quad x_0 = (-2.2, -2.2, 6.5, 6.5, -2.2)^t.$$

**Case 3** Finally, the third case is the system (10.62) for $m = 13$ where $CEI_1 > CEI_4 > CEI_0 > CEI_2 = CEI_3$. The numerical results of this case are in Table 10.9. The root $\alpha$ and the two initial points $x_{-1}$, $x_0$ are

$$\alpha_{1,2,3,5,7,10} = 1.371341671, \quad \alpha_{4,6,8,9,11,12,13} = -.9432774419,$$
$$x_{-1} = (1.3, 1.3, 1.3, -0.9, 1.3, -0.9, 1.3, -0.9, -0.9, 1.3, -0.9, -0.9, -0.9)^t,$$
$$x_0 = (1.4, 1.4, 1.4, -1.0, 1.4, -1.0, 1.4, -1.0, -1.0, 1.4, -1.0, -1.0, -1.0)^t.$$

*Remark 1* In case 1, we can arrange methods $\Phi_2$ and $\Phi_3$ according to the elapsed time $T$ or the time factor TF. The results are different because the final precisions $D_I$ obtained in each method are not comparable. In Sect. 10.5 we explain a better way to compare the elapsed time that is more consistent with the theoretical results of the computational efficiency index $CEI$.

*Remark 2* The first numerical definition of divided difference (10.61) has a counterexample in the following $2 \times 2$ system of nonlinear equations

$$F(x_1, x_2) = \begin{cases} x_1^2 + x_2^2 - 9 = 0, \\ x_1 x_2 - 1 = 0. \end{cases} \tag{10.63}$$

Scheme $\Phi_3$ gives a PCLOC $\check{\rho} = 1 + \sqrt{2}$, instead of the theoretical value $\rho = 1 + \sqrt{3}$. Furthermore, a comparison of the expression (10.43), taking into account the definition of the divided differences operator (10.61), gives the following result

$$
\int_0^1 F'(x+th)\,dt = \begin{pmatrix} 2x_1+h_1 & 2x_2+h_2 \\ x_2+h_2/2 & x_1+h_1/2 \end{pmatrix} \neq [x+h,x;F]^{(1)} = \begin{pmatrix} 2x_1+h_1 & 2x_2+h_2 \\ x_2 & x_1+h_1 \end{pmatrix},
$$

where $x = (x_1,x_2)^t$, $h = (h_1,h_2)^t$ and $t \in \mathbf{R}$. Due to Potra [30, 32] we have the following necessary and sufficient condition to characterize the divided difference operator by means of a Riemann integral.

**Theorem 2**  *If F satisfies the following Lipschitz condition* $\|[x,y;F] - [u,v;F]\| \leq H\,(\|x-u\| + \|y-v\|)$, *then equality (10.61) holds for every pair of distinct points* $(x+h,x) \in D \times D$ *if, and only if, for all* $(u,v) \in D \times D$ *with* $u \neq v$ *and* $2v-u \in D$ *the following relation is satisfied:*

$$
[u,v;F] = 2\,[u,2v-u;F] - [v,2v-u;F]. \tag{10.64}
$$

We can check that the function considered in (10.75) does not hold (10.64). We need a new definition of divided differences instead of the definition given in (10.61) to obtain the local order required when we apply algorithms $\Phi_k$, $k = 0, 1, 2, 3$, in this case. We use the following method to compute the divided difference operator

$$
[y,x;F]^{(2)}_{ij} = \frac{1}{2}\left([y,x;F]^{(1)}_{ij} + [x,y;F]^{(1)}_{ij}\right), \quad 1 \leq i,j \leq m. \tag{10.65}
$$

Notice that the operator defined in (10.65) is symmetric: $[y,x;F] = [x,y;F]$. If we use definition (10.65), we have to double the number of evaluations of the scalar functions in the computation of $[y,x;F]$, and by rewriting (10.65) we have $m^2$ quotients.

If we use (10.65) then method $\Phi_3$ applied to system (10.75), the computational order of convergence is equal to $1 + \sqrt{3}$. Another example with the same behavior as (10.75) is

$$
F_i(x_1,\ldots,x_3) = x_i - \cos\left(\sum_{\substack{j=1 \\ j\neq i}}^{3} x_j - x_i\right), \quad 1 \leq i \leq 3. \tag{10.66}
$$

In Table 10.10, we show the numerical results of method $\Phi_3$ applied to the systems of nonlinear equations (10.75) and (10.66). We denote by $\Phi_3^{(j)}$, $j = 1, 2$, method $\Phi_3$ using the numerical definition of the divided difference operator $[x+h,x;F]^{(j)}$, $j = 1, 2$ respectively. By setting $TF_3^{(j)} = 1/\log CEI_3^{(j)}$ as the time factors of methods $\Phi_3^{(j)}$ and by comparing the two time factors of system (10.75), we can conclude (see Table 10.10) that method $\Phi_3^{(2)}$ is more efficient than $\Phi_3^{(1)}$. This behavior is reversed in example (10.66).

**Table 10.10** Numerical results for the scheme $\Phi_3$ applied to systems (10.75) and (10.66)

| | $\rho_3^{(j)}$ | System (10.75) | | | | System (10.66) | | | |
|---|---|---|---|---|---|---|---|---|---|
| | | $TF_3^{(j)}$ | $I$ | $D_I$ | $\|\Delta\hat{\rho}_I\|$ | $TF_3^{(j)}$ | $I$ | $D_I$ | $\|\Delta\hat{\rho}_I\|$ |
| $\Phi_3^{(1)}$ | $1+\sqrt{2}$ | 65.8 | 7 | 2918 | 1.2E−4 | 2205.4 | 9 | 3386 | 2.3E−4 |
| $\Phi_3^{(2)}$ | $1+\sqrt{3}$ | 63.7 | 6 | 2853 | 4.3E−3 | 2982.3 | 8 | 4199 | 1.7E−4 |

## 10.5  Theoretical Numerical Considerations

Theoretical and experimental studies of numerical applications often move away
from one another. From studies by Ostrowski [28], Traub [41] and Ralston [33],
we have introduced new concepts that allow us to estimate the execution time. We
revisit the time factor [18] and we present a relation between these measures and the
number of iterations, to achieve a given precision in the root computation. In other
words, in the classical comparison between two iterative methods, the following
ratio of efficiency logarithms was introduced

$$\frac{\Theta_1}{\Theta_2} = \frac{\log CEI_2}{\log CEI_1} = \frac{\mathscr{C}_1/\log\rho_1}{\mathscr{C}_2/\log\rho_2}, \qquad (10.67)$$

where $\Theta$ is the total cost of products to obtain the required precision when we apply
a method. That is, if $I$ is the total number of iterations, then

$$\Theta = I \cdot \mathscr{C}. \qquad (10.68)$$

In this section, we also introduce a new factor that provides us with an explicit
expression of the total time $\tilde{\Theta}$.

### 10.5.1  Theoretical Estimations

As an iterative method has local order of convergence $\rho$ and local error $e_n = x_n - \alpha$,
we define $D_n = -\log_{10}\|e_n\|$. That is, $D_n$ is approximately the number of correct
decimal places in the $n$th iteration. From the definition of local order, we have
$\|e_{I+1}\| \approx C\|e_I\|^\rho$ and $D_{I+1} \approx -\log_{10} C + \rho D_I$. The solution of this difference
equation is

$$D_I \approx D_0\rho^I + \log_{10} M, \quad \text{where} \quad M = C^{1/(\rho-1)},$$

and we obtain $D_I \approx D_0\rho^I$. If we apply logarithms to both sides of the preceding
equation and take into account (10.68) we get

$$I = (\log q)/(\log\rho) \quad \text{and} \quad \Theta = \log q\frac{\mathscr{C}}{\log\rho} = \frac{\log q}{\log CEI}, \qquad (10.69)$$

where $q = D_I/D_0$. From (10.68), if we take $\tilde{\Theta}(q) = \Theta t_p$, where $t_p$ is the time required to do one product, then $\tilde{\Theta}(q)$ is the total time. Taking into account (10.55) and (10.69), the total time is $\tilde{\Theta}(q) \approx \log q \dfrac{t_p}{\log CEI}$. If we consider the time in product units, then $1/\log CEI$ will be called the time factor (TF). Notice that the term $\log q$ introduced in (10.69) is simplified in the quotient of Eq. (10.67). In [33], (10.67) is obtained from

$$\frac{\Theta_1}{\Theta_2} = \frac{I_1}{I_2} \frac{\mathscr{C}_1}{\mathscr{C}_2}. \tag{10.70}$$

Then, considering $\Theta = \mathscr{C}/\ln \rho$, the efficiency index is

$$EI = \frac{1}{\Theta} = \frac{\ln \rho}{\mathscr{C}} = \ln \rho^{1/\mathscr{C}}.$$

Not only have we taken $CEI$ as defined in (10.55), but we have also expressed the factor that is simplified in (10.70) as

$$I_k = \frac{\log q}{\log \rho_k} \quad k = 1, 2,$$

as we inferred and deduced in (10.69). We are now in a position to state the following theorem.

**Theorem 3** *In order to obtain $D_I$ correct decimals from $D_0$ correct decimals using an iterative method, we can estimate*

- *the number of iterations $I \approx \dfrac{\log q}{\log \rho}$,*
- *the necessary time $\tilde{\Theta}(q) \approx \log q \dfrac{t_p}{\log CEI}$,*

*where $q = D_I/D_0$ and $t_p$ is the time needed for one product and $\mathscr{C}$ is the computational cost per iteration.*

When we are faced with a numerical problem we rewrite the estimation of time as in the following equation of the straight line in variables $(\log D_I, \tilde{\Theta})$:

$$\tilde{\Theta}(D_I) = \frac{t_p}{\log CEI}(\log D_I - \log D_0) = \kappa \log \frac{D_I}{D_0} = \kappa \log q,$$

where $\kappa = t_p/\log(CEI)$. That is, $\kappa$ is a coefficient that measures the time of execution in function of the approximate number of correct decimals. In order to study and analyze an iterative method, we can consider the straight line $\tilde{\Theta}(D_I)$ with slope $\kappa$ in a semi-logarithmic graph. If we approximate the $(\log D_j, \Theta(D_j))$ pairs in a least-squares sense by a polynomial of degree one, we can compute an experimental slope $\tilde{\kappa}$ that is used in Figs. 10.1, 10.2 and 10.3, and Tables 10.11, 10.12 and 10.13.

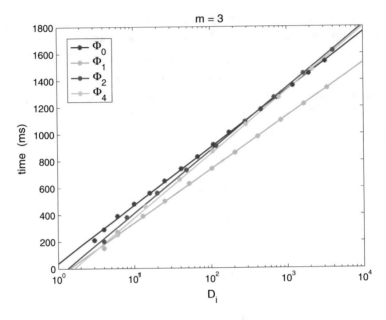

**Fig. 10.1** Time $t$ versus number of correct decimals $D_i$ for $m = 3$

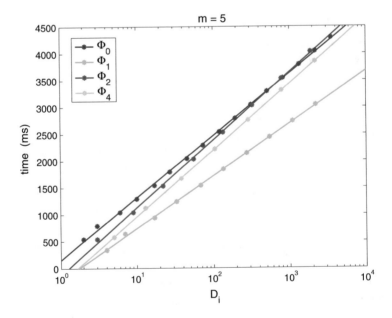

**Fig. 10.2** Time $t$ versus number of correct decimals $D_i$ for $m = 5$

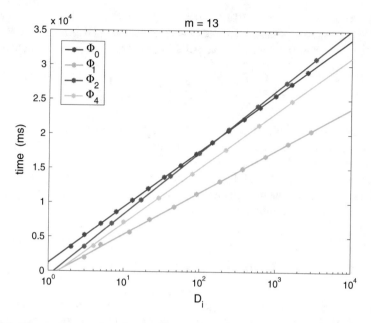

**Fig. 10.3** Time $t$ versus number of correct decimals $D_i$ for $m = 13$

**Table 10.11** Numerical results for case 1, where $m = 3$ and $t_p = 0.1039$

| $\Phi_i$ | TF | I | $D_I$ | $\tilde{\Theta}$ | $\tilde{\kappa}$ | $r_{TF}$ |
|---|---|---|---|---|---|---|
| $\Phi_0$ | 4013.15 | 17 | 3130 | 1292 | 430.7 | 0.997 |
| $\Phi_1$ | 3705.95 | 11 | 3310 | 1123 | 398.1 | 1.024 |
| $\Phi_2$ | 4382.20 | 9 | 3960 | 1388 | 468.8 | 0.994 |
| $\Phi_3$ | 4453.27 | 7 | 2140 | 1187 | 475.7 | 1.008 |

**Table 10.12** Numerical results for case 2, where $m = 5$ and $t_p = 0.1039$

| $\Phi_i$ | TF | I | $D_I$ | $\tilde{\Theta}$ | $\tilde{\kappa}$ | $r_{TF}$ |
|---|---|---|---|---|---|---|
| $\Phi_0$ | 11220.74 | 17 | 3470 | 3776 | 1159.8 | 0.985 |
| $\Phi_1$ | 9356.20 | 10 | 2180 | 2660 | 978.6 | 1.008 |
| $\Phi_2$ | 12252.59 | 8 | 1880 | 3561 | 1262.2 | 0.989 |
| $\Phi_3$ | 11762.06 | 7 | 2140 | 3216 | 1239.2 | 1.003 |

**Table 10.13** Numerical results for case 3, where $m = 13$ and $t_p = 0.1039$

| $\Phi_i$ | TF | I | $D_I$ | $\tilde{\Theta}$ | $\tilde{\kappa}$ | $r_{TF}$ |
|---|---|---|---|---|---|---|
| $\Phi_0$ | 77967.67 | 17 | 2630 | 25266 | 8125.6 | 1.008 |
| $\Phi_1$ | 58546.41 | 11 | 2950 | 18205 | 6117.1 | 1.010 |
| $\Phi_2$ | 85137.03 | 9 | 3380 | 26995 | 8950.9 | 1.003 |
| $\Phi_3$ | 77305.42 | 7 | 1650 | 21007 | 8040.0 | 1.006 |

Tables 10.11, 10.12 and 10.13 show the time factor (*TF*); the last iteration reached (*I*); the approximated number of correct decimal places in the *I*th iteration ($D_I$); the elapsed time $\tilde{\Theta}$; the slope $\tilde{\kappa}$; and the computed time factor $\widetilde{TF}$ defined by

$$\widetilde{TF} = \frac{\tilde{\Theta}(D_I)}{t_p \log q} = \frac{\tilde{\kappa}}{t_p} \approx TF = \frac{1}{\log CEI}.$$

Furthermore, the last column shows the percentage of relative error $r_{TF}$ between *TF* and $\widetilde{TF}$. Note that the ordering of the methods according to time factor (*TF*) (or *CEI*) matches the ordering according to $\tilde{\kappa}$.

Figures 10.1, 10.2 and 10.3 show in a semi-logarithmic graph the $(\log D_j, \Theta(D_j))$ pairs and the straight line of each method. Note that the smaller the slope of the line, the more efficient the method.

## 10.6   Ball of Local Convergence

In this section, we use the convergence ball to study an aspect of local convergence theory. For this purpose, we present an algorithm devised by Schmidt and Schwetlick [34] and subsequently studied by Potra and Pták [31]. The procedure consists of fixing a natural number $k$, and keeping the same linear operator of the Secant method for sections of the process consisting of $k$ steps each. It may be described as follows: starting with $x_n, z_n \in D$, for $0 \leq j \leq k - 1$, $n \geq 0$ and $x_n^{(0)} = z_n$,

$$x_n^{(j+1)} = \Phi_{4,j+1}(x_n; x_{n-1}) = x_n^{(j)} - [x_n, z_n; F]^{-1} F(x_n^{(j)}). \qquad (10.71)$$

In the two last steps we take $x_{n+1} = \Phi_{4,k-1}(x_n; x_{n-1}) = x_n^{(k-1)}$ and finally $z_{n+1} = \Phi_{4,k}(x_n; x_{n-1}) = x_n^{(k)}$.

The iterative method $\Phi_{4,k}$ defined in (10.71) has a local convergence order equal to at least $\rho_{4,k} = \frac{1}{2}\left(k + \sqrt{k^2 + 4}\right)$ [21].

We introduce a theorem (see [21]) on the local convergence of sequences defined in (10.71), following the ideas given in [8, 36]. We denote $B(\alpha, r)$ the open ball $\{x \in \mathbf{R}^m; \|x - \alpha\| < r\}$.

**Theorem 4** *Let $\alpha$ be a solution of $F(x) = 0$ such that $[F'(\alpha)]^{-1}$ exists. We suppose that there is a first order divided difference $[x, y; F] \in \mathscr{L}(D, \mathbf{R}^m)$, for all $x, y \in D$, that satisfies*

$$\left\| [F'(\alpha)]^{-1} ([x, y; F] - [u, v; F]) \right\| \leq K (\|x - u\| + \|y - v\|), \ x, y, u, v \in D, \qquad (10.72)$$

*and $B(\alpha, r) \subset D$, where $r = \dfrac{1}{5K}$. Then, for $x_0, z_0 \in B(\alpha, r)$, the sequence $\{x_n^{(j+1)}\}$, $0 \leq j \leq k - 1$, $n \geq 0$, given in (10.71) is well-defined, belongs to $B(\alpha, r)$*

*and converges to α. Moreover,*

$$\left\| x_n^{(j+1)} - \alpha \right\| = \frac{K\left( \|x_n - \alpha\| + \|z_n - \alpha\| + \|x_n^{(j)} - \alpha\| \right)}{1 - K\left( \|x_n - \alpha\| + \|z_n - \alpha\| \right)} \left\| x_n^{(j)} - \alpha \right\|. \qquad (10.73)$$

Theorem 4 can be seen as a result of the accessibility of the solution in the following way, if $x_0, z_0 \in B(\alpha, r)$, where $r = \dfrac{1}{5K}$, the sequence $\{x_n^{(j+1)}\}$, given in (10.71), converges to $\alpha$. The radius $r$ could be slightly increased if we consider center-Lipschitz conditions, as in [3] or [4], for instance. In fact, let us assume that, together with (10.72), the following condition holds:

$$\left\| [F'(\alpha)]^{-1} \left( [\alpha, \alpha; F] - [u, v; F] \right) \right\| \leq K_1 \left( \|\alpha - u\| + \|\alpha - v\| \right), \quad u, v \in D. \qquad (10.74)$$

Obviously $K_1$ is always less than or equal to $K$. Then, we can mimic the proof of Theorem 4 (see [21]) to obtain the radius $r_1 = 1/(2K_1 + 3K) \geq r$.

**Application** Now, we consider an application of the previous analysis to the nonlinear integral equation of mixed Hammerstein type of the form:

$$x(s) = 1 + \frac{1}{2} \int_0^1 G(s, t) \, x(t)^3 \, dt, \quad s \in [0, 1],$$

where $x \in C[0, 1]$, $t \in [0, 1]$, and the kernel $G$ is the Green function

$$G(s, t) = \begin{cases} (1 - s)\, t, \ t \leq s, \\ s\, (1 - t), \ s < t. \end{cases}$$

Using the following Gauss-Legendre formula to approximate an integral, we obtain the following nonlinear system of equations

$$x_i = 1 + \frac{1}{2} \sum_{j=1}^8 b_{ij} x_j^3, \quad b_{ij} = \begin{cases} \varpi_j t_j (1 - t_i), \ \text{if } j \leq i, \\ \varpi_j t_i (1 - t_j), \ \text{if } j > i, \end{cases} \quad i = 1, \ldots, 8, \qquad (10.75)$$

where the abscissas $t_j$ and the weights $\varpi_j$ are determined for $m = 8$ (see [6]). We have denoted the approximation of $x(t_i)$ by $x_i$ ($i = 1, 2, \ldots, 8$). The solution of this system is $\alpha = (\alpha_1, \alpha_2, \alpha_3, \alpha_4, \alpha_4, \alpha_3, \alpha_2, \alpha_1)^t$, where

$$\alpha_1 \approx 1.00577, \ \alpha_2 \approx 1.02744, \alpha_3 \approx 1.05518 \text{ and } \alpha_4 \approx 1.07441.$$

The nonlinear system (10.75) can be written in the form

$$F(x) = x - \overline{1} - \frac{1}{2} B \hat{x} = 0,$$

for $x = (x_1, \ldots, x_m)^t$, $\overline{1} = (1, 1, \ldots, 1)^t$, $\hat{x} = (x_1^3, \ldots, x_m^3)^t$ and $B = (b_{ij})$.

Taking the definition of the divided difference operator (10.61) we have

$$[x, y; F] = I - \frac{1}{2} B \operatorname{diag}(\tilde{p}),$$

where $\tilde{p} \in \mathbf{R}^m$ and $\tilde{p}_i = x_i^2 + x_i y_i + y_i^2$, $1 \le i \le m$. The Fréchet-derivative of operator $F$ is given by

$$F'(X) = I - \frac{3}{2} B \operatorname{diag}(q),$$

where $q \in \mathbf{R}^m$ and $q_i = X_i^2$, $1 \le i \le m$, and, in addition, we have $F'(X) - F'(Y) = -\frac{3}{2} B \operatorname{diag}(r)$, where $r \in \mathbf{R}^m$ and $r_i = X_i^2 - Y_i^2$, $1 \le i \le m$. Setting

$$\Omega = \{X \in \mathbf{R}^m | \|X\|_\infty \le \delta\} \tag{10.76}$$

and taking norms we obtain

$$\|F'(X) - F'(Y)\| \le \frac{3}{2} \|B\| \max_{1 \le i \le m} |2c_i| \, \|X - Y\|,$$

where $c \in \Omega$, and we get

$$\|F'(X) - F'(Y)\| \le 3\delta \|B\| \, \|X - Y\|. \tag{10.77}$$

The divided difference operator can also be written as follows (see [32]):

$$[x, y; F] = \int_0^1 F'(\tau x + (1 - \tau)y) \, d\tau.$$

Then we have

$$\|[x, y; F] - [u, v; F]\| \le \int_0^1 \|F'(\tau x + (1 - \tau)y) - F'(\tau u + (1 - \tau)v)\| \, d\tau$$

$$\le 3\delta \|B\| \int_0^1 (\tau \|x - u\| + (1 - \tau)\|y - v\|) \, d\tau$$

$$= \frac{3}{2} \delta \|B\| (\|x - u\| + \|y - v\|).$$

**Table 10.14** Numerical
results for the radius of the
existence ball in the nonlinear
system (10.75) for different
values of $\delta$ defined in (10.76)
and the corresponding $K$
introduced in (10.78)

| $\delta$ | $K$ | $r$ |
|-----|---------|--------|
| 1.3 | 0.29708 | 0.6732 |
| 1.2 | 0.27423 | 0.7293 |
| 1.1 | 0.25137 | 0.7956 |

Here $r$ is the radius for the
frozen Schmidt-Schwetlick
method (Theorem 4)

Next, we compute an upper bound for the inverse of $F'(\alpha)$, $\|F'(\alpha)^{-1}\| \leq 1.233$,
and finally, taking into account

$$\|F'(\alpha)^{-1}\left([x, y; F] - [u, v; F]\right)\| \leq \|F'(\alpha)^{-1}\| \frac{3}{2} \delta \|B\| \left(\|x - u\| + \|y - v\|\right),$$

we deduce the following value for the parameter $K$ introduced in (10.72):

$$K = \frac{3}{2} \delta \|F'(\alpha)^{-1}\| \|B\|. \tag{10.78}$$

Table 10.14 shows the size of the balls centered in $\alpha$, $r = 1/(5K)$, for the frozen
Schmidt-Schwetlick method and for $x_{-1} = \bar{\delta}$ and $x_0 = \bar{1}$ (see Theorem 4).

The numerical computations were performed on an MPFR library of C++ multi-
precision arithmetics with 4096 digits of mantissa. All programs were compiled by
g++ (4.2.1) for i686-apple-darwin1 with libgmp (v.4.2.4) and libmpfr
(v.2.4.0) libraries on an Intel® Core i7, 2.8 GHz (64-bit machine) processor.
For this hardware and software, the computational cost of the quotient respect to the
product is $\ell = 1.7$. Within each example, the starting point is the same for all the
methods tested. The classical stopping criteria $\|e_{I+1}\| = \|x_{I+1} - \alpha\| < 10^{-\varepsilon}$ and
$\|e_I\| > 10^{-\varepsilon}$, where $\varepsilon = 4096$, is replaced by $\|\tilde{e}_{I+1}\| = \|x_{I+1} - \tilde{\alpha}_{I+1}\| < \varepsilon$ and
$\|\tilde{e}_I\| > \varepsilon$, where $\tilde{\alpha}_n$ is obtained by the $\delta^2$-Aitken procedure, that is

$$\tilde{e}_n = \left(\frac{(\delta x_{n-1}^{(r)})^2}{\delta^2 x_{n-2}^{(r)}}\right)_{r=1 \div m} \tag{10.79}$$

where $\delta x_{n-1} = x_n - x_{n-1}$ and the stopping criterion is now $\|\tilde{e}_{I+1}\| < 10^{-\eta}$, where
$\eta = \left[\varepsilon(2\rho - 1)/\rho^2\right]$. Note that this criterion is independent of the knowledge of the
root (see [14]).

In this case the concrete values of the parameters for the method $\Phi_{4,k}$ are
$(m, \mu) = (8, 11)$. Taking as initial approximations $x_{-1} = \bar{1}$ and $x_0 = \overline{1.1}$, which
satisfy the conditions of Theorem 4 (see Table 10.14), we compare the convergence
of the methods $\Phi_{4,k}$, towards the root $\alpha$. We get the results shown in Table 10.15.

**Table 10.15** Numerical results for the nonlinear system (10.75), where we show $I$, the number of iterations, $\rho_{4,k}$, the local order of convergence, $CEI$, the computational efficiency index, $TF$, the time factor, $D_I$ and the estimation of the corrected decimal number in the last iteration $x_I$

|            | I | $\rho_{4,k}$ | CEI         | TF     | $D_I$ |
|------------|---|--------------|-------------|--------|-------|
| $\Phi_{4,5}$ | 4 | 5.193        | 1.000969201 | 1032.3 | 1477  |
| $\Phi_{4,6}$ | 4 | 6.162        | 1.000979191 | 1021.8 | 2973  |
| $\Phi_{4,7}$ | 3 | 7.140        | 1.000975729 | 1025.4 | 757   |

Finally, in the computations we substitute the computational order of convergence (COC) [43] by an extrapolation (ECLOC) denoted by $\tilde{\rho}$ and defined as follows

$$\tilde{\rho} = \frac{\ln\|\tilde{e}_I\|}{\ln\|\tilde{e}_{I-1}\|},$$

where $\tilde{e}_I$ is given in (10.79). If $\rho = \tilde{\rho} \pm \Delta\tilde{\rho}$, where $\rho$ is the local order of convergence and $\Delta\tilde{\rho}$ is the error of ECLOC, then we get $\Delta\tilde{\rho} < 10^{-3}$. This means that in all computations of ECLOC we obtain at least three significant digits. Therefore, it is a good check of the local convergence orders of the family of iterative methods presented in this section.

## 10.7 Adaptive Arithmetic

In this section, we present a new way to compute the iterates. It consists of adapting the length of the mantissa to the number of significant figures that should be computed in the next iteration. The role of the local convergence order $\rho$ is the key concept to forecast the precision of the next iterate. Furthermore, if the root is known, then an expression of the forecast of digits in terms of $x_n$ and $e_n$ is

$$\Delta_{e_n} = \lceil \rho\left(-\log_{10}\|e_n\| + 4\right) + \log_{10}\|x_n\| \rceil,$$

where 4 is a security term that is empirically obtained. When the root is unknown according to

$$D_k \approx -\log_{10}\|e_k\| \approx -\frac{\rho}{\rho-1}\log_{10}\|\breve{e}_k\|. \tag{10.80}$$

(see Sect. 10.2 for more details) we may use the following forecast formulae

$$\Delta_{\breve{e}_n} = \left\lceil \frac{\rho^2}{\rho-1}\left(-\log_{10}\|\breve{e}_n\| + 2\right) + \log_{10}\|x_n\| \right\rceil,$$

where $\Delta_{e_n}$ and $\Delta_{\breve{e}_n}$ are the lengths of the mantissa for the next iteration.

## 10.7.1  Iterative Method

The composition of two Newton's iteration functions is a well-known technique that allows us to improve the efficiency of iterative methods in the scalar case. The two-step iteration function obtained in this way is

$$\begin{cases} y = \mathcal{N}(x) = x - \dfrac{f(x)}{f'(x)}, \\ X = \mathcal{N}(y). \end{cases} \tag{10.81}$$

In order to avoid the computation and the evaluation of a new derivative function $f'(y)$ in the second step, some authors have proposed the following variant:

$$X = y - \frac{f(y)}{f'(x)}. \tag{10.82}$$

In this case, the derivative is "frozen" (we only need to calculate $f'(x)$ at each step). Note that the local order of convergence decreases from $\rho = 4$ to $\rho = 3$, but the efficiency index in (10.81) is $EI = 4^{1/4} = 1.414$, while in (10.82) we have an improvement: $EI = 3^{1/3} = 1.441$.

Chung in [5] considers a different denominator in the second step of (10.81); namely,

$$X = y - \frac{f(y)}{f'(x)\,h(t)}, \tag{10.83}$$

where $t = f(y)/f(x)$ and $h(t)$ is a real valued function. The conditions established by Chung [5] in order to obtain a fourth-order method with only three evaluations of functions, $f(x), f(y), f'(x)$, are the following

$$h(0) = 1, \quad h'(0) = -2 \quad \text{and} \quad |h''(0)| < \infty.$$

In particular, we consider a specific member of the King's family [25] defined by $h(t) = \dfrac{1}{1+2t}$ (see also [2]) and the second step of (10.81) is

$$X = y - (1 + 2t)\frac{f(y)}{f'(x)}.$$

Taking into account that

$$t = \frac{f(y)}{f(x)} = \frac{f(x) + f(y) - f(x)}{f(x)} = 1 + \frac{f(y) - f(x)}{f(x)},$$

**Table 10.16** Computational cost of iterative methods $\Phi_5$

|          | $a_0(m)$   | $a_1(m)$ | $\omega(m, \ell)$                              |
|----------|------------|----------|-----------------------------------------------|
| $\Phi_5$ | $m^2 + m$  | $m^2$    | $\dfrac{m}{6}(2m^2 + 3(3\ell + 7)m + (15\ell - 17))$ |

and $f(x) = -f'(x)(y - x)$, we have

$$t = 1 - \frac{f(y) - f(x)}{f'(x)(y - x)} = 1 - \frac{[x, y]_f}{f'(x)}, \tag{10.84}$$

we generalize to the operator $T$ defined by

$$T = I - F'(x)^{-1}[x, y; F].$$

So we consider the iterative method

$$\begin{cases} y_n = x_n - F'(x_n)^{-1} F(x_n), \\ x_{n+1} = \Phi_5(x_n) = y_n - (I + 2T) F'(x)^{-1} F(y) \\ \qquad = y_n - \left(3I - 2F'(x_n)^{-1}[x_n, y_n; F]\right) F'(x)^{-1} F(y). \end{cases} \tag{10.85}$$

This algorithm has been deduced independently by Sharma and Arora [37], who prove that the $R$-order is at least four. In Table 10.16 we present the parameters of the computational cost of this scheme.

### 10.7.2 Numerical Example

We have applied the method (10.85) to the system (10.62) for $m = 11$ and the initial vector $x_0$ with all its components equal to 0.1. The solution reached with this initial point is the vector $\alpha$ with their eleven components approximately equal to $0.09127652716\ldots$ In this case, taking into account $\mu_1 = \mu_0/m$, we have $(m, \mu_0, \mu_1) = (11, 76.4, 6.95)$.

Table 10.17 shows the correct decimals and forecast for this function using the $\Phi_5$ method. In the first two rows, we consider the correct decimals number $D_n$ using fixed arithmetic (FA) and adaptive arithmetic (AA). In the third and fourth rows, the forecasts of lengths of the mantissa are obtained in adaptive arithmetic when we know the root $\Delta_{e_n}$, or not $\Delta_{\breve{e}_n}$, respectively.

As can be seen, all the forecasts overestimate the real values. Note that the values of $\Delta_{e_n}$ and $\alpha$ are very similar.

The two first rows in Table 10.18 show the partial and total elapsed time ($t_e$ and $T_e$) when the root is known. The third and fourth rows show these times ($t_{\breve{e}}$ and $T_{\breve{e}}$) when the root is unknown. Moreover, using fixed arithmetic, the total elapsed time

**Table 10.17** Number of correct decimals and forecasts for each iteration $1 \leq n \leq 5$

| $n$ | 1 | 2 | 3 | 4 | 5 |
|---|---|---|---|---|---|
| $D_n$ FA | 12 | 51 | 209 | 839 | 3358 |
| $D_n$ AA | 12 | 51 | 209 | 837 | 3346 |
| $\Delta_{e_n}$ | 32 | 63 | 221 | 850 | 3363 |
| $\Delta_{\check{e}_n}$ | 32 | 63 | 220 | 850 | 3369 |

**Table 10.18** Partial and total elapsed time in ms for each iteration $1 \leq n \leq 5$

| $n$ | 1 | 2 | 3 | 4 | 5 |
|---|---|---|---|---|---|
| $t_e$ | 1.112 | 1.591 | 3.639 | 23.598 | 209.953 |
| $T_e$ | 1.112 | 2.703 | 6.342 | 29.940 | 239.893 |
| $t_{\check{e}}$ | 1.215 | 1.722 | 3.902 | 25.352 | 226.234 |
| $T_{\check{e}}$ | 1.215 | 2.937 | 6.839 | 32.191 | 258.425 |

**Table 10.19** Elapsed time in ms for 100,000 products

| Digits | 32 | 64 | 128 | 256 | 512 | 1024 | 2048 | 4096 | 8192 | 16,384 | 32,768 |
|---|---|---|---|---|---|---|---|---|---|---|---|
| Time (ms) | 5.0 | 9.1 | 16.0 | 33.1 | 89.1 | 274.6 | 820.1 | 2419.0 | 7190.9 | 19,019.7 | 53,083.7 |

to obtain the same solution is 1364.151 ms. Henceforth, the time in the forecast of the mantissa length is sufficiently short that we can state that the use of adaptive arithmetic is a technique five times faster than the fixed arithmetic technique (in fact, 5.7 using $e$ and 5.3 using $\check{e}$).

### 10.7.3 Practical Result

Note that, if the number of digits is increased, then the time needed to compute a product is also increased. In particular, when the number of digits is doubled from 256 digits the computational time is tripled (see Table 10.19). Following Karatsuba's algorithm [24], we have

$$t_n = a\, \Delta_n^\lambda,$$

where $\lambda = \log_2 3 = 1.58496\ldots$ Note that $\Delta_n = D_n + \Lambda$ where $D_n$ is the number of correct decimals and $\Lambda$ is the number of integer digits of $x_n$. For the range $256 \leq \Delta_n \leq 4096$ we have $\Delta_n^\lambda \approx D_n^\lambda$. Therefore, from $D_{n+1} \approx \rho\, D_n$ we obtain

$$\mathscr{C}_{n+1} = a\, \Delta_{n+1}^\lambda \approx a\, D_{n+1}^\lambda \approx a\, \rho^\lambda\, D_n^\lambda \approx \rho^\lambda\, \mathscr{C}_n.$$

Denoting by $\mathscr{C}_I$ the computational cost of the last iteration, if we consider an infinite number of iterations, the total cost is

$$\widetilde{\mathscr{C}} = \mathscr{C}_I \left( 1 + \frac{1}{\rho^\lambda} + \frac{1}{\rho^{2\lambda}} + \cdots \right) = \mathscr{C}_I \frac{\rho^\lambda}{\rho^\lambda - 1}.$$

If we only consider the last iterate then we have

$$r = \frac{\mathscr{C}_I}{\widetilde{\mathscr{C}}} = \frac{\rho^\lambda - 1}{\rho^\lambda}.$$

Notice that for $\rho = 4$ we have $r = 0.889$. From Table 10.18 we can deduce that in the two cases (knowledge of the root and no knowledge of the root) in adaptive arithmetic, the computational cost of the last iteration is 87.5 % of the total elapsed time. Actually, we can assert that for iterative methods of an order equal to or greater than the fourth order, we only need to consider the cost of the last iteration to obtain a first approximation.

# References

1. Aitken, A.: On Bernoulli's numerical solution of algebraic equations. Proc. R. Soc. Edinb. **46**, 289–305 (1926)
2. Amat, S., Busquier, S., Plaza, S.: Dynamics of the King and Jarrat iterations. Aequationes Math. **69**, 212–223 (2005)
3. Argyros, I.K., Gutiérrez, J.M.: A unified approach for enlarging the radius of convergence for Newton's method and applications. Nonlinear Funct. Anal. Appl. **10**, 555–563 (2005)
4. Argyros, I.K., Gutiérrez, J.M.: A unifying local and semilocal convergence analysis of Newton-like methods. Adv. Nonlinear Var. Inequal. **10**, 1–11 (2007)
5. Chung, C.: Some fourth-order iterative methods for solving nonlinear equations. Appl. Math. Comput. **195**, 454–459 (2008)
6. Ezquerro, J.A., Grau-Sánchez, M., Grau, A., Hernández, M.A., Noguera, M., Romero, N.: On iterative methods with accelerated convergence for solving systems of nonlinear equations. J. Optim. Theory Appl. **151**, 163–174 (2011)
7. Ezquerro, J.A., Grau-Sánchez, M., Grau, A., Hernández, M.A., Noguera, M.: Analysing the efficiency of some modifications of the secant method. Comput. Math. Appl. **64**, 2066–2073 (2012)
8. Ezquerro, J.A., Grau-Sánchez, M., Grau, A., Hernández, M.A.: Construction of derivative-free iterative methods from Chebyshev's method. Anal. Appl. **11**(3), 1350009 (16 pp.) (2013)
9. Ezquerro, J.A., Grau-Sánchez, M., Hernández, M.A., Noguera, M.: Semilocal convergence of secant-like methods for differentiable and nondifferentiable operator equations. J. Math. Anal. Appl. **398**, 100–112 (2013)
10. Fousse, L., Hanrot, G., Lefèvre, V., Pélissier, P., Zimmermann, P.: MPFR: a multiple-precision binary floating-point library with correct rounding. ACM Trans. Math. Softw. **33** (2007). doi:10.1145/1236463.1236468
11. Grau, M., Díaz-Barrero, J.L.: A weighted variant family of Newton's method with accelerated third-order convergence. Appl. Math. Comput. **186**, 1005–1009 (2007)

12. Grau-Sánchez, M., Gutiérrez, J.M.: Zero-finder methods derived from Obreshkov's techniques. Appl. Math. Comput. **215**, 2992–3001 (2009)
13. Grau-Sánchez, M., Noguera, M.: A technique to choose the most efficient method between secant method and some variants. Appl. Math. Comput. **218**, 6415–6426 (2012)
14. Grau-Sánchez, M., Noguera, M., Gutiérrez, J.M.: On some computational orders of convergence. Appl. Math. Lett. **23**, 472–478 (2010)
15. Grau-Sánchez, M., Grau, A., Díaz-Barrero, J.L.: On computational order of convergence of some multi-precision solvers of nonlinear systems of equations. ArXiv e-prints (2011). Available at http://arxiv.org/pdf/1106.0994.pdf
16. Grau-Sánchez, M., Grau, A., Noguera, M.: On the computational efficiency index and some iterative methods for solving systems of nonlinear equations. J. Comput. Appl. Math. **236**, 1259–1266 (2011)
17. Grau-Sánchez, M., Grau, A., Noguera, M.: Ostrowski type methods for solving systems of nonlinear equations. Appl. Math. Comput. **218**, 2377–2385 (2011)
18. Grau-Sánchez, M., Grau, A., Noguera, M.: Frozen divided differences scheme for solving systems of nonlinear equations. J. Comput. Appl. Math. **235**, 1739–1743 (2011)
19. Grau-Sánchez, M., Grau, A., Noguera, M., Herrero, J.R.: On new computational local orders of convergence. Appl. Math. Lett. **25**, 2023–2030 (2012)
20. Grau-Sánchez, M., Grau, A., Noguera, M., Herrero, J.R.: A study on new computational local orders of convergence. ArXiv e-prints (2012). Available at http://arxiv.org/pdf/1202.4236.pdf
21. Grau–Sánchez, M., Noguera, M., Gutiérrez, J.M.: Frozen iterative methods using divided differences "à la Schmidt-Schwetlick". J. Optim. Theory Appl. **160**, 93–948 (2014)
22. Hueso, J.L., Martínez, E., Torregrosa, J.R.: Third and fourth order iterative methods free from second derivative for nonlinear systems. Appl. Math. Comput. **211**, 190–197 (2009)
23. Jay, L.O.: A note on Q-order of convergence. BIT **41**, 422–429 (2001)
24. Karatsuba A., Ofman, Y.: Multiplication of many-digital numbers by automatic computers. Proc. USSR Acad. Sci. **145**, 293–294 (1962). Transl. Acad. J. Phys. Dokl. **7**, 595–596 (1963)
25. King, R.F.: A family of fourth-order methods for nonlinear equations. SIAM J. Numer. Anal. **10**, 876–879 (1973)
26. Kurchatov, V.A.: On a method of linear interpolation for the solution of functional equations. Dokl. Akad. Nauk SSSR **198**, 524–526 (1971). Transl. Sov. Math. Dokl. **12**, 835–838 (1971)
27. Ortega, J.M., Rheinboldt, W.C.: Iterative Solution of Nonlinear Equations in Several Variables. Academic, New York (1970)
28. Ostrowski, A.M.: Solutions of Equations and System of Equations. Academic, New York (1960)
29. Petković, M.S.: Remarks on "On a general class of multipoint root-finding methods of high computational efficiency". SIAM J. Numer. Anal. **49**, 1317–1319 (2011)
30. Potra, F.A.: A characterisation of the divided differences of an operator which can be represented by Riemann integrals. Revue d'analyse numérique et de la théorie de l'approximation **2**, 251–253 (1980)
31. Potra, F.A., Pták, V.: A generalization of Regula Falsi. Numer. Math. **36**, 333–346 (1981)
32. Potra, F.A., Pták, V.: Nondiscrete Induction and Iterative Processes. Research Notes in Mathematics, vol. 103. Pitman Advanced Publishing Program, Boston (1984)
33. Ralston, A.: A First Course in Numerical Analysis. McGraw-Hill, New York (1965)
34. Schmidt, J.W., Schwetlick, H.: Ableitungsfreie Verfahren mit höherer Konvergenzgeschwindigkeit. Computing **3**, 215–226 (1968)
35. Schröder, E.: Über unendlich viele Algorithmen zur Auflösung der Gleichungen. Math. Ann. **2**, 317–365 (1870). Translated by G.W. Stewart, On Infinitely Many Algorithms for Solving Equations (1998). Available at http://drum.lib.umd.edu/handle/1903/577
36. Shakno, S.M.: On an iterative algorithm with superquadratic convergence for solving nonlinear operator equations. J. Comput. Appl. Math. **231**, 222–335 (2009)
37. Sharma, J.R., Arora, H.: On efficient weighted-Newton methods for solving systems of nonlinear equations. Appl. Math. Comput. **222**, 497–506 (2013)
38. The MPFR library 3.0.0. (2010). Timings in http://www.mpfr.org/mpfr-3.0.0/timings.html

39. The MPFR library 3.1.0. (2011). Available in http://www.mpfr.org
40. Tornheim, L.: Convergence of multipoint iterative methods. J. ACM **11**, 210–220 (1964)
41. Traub, J.F.: Iterative Methods for the Solution of Equations. Prentice-Hall, Englewood Cliffs (1964)
42. Wall, D.D.: The order of an iteration formula. Math. Tables Aids Comput. **10**, 167–168 (1956)
43. Weerakoon, S., Fernando, T.G.I.: A variant of Newton's method with accelerated third-order convergence. Appl. Math. Lett. **13**, 87–93 (2000)

Printed in the United States
By Bookmasters